Advanced Materials & Demanding Applications (AMDA 2020)

IOP Conference Series: Materials Science and Engineering
Volume 1060

Online
22 - 26 June 2020

Editors:

A.J. McMillan
A.L. Galinovsky

ISBN: 978-1-7138-3339-0
ISSN: 1757-8981

Printed from e-media with permission by:

Curran Associates, Inc.
57 Morehouse Lane
Red Hook, NY 12571

CURRAN ASSOCIATES INC.
proceedings
.com

Some format issues inherent in the e-media version may also appear in this print version.

Institute of Physics
Dirac House, Temple Back
Bristol BS1 6BE UK

Phone: 44 1 17 929 7481
Fax: 44 1 17 920 0979

techtracking@iop.org

Additional copies of this publication are available from:

Curran Associates, Inc.
57 Morehouse Lane
Red Hook, NY 12571 USA
Phone: 845-758-0400
Fax: 845-758-2633
Email: curran@proceedings.com
Web: www.proceedings.com

TABLE OF CONTENTS

IOP Conf. Series: Materials Science and Engineering 1060 (2021) 011001 doi:10.1088/1757-899X/1060/1/011001

Advanced Materials & Demanding Applications

Co-Editors: A J McMillan[1] and A L Galinovsky[2]

Co-Chairs: A J McMillan[1] and S V Reznik[2]

[1] Wrexham Glyndŵr University, UK
[2] Bauman Moscow State Technical University, Russia

E-mail: a.mcmillan@glyndwr.ac.uk

1. Proceedings

This conference was originally planned to be held at Wrexham Glyndŵr University in April 2020. It was one of a series of conferences and meetings held as part of the collaborative relationship between Wrexham Glyndŵr University and Bauman Moscow State Technical University, and with our wider research partners. The purpose of the conferences being to build research collaborations between the more senior participants, and also to provide an opportunity for our students and junior colleagues to participate in a conference and publish their work in the proceedings.

Most of the plans for the conference were being finalized just as the world became locked down by the COVID-19 pandemic. Accordingly, the conference team was reluctant to lose the opportunity that we had been working towards, and it was decided to hold the conference online instead. This was replanned for 22-26 June 2020, and held as an asynchronous conference, with participants joining from time zones spanning the globe: Russia (Yakutsk, Samara, Moscow), Australia, China, Myanmar, Ukraine, UK, and the USA. This was facilitated by presentations being prerecorded and made available as video links, and social media tools used for questions and answers. Social media was also used to enable some conference community and relationship building, with participants writing and sharing pictures of their local area. In place of the usual conference dinner we held an online quiz night.

As a result of our early experimentation with online tools and conferencing, we found that we were able to attract a somewhat larger participation than had been previously anticipated. This is probably a reflection of the costs in travel and accommodation that would usually present a barrier to attending a conference. Furthermore, many of the logistical issues of running a conference became simpler as a result of conferencing online. We now recognize online conferencing as an attractive and very feasible option for the future.

2. Conference Team

There were very many people involved in the organisation and smooth running of the conference. I would particularly like to give thanks to

- Co-Chair of the Conference: Professor Reznik
- Co-Editor of the Proceedings: Professor Galinovsky

I would also like to make a special mention of Maria Prokhorova for her support with the logistical aspects of running a conference between multiple countries, and to Professor Alexander Chernikov for his support to colleagues at Bauman Moscow State Technical University and his steadfast dedication to the continuing relationship between the Glyndŵr and Bauman universities.

2.1. "Local" Organising Committee

2.2. International Scientific Committee

3. In Memoriam

This Advanced Materials & Demanding Applications conference would never have taken place without the work of many people who have contributed to the developing of research collaboration and relationship building between the participants. I would like to take the opportunity to mark the passing of three very significant people.

Brian Foxon, retired Director of International Operations, Wrexham Glyndŵr University, died in November 2019. It was his work, initially with Bauman Moscow State Technical University, and later with Samara University, Russia, and universities in Ukraine, including Lutsk National Technical University, that built the foundation for the collaboration today.

Valeriy Elenev, retired Director of the Institute of Aeronautical Engineering, Samara University, Russia, died in July 2020. Valeriy initiated the partnership and research cooperation between Samara University and Wrexham Glyndŵr University, and later worked together with Wrexham Glyndŵr University staff to extend the collaboration worldwide.

Peter Excell, retired Deputy Vice Chancellor and Professor Emeritus, Wrexham Glyndŵr University, died in August 2020. Peter gave great encouragement to the engineering staff and students at Wrexham Glyndŵr University and supported the research relationships between the universities as a frequent visitor to collaborators in Moscow.

Professor Alison J McMillan
Wrexham Glyndŵr University, UK

Professor Andrey L Galinovsky
Bauman Moscow State Technical University, Russia

Professor Sergey V Reznik
Bauman Moscow State Technical University, Russia

IOP Conf. Series: Materials Science and Engineering 1060 (2021) 011002 doi:10.1088/1757-899X/1060/1/011002

Peer review declaration

All conference organisers/editors are required to declare details about their peer review. Therefore, please provide the following information:

- **Type of peer review: Single-blind**

Single-anonymous: authors' identities are known to the reviewers, reviewers' identities are hidden from authors

- **Describe criteria used by Reviewers when accepting/declining papers. Was there the opportunity to resubmit articles after revisions?**

We advertised the conference with the expectation that each participant would author or co-author a presentation, and that most presentations would lead to a paper in the proceedings. Peer reviewers were directed to the IOPP Conference Series guidance, including the Proceedings peer review policy. Two peer reviewers were assigned to each paper: care was taken to ensure good mixing of pairings of reviewers.

As many of the authors were students or early career researchers, we anticipated that there would be a need for multiple review iterations. Reviewer feedback was intended to be helpful and focussed on the technical content and the clarity of exposition. In addition to review of the technical content, there was also support for using the template, improving the layout of content, and referencing. We encouraged resubmissions and encouraged the authors to make as many improvements as would be feasible within reasonable amount of time.

- **Conference submission management system:**

The papers were numbered, and resubmitted versions were also numbered. Progress with the submission and the review process was monitored using a spreadsheet.

- **Number of submissions received:**

Initially, there were expected to be 35 papers, but one missed the abstract deadline, two did not register for the conference, and one withdrew after the conference. As a result, 31 paper submissions were received.

- **Number of submissions sent for review:** 31 submissions were sent for review.
- **Number of submissions accepted:** 27 submissions were accepted.
- **Acceptance Rate (Number of Submissions Accepted / Number of Submissions Received X 100):** Acceptance rate = 87%
- **Average number of reviews per paper:** Each paper was reviewed by two reviewers
- **Total number of reviewers involved:** 13 reviewers
- **Any additional info on review process (ie plagiarism check system):**

Some support was provided for use of Word templates, English language grammar, checking accuracy of reference information. The two proceedings Editors provided support to reviewers and authors when required.

- **Contact person for queries:** Alison McMillan

IOP Conf. Series: Materials Science and Engineering 1060 (2021) 012001 doi:10.1088/1757-899X/1060/1/012001

Fractal Graphene Patch Antennas and the THz Communications Revolution

J M Blackledge[1], A Boretti[2], L Rosa[3] and S Castelletto[4]

[1]School of Electrical and Electronic Engineering, Technological University Dublin, Ireland; Department of Computer Science, University of Western Cape, South Africa; School of Mathematics, Statistics and Computer Science, University of KwaZulu-Natal, South Africa; Faculty of Arts, Science and Technology, Wrexham Glyndwr University of Wales, UK.
[2]Department of Mechanical Engineering, College of Engineering, Prince Mohammad Bin Fahd University, Al Khobar, Saudi Arabia.
[3]Department of Engineering 'Enzo Ferrari', University of Modena and Reggio Emilia, Modena, Italy and Applied Plasmonics Laboratory, Center for Micro-Photonics, Swinburne University of Technology, Australia.
[4]School of Engineering, Royal Melbourne Institute of Technology, Australia.

E-mail: [1]`jonathan.blackledge@tudublin.ie`

Abstract.
Fractal antennas have and are continuing to receive attention in regard to the future of wireless communications. This is because of their wide- and multi-band capabilities, the opportunity of fractal geometries to drive multiple resonances, and, the ability to make smaller and lighter antennas with fewer components and radiative elements with higher gains. Small scale (i.e. on the micro- and nano-scale) and ultra high frequency (in the Terahertz or THz range) fractal antennas composed of Graphene have the potential to enhance wireless communications at a data rate that is unprecedented, i.e. $\sim 10^{12}$ bits per second. A Fractal Graphene antenna is a high-frequency tuneable antenna for radio communications in the THz spectrum, enabling unique applications such as wireless nano-networks. This is because (mono-layer) Graphene is a one-atom-thick two-dimensional allotrope of Carbon with the highest known electrical conductivity that is currently unavailable in any other material, including metals such as Gold and Silver. Thus, combining the properties of Graphene with the self-affine characteristics of a fractal at the micro- and nano-scale, provides the potential to revolutionise communications, at least in the near field (the order of a few metres) for low power systems. In this paper, we consider the basic physics and some of the principle mathematical models associated with the development of this new disruptive technology in order to provide a guide to those engaged in current and future research, a fractal Graphene antenna being an example of an advanced material for demanding applications. This includes some example simulations on the THz field patterns generated by a fractal patch antenna composed of Graphene whose conductivity is taken to scale with the inverse of the frequency according to a 'Drude' model. The approach to generating THz sources using Graphene is also explored based on Infrared laser pumping to induce a THz photo-current.

1. Introduction

The theoretical principles of what is now known as Fractal Geometry can be traced back to the work of mathematician Gottfried Leibniz in the 17th century. The theoretical properties of Graphene were first explored in the late 1940s by Philip Wallace [1] as a starting point for understanding the electronic properties of Graphite. Similarly, the design and implementation of radio antennas has been ongoing since the pioneering work of Heinrich Hertz [2] in the late 1880s following the theoretical prediction of Electromagnetic (EM) wave propagation by James Clark Maxwell in the 1860s. This paper investigates the coupling of fractal structures and the (electrical) properties of Graphene to design a new type of antenna operating in the THz region of the electromagnetic spectrum. The paper provides a contribution in the current quest to fill the 'THz gap' as illustrated in Figure 1, [4]. This gap is arguably the last of the frequency bands to be considered in the field of wireless communications. It is desirable to 'fill this gap' because communications system that operates at frequencies in the THz band will facilitate the wireless communication of binary information at a rate of Tera bits per second (Tbps), provided a suitable modulation method is available, which is not possible at lower frequency bands.

Technological solution to filling the THz gap represents a transition between the field of electronics for the generation of kHz, MHz and GHz radiation and the field of photonics in which binary information is transmitted through a sequence of laser light pulses in the optical frequency band $\sim 10^{15}$ Hz. The THz frequency band is therefore a band (the last band of it type) where it is physically possible to communicate using a wireless system. To realise this requires an antenna to be fabricated which is on the scale of the wavelength (i.e. the μm-nm scale) given that the physical size of an antenna generally reflects the wavelength at which the antenna radiates an electromagnetic field. For this reason, the power rating of such THz systems is relatively low, constraining the effective scale over which information can be communicated to the near field ($\sim$ meters). This is because there are no efficient methods to generate powerful emissions in the THz band at room temperature compared to lower and higher frequency sources as illustrated in Figure 2. It therefore becomes imperative to construct the antenna from a material that provides maximal current densities over the scale of the antenna. The current density is limited by the conductivity of the material and this is why Graphene is considered, i.e. it is the most conductive material known. Moreover, as shall be discussed in this paper, if the conductivity of Graphene is taken to scale with the inverse of the frequency (in the THz range), then the strength of the electric field emitted by such a conductor becomes independent of the frequency.

The application of Graphene enables wireless communication for nano-scale electronic devices but with a significant reduction in the emission energies. As de-localised electrons on the surface of Graphene (known as 'Graphon's') oscillate, they create an electromagnetic wave which is emitted from the surface of the Graphene. This is referred to as a Surface Plasmon Polariton (SPP) wave which is an electromagnetic wave that travel along a metal-dielectric interface and enables an antenna to operate at the lower end of the THz frequency spectrum with significantly greater efficiencies than are available with copper and other metal-based antennas, for example. Nano-antennas based on SPP waves enable the conversion of light from free-space into sub-wavelength volumes establishing a way of communication using free electron propagation within networks of nano-sized devices. This is impacting on many nano-applications, including biochemical sensors, reconfigurable meta-surfaces, compact optoelectronic devices and wireless nano-sensor networks (nano-networks). Due to its unique electronic properties, Graphene has been identified as a promising platform for building integrated active plasmonic nano-antennas for wavelengths in the mid-infrared range, i.e. 10 - 100 THz.

The electrical properties of (mono-layer) Graphene are such that a Fermionic or Dirac electron fluid [5] is present on the surface of the material. Such a fluid is a form of condensed matter that can be effectively described by the Dirac equations. Although the Dirac equations describe fermions, the quasi-particles present within Dirac matter can have variable statistics. By using infrared laser pumping (i.e. irradiating the electron fluid with a coherent oscillating electric field in the infrared spectrum) to generate a resonant wave pattern in the electron fluid, a Graphene ribbon can be tuned to operate as a THz resonator [6]. The output of this resonator is then coupled to a fractal Graphene patch antenna, thereby producing a THz radiator in the near field.

In this context, the basic principles associated with the design of a fractal Graphene antenna are as follows:

(i) Graphene can be used to generate a source of THz radiation;

(ii) Graphon's can propagate over a surface without incurring a significant number of collisions leading to a loss of energy and thus, the oscillation of Graphon's generate EM fields with relatively high energy densities for the scale of the antenna;

(iii) Fractal geometric shapes can generate EM fields with multiple bands due to the spatial frequency characteristics of self-affine structures.

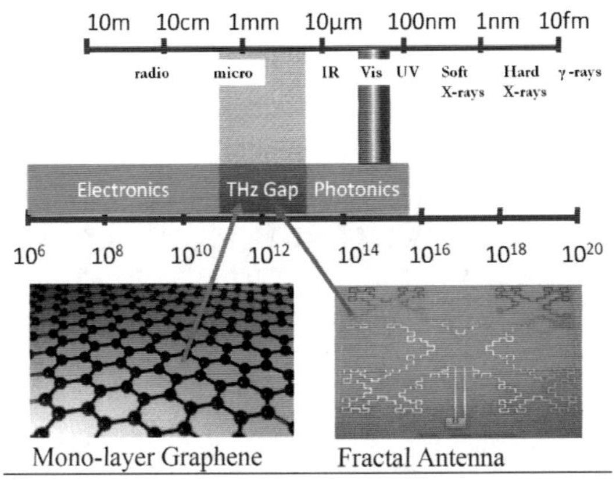

Figure 1. Illustration of the 'THz gap' and the role that Graphene and fractal antennas (composed from Graphene) can play in filling this gap.

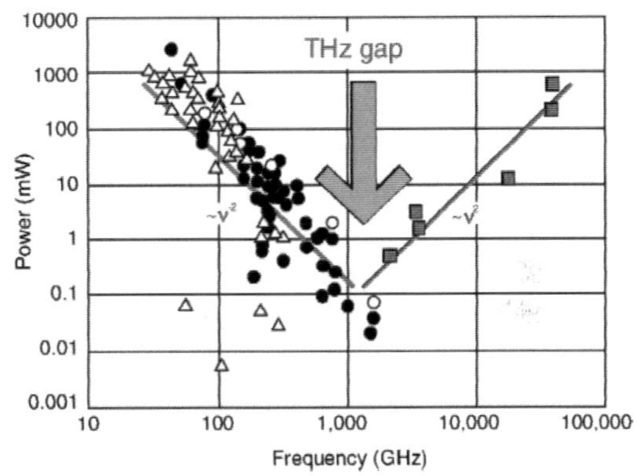

Figure 2. The 'THz gap' with respect to the output power and the frequency ν associated with different technologies: quantum cascade lasers (■), frequency multipliers (•), other electronic devices (△) and cryogenic sources (○) [4].

1.1. About this Paper

After a brief overview of electromagnetic theory and the electrical properties of Graphene, this paper presents some results of simulating the emission field patterns of a fractal Graphene antenna. The principal focus of the material presented is on THz wireless communications using nano-scale antennas. We consider the basic physical models that are the basis for the development of THz technology at this scale, models that include the far field emission of THz radiation and electron transport in Graphene. The combination of metallic structures and Graphene with gate-voltage dependent properties has already provided tuneable antennas with narrower frequency operations, e.g. [7] and [8]. In this paper, the aim is to design a compact structure while increasing the order of fractal structure to a high level by engineering nano-metric fractal Graphene. This fabrication can be achieved in practice using Electron Beam Lithography (EBL) or 3D nano-patterning using focused ion beam lithography, the Graphene being transferred on to a SiO_2 substrate to control the size and features at the nano-scale [9].

1.2. Structure of the Paper

The paper is structured in to the following sections.

- Section 2 provides a short overview and tutorial on the basic electromagnetic theory associated with antennas for harmonic frequency sources in order to introduce two basic antenna models for the intermediate and far fields which are used later in the paper.
- Section 3 uses the theory presented in Section 2 to introduce an idealised model for a fractal patch antenna and considers a new approach to quantifying the electric field generated by such an antenna in the far field based on the principles of scale invariant functions.
- Section 4 provides a short introduction to Graphene to contextualise Section 5, which focuses on the electrical conductivity of Graphene, in particular, its scaling relationship with frequency using a quantum mechanical model in different dimensions. The results show that the classical Drude model for the conductivity is only compatible with the quantum conductivity for a one-dimensional model.
- Section 6 presents some examples of a new Finite Element Analysis for a nano-metric scale fractal Graphene antenna.
- Section 7 discusses the physical principles associated with using a Graphene ribbon for generating a THz source and the optimal modulation method that should be applied.
- Sections 8, 9 and 10 present, respectively, a summary of the work, its principle conclusions and some future directions in regard to further development of the physical models considered, engineering issues and applications.

The principal original contributions are compounded in the material presented in Sections 3-7.

2. Physical Principles of EM Emissions from an Antenna

In this section we briefly review the physical models for the emission of EM radiation from an antenna focusing on a scalar model for a patch antenna. We begin with the basic electromagnetic equations for an antenna.

2.1. Basic Antenna Theory

For a three-dimensional continuum with a spatial vector $\mathbf{r}$ that is linear and isotropic, Maxwell's time t dependent equations (using the International System of units) can be written in the form

$$\nabla \cdot \epsilon \mathbf{E} = \rho \qquad \nabla \cdot \mu \mathbf{H} = 0$$
$$\nabla \times \mathbf{E} = -\mu \frac{\partial \mathbf{H}}{\partial t} \qquad \nabla \times \mathbf{H} = \epsilon \frac{\partial \mathbf{E}}{\partial t} + \mathbf{J}$$

where $\mathbf{E}(\mathbf{r}, t)$ is the electric field, $\mathbf{H}(\mathbf{r}, t)$ is the magnetic field, $\mathbf{J}(\mathbf{r}, t)$ is the current density, $\rho(\mathbf{r}, t)$ is the charge density, $\epsilon(\mathbf{r})$ is the permittivity and $\mu(\mathbf{r})$ is the permeability. The material parameters ϵ and μ can be written in the form $\epsilon = \epsilon_r \epsilon_0$ and $\mu = \mu_r \mu_0$ where $\epsilon_r \geq 1$ and $\mu_r \geq 1$ are the relative permittivity and relative permeability, respectively. In a perfect vacuum (where $\epsilon_r = 1$ and $\mu_r = 1$), $\epsilon = \epsilon_0 = 8.854 \times 10^{-12}$ Farads/meter and $\mu = \mu_0 = 4\pi \times 10^{-7}$ Henrys/meter. The equations above are the macroscopic form of Maxwell's equations for material properties characterised by ϵ and μ.

A conventional antenna is composed of conductive elements and can be considered to a 'conductive continuum' that is a good conductor in which a current (due to flow and oscillation of electrons) can be induced depending on the magnitude of an applied electric field and the conductivity σ (in Siemens/meter or S/m). The relationship between the electric field and the current density is given by Ohm's law, i.e.

$$\mathbf{J} = \sigma \mathbf{E} \tag{1}$$

IOP Conf. Series: Materials Science and Engineering 1060 (2021) 012001 doi:10.1088/1757-899X/1060/1/012001

Noting that $\nabla \cdot (\nabla \times \mathbf{H}) = 0$, for constant values of ϵ and σ,

$$\frac{\partial \rho}{\partial t} + \frac{\sigma}{\epsilon}\rho = 0$$

whose solution is

$$\rho(t) = \rho_0 \exp(-\sigma t/\epsilon), \ \rho_0 = \rho(t=0)$$

This result shows that the charge density decays exponentially with time. Thus, with values of $\epsilon \sim 10^{-12} - 10^{-10}$ Farads/meter, and, provided that σ is not a semi-conductor, the dissipation of charge is very rapid. For conductive materials, it is therefore physically reasonable to set the charge density to zero, and, for problems involving the radiation and/or interaction of electromagnetic waves with good conductors, we can consider the case when

$$\nabla \cdot \epsilon \mathbf{E} = 0 \tag{2}$$

This is a useful result as it allows a relatively simple scalar model to be developed for the emission of an electric field from an antenna (at least for the case when ϵ is taken to be a constant).

2.2. *Model for a Scalar Wave Field*

From Maxwell's equations given above, we can write

$$\nabla \times \nabla \times \mathbf{E} = -\mu \frac{\partial}{\partial t} \nabla \times \mathbf{H} = -\mu \frac{\partial}{\partial t}\left(\epsilon \frac{\partial \mathbf{E}}{\partial t} + \mathbf{J}\right) \quad \text{or} \quad \nabla \times \nabla \times \mathbf{E} + \epsilon\mu \frac{\partial^2}{\partial t^2}\mathbf{E} = -\mu \frac{\partial}{\partial t}\mathbf{J}$$

Thus, noting that $\nabla \times \nabla \times \mathbf{E} = \nabla(\nabla \cdot \mathbf{E}) - \nabla^2 \mathbf{E}$, and, given Equation (2), for constant ϵ, we obtain a wave equation for the electric field with a wave velocity c given by

$$\nabla^2 \mathbf{E} - \frac{1}{c^2}\frac{\partial^2}{\partial t^2}\mathbf{E} = \mu \frac{\partial}{\partial t}\mathbf{J} \ \text{ where } \ c = \frac{1}{\sqrt{\epsilon\mu}}$$

In this case, the electric field is the same for any vector component of the current density. There is no field component coupling and we can therefore work with the scalar wave equation

$$\nabla^2 E - \frac{1}{c^2}\frac{\partial^2}{\partial t^2}E = \mu \frac{\partial}{\partial t}J \tag{3}$$

where E and J are any one of the three components of the corresponding vector fields. The interpretation of Equation (3) is that a rate of change in time of the current density J generates an electric wave field E. This is is a scalar antenna model. It is predicated on Equation (3) and consequently, does not take into account polarisation effects.

For a three-dimensional model of an antenna, the current density is of compact support $\mathbf{r} \in \mathbb{R}^3$. The electric field generated by a rate of change in the current density is taken to exit both inside and outside the domain $\mathbf{r} \in \mathbb{R}^3$. For $\mathbf{r} \notin \mathbb{R}^3$ where $\mathbf{J}(\mathbf{r}) = \mathbf{0}$ this electric field induces a magnetic field. Hence

$$\nabla \times \mathbf{H} = \epsilon \frac{\partial \mathbf{E}}{\partial t}$$

from which is follows (form Maxwell's equations) that the magnetic field $\mathbf{H}$ also adheres to a wave equation, i.e.

$$\nabla^2 \mathbf{H} - \frac{1}{c^2}\frac{\partial^2}{\partial t^2}\mathbf{H} = 0$$

Thus, a rate of change in current density induces an electric wave field which in turn induces a magnetic wave field. The combined fields generate an electromagnetic wave which is a self-propagating wave requiring no medium of support. This is because the oscillation of an electric field yields an oscillating magnetic field which in turn, generates another cycle of the electric field and so on, i.e. each electric and magnetic field component creates the other in a succession that leads to the propagation of an EM field in 'free space' where, for a vacuum

$$c_0 = \frac{1}{\sqrt{\epsilon_0 \mu_0}} \simeq 3 \times 10^8 \text{ ms}^{-1}, \quad c \leq c_0$$

The 'arrival' of an EM field at some 'receiver' antenna with conductivity σ then induces a current density through the electric field component given by Equation (1).

2.3. *Model for Harmonic Oscillation*

From Equation (3), it is clear that to produce an electric field, the current density (the source function) can not be a constant but must undergo some change in time, the characteristic rate of change determining the characteristics of the electric field that is generated. Most antennas utilise a current density source that is oscillatory, in particular, simple harmonic oscillators. In this case we can consider the transformations $J := J\exp(-i\omega t)$ and $E := E\exp(-i\omega t)$ for angular frequency ω. With $\epsilon = \epsilon_0$ and $\mu = \mu_0$, Equation (3) can then be reduced to the inhomogeneous Helmholtz equation

$$(\nabla^2 + k^2)E(\mathbf{r}, k) = -ikz_0 J(\mathbf{r}) \tag{4}$$

where $k = \omega/c_0$ and $z_0 = \mu_0 c_0 \simeq 377$ Ohms is the impedance of free space. Note, that a transformation using $\exp(i\omega t)$ instead of $\exp(-i\omega t)$ is equally valid and only changes the polarity of the source term in Equation (4) and thus, the polarity of a solution to Equation (4).

The harmonic transformation to Equation (4) allows a solution to be formulated through application of the three-dimensional Green's function [10]. For homogenous boundary conditions, when $E = 0$ and $\nabla E = \mathbf{0}$ on the surface of J, this solution is given by [10]

$$E(\mathbf{r}, k) = ikz_0 \int\limits_{\mathbf{s}\in\mathbb{R}^3} \frac{\exp(ik \mid \mathbf{r} - \mathbf{s} \mid)}{4\pi \mid \mathbf{r} - \mathbf{s} \mid} J(\mathbf{s}) d^3\mathbf{s} \equiv ikz_0 \frac{\exp(ikr)}{4\pi r} \otimes J(\mathbf{r}) \tag{5}$$

where $\otimes$ denotes the (three-dimensional) convolution integral over the spatial domain of the current density which is of compact support, i.e. $J(\mathbf{r}) \exists \forall \mathbf{r} \in \mathbb{R}^3$.

Equation (5) provides a general solution for the electric field generated by a (current density) source. It can be appreciated physically by considering a complex of oscillating electrons undergoing a harmonic acceleration given that a free charged particle undergoing an acceleration radiates an EM field. A conductive material is composed of delocalised electrons which are relatively free to oscillate within the crystal lattice of the material. A single electron oscillating at frequency ω represents a (three-dimensional) point source $\delta^3(\mathbf{r})$ which radiates a field described by the Green's function itself, i.e.

$$g(\mathbf{r}, k) = \frac{\exp(ikr)}{4\pi r}, \quad r \equiv \mid \mathbf{r} \mid \tag{6}$$

which is the solution of

$$(\nabla^2 + k^2) g(\mathbf{r}, k) = -\delta^3(\mathbf{r})$$

Thus, the field generated by a complex of oscillating electrons at different points in space (determined by the spatial extent of an antenna) is the sum of all such fields emanating from each point. The sum must therefore be taken over the displaced fields according to the position of all such sources which is the reason why Equation (5) is characterised by a convolution integral. In this context, the current density, in particular, the source function ikz_0J, is that function which represents the spatial source of all such delocalised oscillating electrons in a conductive material as derived from Maxwell's macroscopic equations given in Section 2.1 subject to the condition that $\nabla \cdot \mathbf{E} = 0$. Note that, without this condition, we are required to solve the equation for the electric vector given by

$$\nabla \times \nabla \times \mathbf{E} - k^2\mathbf{E} = -ikz_0\mathbf{J}$$

which requires application of the Green's Dyadic. This model is considered later on in regard to the simulation of the electric field generated by a THz fractal Graphene antenna.

2.4. Intermediate and Far Field Solutions

Given Equation (5), our purpose is to generate an expression for the electric field E that is outside the spatial domain $\mathbf{r} \in \mathbb{R}^3$ of the current density J. There are two standard results that can be formulated in this respect. Both of these results come from applying an approximation to the convolution kernel - the Green's function - in Equation (5).

It is shown in Appendix A, that in the intermediate or Fresnel zone,

$$E(\mathbf{r},k) = \frac{ikz_0}{4\pi r}\exp(ikr/2)\int\limits_{\mathbf{s}\in\mathbb{R}^3}\exp(i\alpha \mid \mathbf{r}-\mathbf{s}\mid^2)J(\mathbf{s})d^3\mathbf{s} = \frac{ikz_0}{4\pi r}\exp(ikr/2)K(\mathbf{r},k)$$

where

$$K(\mathbf{r},k) = \exp(i\alpha r^2) \otimes J(\mathbf{r}) \tag{7}$$

and $\alpha = k/2r,\ r \notin \mathbb{R}^3$ and in the far field,

$$E(\mathbf{r},k) = \frac{ikz_0}{4\pi r}\exp(ikr)\widetilde{J}(\mathbf{k})$$

where, with $\mathcal{F}_3$ denoting the three-dimensional Fourier transform operator,

$$\widetilde{J}(\mathbf{k}) = \mathcal{F}_3[J(\mathbf{s})] = \int\limits_{\mathbf{s}\in\mathbb{R}^3}\exp(-i\mathbf{k}\cdot\mathbf{s})J(\mathbf{s})d^3\mathbf{s},\ \mathbf{k} = k\hat{\mathbf{n}},\ \hat{\mathbf{n}} = \frac{\mathbf{r}}{r} \tag{8}$$

The electric field in the intermediate zone is determined by a convolution integral - Equation (7) - and the electric field in the far zone is determined by a Fourier transform - Equation (8). Thus, in the far field, which is the more usual case in antenna theory, the power distribution of the electric field is given by

$$\mid E(\mathbf{r},\omega)\mid^2 = \frac{k^2z_0^2}{16\pi^2r^2}\mid \mathcal{F}_3[J(\mathbf{r})]\mid^2 \sim 10^{-14}\frac{\omega^2}{r^2}\mid \mathcal{F}_3[J(\mathbf{r})]\mid^2$$

showing that the electric field intensity is limited by the inverse square law but varies as the square of the frequency.

The far field model for radiation emission from an antenna compounded in Equation (8) is the basic model used for analysing antenna designs in order to evaluate the spatial characteristics of the field pattern that they generate. There is a synergy between this model and the principles of Fourier optics where a well corrected and diffraction limited lens is taken to perform a Fourier transform in the focal plane. The Fourier transformation which takes place is of light that has been scattered from an object, the model for the far field scattering being given by the Fourier transform of the object. Although near field analysis of an antenna is possible through application of Equation (7), in most cases, we are interested in antennas that emit and receive an EM field that is some distance away from the source of emission. Hence, a far field analysis is most common.

2.5. Idealised Model for a Patch Antenna

A patch antenna is an antenna with a low profile, mounted on a flat surface. It typically consists of a flat rectangular sheet or 'patch' of conductive material (typically a metal), mounted on a larger dielectric substrate (with a relative permittivity ϵ_r of depth h and dimensions $L \times W$) as illustrated schematically in Figure 3. A second ground plate is then mounted on the same substrate below the patch which produces a resonance generated by what is in effect a micro-transmission line with a length of approximately one-half the wavelength of the EM wave to be radiated. The dielectric substrate must be a stable homogenous insulator, ideally, with as lower relative permittivity as possible. This is because the bandwidth is inversely proportional to relative permittivity [11]. In this context, Silicon Dioxide SiO_2 has and remains the most commonly used insulator. It is the primary gate dielectric material in metal-oxide-semiconductor field effect transistors, for example, and is used in the fabrication of many patch antennas.

If we consider the patch to be infinitely thin in the z-plane, then we can take a patch antenna to represent a source where the current density is given by $J(\mathbf{r})\delta(z)$, $\mathbf{r} \in \mathbb{R}^2$. The effect of this is to reduce Equation (7) to the form,

$$K(\mathbf{r}) = \exp(i\alpha r^2) \otimes J(\mathbf{r}) \tag{9}$$

where $\otimes$ now denotes the two-dimensional convolution integral [ignoring scaling by the phase-only function $\exp(i\alpha z^2)$], and Equation (8) to

$$\widetilde{J}(\mathbf{k}) = \mathcal{F}_2[J(\mathbf{r})] \tag{10}$$

where $\mathcal{F}_2$ denotes the two-dimensional Fourier transform with spatial frequencies $k_x = kx/z$ and $k_y = ky/z$. Thus, for an infinitely thin patch antenna, the electric field pattern in the two-dimensional plane, far from the plane of the patch, is determined by the two-dimensional Fourier transform of the current density. Since x/z and y/z define the tangents in the x and y planes, respectively, it is common to visualise the electric field using a polar plot in which the electric field 'strength' (the power spectrum given by $\mid \widetilde{J}(\mathbf{k}) \mid^2$) is plotted as a function of the angle from the patch plane to the Fourier plane as illustrated in Figure 4. This provides a quasi-physical visualisation of the radiating field pattern which is typically viewed on a logarithmic scale in order to enhance the relative amplitudes of the higher frequency components of the pattern - the 'side-lobes'.

In the case of Figure 4, the field pattern is a colour coded surface representation of the field intensity and includes the current density distribution in the patch plane. This is typical of the graphical representations used to visualise both the current density and the electric field pattern compounded in Equation (10). If the electric field pattern $\widetilde{J}(kx/z, ky/z)$ for a given value of k is known at a distance z from the source, then the current density distribution induced at a receiver is taken to be given by (from Ohm's law) $J(x,y) = \sigma(x,y)\widetilde{J}(x,y)$, $x := kx/z, y := ky/z$. This result does not of course take into account the scaling constant $\sim ikz_0/4\pi z$ associated with the electric field but it does provides a relationship between the electric far field pattern and the induced current density in terms of the two-dimensional conductivity function $\sigma(x,y)$ which defines the 'shape' of a receiver (infinitely thin patch antenna).

Given Equations (9) and (10) where $J = \sigma E$, it is clear that, for a fixed geometry, the greater the conductivity of the materials from which the antenna is made, the larger the amplitude of the electric field that is emitted. Similarly, for a given electric field amplitude at a distance z, the greater the conductivity of the receiving antenna, the higher the induced current density. Given that the electric field amplitude is proportional to the frequency, it is clear that making antennas from highly conductive materials operating at very high frequencies provides for greater electric field strengths and induced current densities. On the other hand, since the physical size of a patch antenna has to be of the order of the wavelength, operating at higher frequencies requires a reduction in their physical size and therefore the power that can be applied - essentially the amplitude of the current density - subject to the conductivity of the material from which the antenna is composed. This leads to the application of Graphene as a suitable material or constructing high frequency antennas in the THz region because Graphene is the most conductive of any room temperature material currently known. While the precise definitions and experimental determinations very, its conductivity is the order of 10^8 S/m which is an order of magnitude greater than the most highly conductive of the elements such a Copper, Silver and Gold, for example, from which conventional antennas are composed - primarily pure copper and copper based amalgam's due to the price of copper being lower.

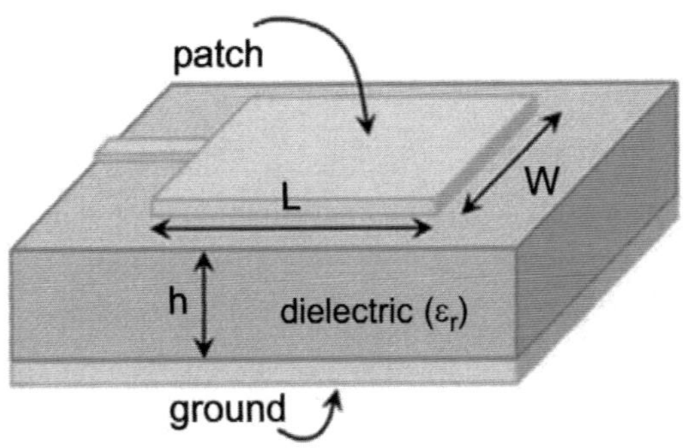

Figure 3. Example of a patch antenna.

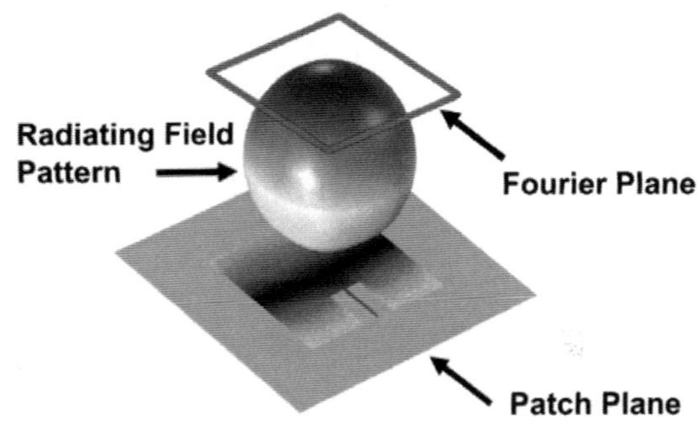

Figure 4. Example visualization of the radiating electric field patterns generated by a patch antenna.

3. Fractal Patch Antennas

A fractal is a self-similar or self-affine function that may be deterministic or stochastic (a statistically self-affine field); it is a field that looks the same or similar at different scales and maybe exactly self-similar at different scales or have an affinity at different scales [12]. In either case, fractals are objects that have a specific fractal dimension, a value that is strictly greater than the topological dimension upon which the fractal is based. Thus, a fractal curve is one with a fractal dimension $D \in (1, 2)$ and a fractal surface has a fractal dimension $D \in (2, 3)$ and so on. As a fractal dimension approaches the upper bound, it becomes more 'space filling' and the 'texture' of the object exhibits greater 'complexity'. This is why the fractal dimension of a field can be used to classify texture which is an otherwise rather elusive term. However, the fractal dimension alone is not a unique measure as different fractals that exhibit different textures can have the same fractal dimension. In this section, we provide a short study of the effect on introducing patch antennas with a fractal structure.

3.1. Scale Invariant Functions and the Characteristic Scaling Properties of Fractals

For some scaling factor L, say, a scale invariant function $f(\mathbf{r})$, $\mathbf{r} \in \mathbb{R}^n$ is one that conforms to the equation

$$L^q f(\mathbf{r}) = f(L\mathbf{r}),\ q > 0 \tag{11}$$

for some choice of exponent q and for any dilation L. A fractal is usually scale invariant but only for a discrete set of values L with translations and rotations often being required to match the fractal with itself at different (discrete) scales. An exact self-similar fractal is one for which $q = 1$.

Deterministic fractals are mostly based on iterative sequences that produce a consecutive repetition of an original object - its shape - to provide a number of scaled down copies of itself.

If, at each iteration, the scale length of an object is reduced by a factor L, and, at each iteration N, copies of the object are made and re-assembled in a coordinated way that is the same at each iteration, then the resulting pattern can be taken to adhere to the equation $NL^D = 1$. The fractal dimension is then given by

$$D = -\frac{\log(N)}{\log(L)} \tag{12}$$

Figure 5 provides an example of this iterative process to produce what is known as a Cantor set. In this case, starting at the top the graphic, we consider a line of length 1 unit. The line is cut into three lines of equal length, one of which is discarded (in the case of Figure 5, this is the central line but it could be the lines to the left or right hand side). If this process is then repeated again and again, then at each iteration, it is clear that $N = 2$ and $L = 1/3$ from which is follows that $D = \log(2)/\log(3) \simeq 0.6309$, i.e. $D \in (0, 1)$. The Cantor set is an example of so called 'fractal dust'. This is because it has a fractal dimension that is less than a topological dimension of 1. Note, that if the line at each iteration is retained instead of deleted, then at each iteration $D = \log(3)/\log(3) = 1$ and the original topological dimension is retained. This is why fractals are referred to as space filling objects as $D \to n$ for a topological dimension $n = 1, 2, 3$.

An important relationship between q, n and D is given by [12]

$$D = \frac{3}{2}n + 1 - q$$

where the range of the values of q is strictly determined by the value of n and the corresponding range of values of D, i.e. $n = 0 \Rightarrow D \in (0, 1)$, $n = 1 \Rightarrow D \in (1, 2)$, ... This relationship applies, for example, to the scaling characteristics associated with the amplitude spectrum $F(\mathbf{k})$ of a fractal $f(\mathbf{r})$ given by [12]

$$F(\mathbf{k}) \sim \frac{1}{|\ \mathbf{k}\ |^q},\ q > 0 \tag{13}$$

Ignoring scaling coefficients, the fractal then has a scaling relationship characterised by by $1/\mid \mathbf{r} \mid^{n-q}$, $n - q > 0$, $\mathbf{r} \in \mathbb{R}^n$ so that for some object function $o(\mathbf{r})$, say,

$$f(\mathbf{r}) = o(\mathbf{r}) \otimes \frac{1}{|\ \mathbf{r}\ |^{n-q}}$$

where the fractal is taken to conform to a linear stationary process, i.e. the sum of all points over which the object function is constructed subject to the scaling property $1/\mid \mathbf{r} \mid^{n-q}$. This is a fundamental 'signature' of a self-affine function, given that

$$f(L\mathbf{r}) = \int_{-\infty}^{\infty} \frac{o(\mathbf{s})}{|\ L\mathbf{r} - \mathbf{s}\ |^{n-q}} d^n\mathbf{s} = L^n \int_{-\infty}^{\infty} \frac{o(L\mathbf{s})}{|\ L\mathbf{r} - L\mathbf{s}\ |^{n-q}} d^n\mathbf{s} = L^q f(\mathbf{r})$$

which recovers Equation (11).

3.2. Why use a Fractal Patch Antenna?

A conventional patch antenna (modeled as a thin and primarily rectangular conductor) has a topological dimension of $n = 2$. The wavelength of the emission/reception is essentially on the same physical scale of the patch. For communications in the GHz or microwave range, this is on the scale of a few centimeters. A wire antenna (with same same electrical conductivity as the patch antenna) with a topological dimension of $n = 1$ and of the same scale will also radiate and EM field but with significant less energy. This is due to the following:

(i) the population density of delocalised electrons is lower;

Figure 5. The Cantor Set with a Fractal Dimension of 0.6309.

(ii) it is not possible to develop a 'line resonator' to produce a comparable power output.

Thus, the first point to note about a fractal patch antenna where $D \in (1, 2)$ is that the energy of an emission will be less; there is, by the very nature of a 'fractal patch', less conductive material to source an electric field. On the other hand, the self-affine structure of a fractal antenna provides, what is in effect, a set of multiple antennas in one, supporting a multitude of frequency bands. To explain this, we refer to Figure 6. This figure shows a Sierpinski and a Tree antenna fabricated from copper. In the former case, the figure illustrates the correlation between the physical size of the Sierpinski triangle and it relationship to a single wavelength of an electric field (emitted or received) and how electric wave fields consisting of $2, 4, 8, 16$ and 36 periods (from left to right of the central graphic) relate to the size of the 'triangles' from which this self-similar structure is composed. The Sierpinski triangle is a deterministic fractal obtained by iteratively replicating an equilateral triangle four times at half the scale length of the original and deleting one of the four resulting triangles (the central component) to leave just three. Hence, at each iteration $N = 3$ and $L = 1/2$, giving a fractal dimension of $D = \log(3)/\log 2 \simeq 1.5850$.

There are a multitude of fractals that can, in principle, be used as patch antennas but, strictly speaking, for them to be patch antennas necessitates the use of fractals such that $D \in (1, 2)$. On the other hand, it is possible to fabricate patch antennas of the type shown in Figure 5 for $D \in (0, 1)$ but only when the current density is taken to exist in the background space to the fractal structure. This is because a continuous current density is required within and over the extent of the antenna. Thus, in the case of Figure 5, the Cantor set describes the 'geometry' of the dielectric substrate upon which the conductor is mounted and not the conductor itself.

In regard to patch antennas for which $D \in (1, 2)$, the choice of which components of the pattern are used to support a current density depends critically on the continuity of the structure so that electrons can oscillate and flow continuously over the fractal structure as a whole. This leads to some important issues that need to be considered in the practical fabrication of such antennas which include:

(i) the number of iterations that can be applied in practice given the material from which the antenna is made, i.e. what are the engineering limits associated with cutting a material into a fractal patch antenna at a specific physical scale?

(ii) the extent to which the fine structure elements of the fractal antenna do not over heat and/or possibly melt for the current densities that are applied.

Points (i) and (ii) are interrelated because in order to maximise the number of multiple bands that a fractal antenna can operate over, the structure must be as fine as possible. However, to optimise the power output from such an antenna, it must be fabricated from a highly conductive material with minimal resistance to the flow and oscillation of electrons, thereby allowing a relatively high current density to be applied which does not lead to over heating and/or burn-out of the fine structure components. These issues lead naturally to the application of Graphene to fabricate such antennas using technologies that are available to cut a Graphene mono-layer to produce a fractal structure using e-beam lithography, for example.

Irrespective of the (conductive) material that is used to fabricate a fractal antenna, such an antenna provides, multi-band support, wide bandwidths and high gains at smaller dimensions. The dimensions of such antennas becomes critical in regard to generating THz emissions since the wavelength of an emitting and receiving antenna is determined by its physical size; the smaller the wavelength, the smaller the antenna. In this context, the fine structure at small scales required to fabricate a THz fractal patch antenna necessitates the use Graphene since it is the most conductive of all known (room temperature) materials.

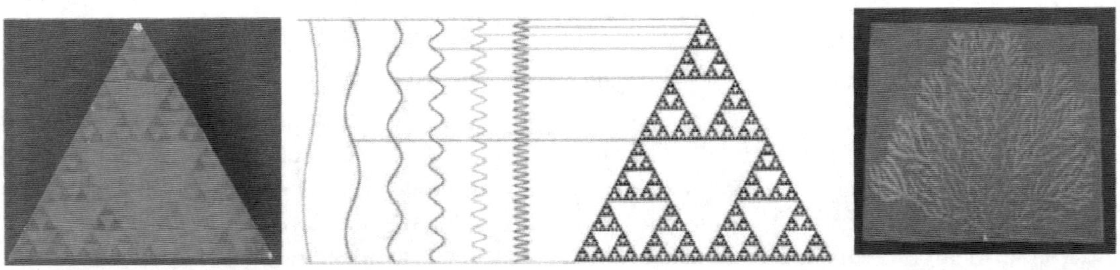

Figure 6. A deterministic Sierpinski triangle antenna (left), a schematic explanation of its multi-band properties (centre) and a further example of a random fractal 'tree' antenna (right).

3.3. Far-Field Patterns Generated by Fractal Patch Antennas

In the far field (specifically the Fourier plane that is parallel to the patch plane), the electric field pattern generated by a patch antenna is characterised by Equation (10), namely, the two-dimensional Fourier transform of the (two-dimensional) current density. It is therefore informative to study the Fourier transform of 'patch fractals' for $D \in (1, 2)$.

Figure 7 shows the radiation far-field patterns (specifically, the logarithmic power spectrum - $\log \mid \widetilde{J}(\mathbf{k}) \mid^2$ - for three patches with a topological dimension of 2 and fractal dimensions of 1.8928 (a Sierpinski Carpet) and 1.5850 (a Sierpinski Triangle). In these examples, it is clear that the power spectra of the fractal structures is broader than the power spectrum associated with a continuous square patch when the (amplitude) spectrum is characterised by a separable sinc function.

Since the power spectrum of a fractal is characterised by the scaling relationship given by Equation (13), for a fractal patch antenna, the log power spectrum is characterised by

$$\log P(\mathbf{k}) = \log \mid F(\mathbf{k}) \mid^2 \sim (2D - 5) \log \mid k \mid, \; D \in (1, 2)$$

Thus as $D \to 2$ the power spectrum is proportionately greater over the equivalent range of $\log \mid k \mid$ indicating that greater power will be emitted by a fractal patch antenna for higher values of the fractal dimension. However, as discussed in the latter part of the previous subsection, this requirement has to take into account the engineering associated with fabricating high fractal dimension fractals, especially those that are to be composed of mono-layer Graphene.

The far-field radiation pattern generated by a fractal patch antenna can be quantified theoretically by analysing the Fourier transformation of a scale invariant function. On the basis of Equation (10), we consider the geometry of the current density of the antenna to be characterised by the equation

$$L^q J(\mathbf{r}) = J(L\mathbf{r}), \; \mathbf{r} \in \mathbb{R}^2, \; q = \frac{5}{2} - D, \; D \in (1, 2)$$

IOP Conf. Series: Materials Science and Engineering 1060 (2021) 012001 doi:10.1088/1757-899X/1060/1/012001

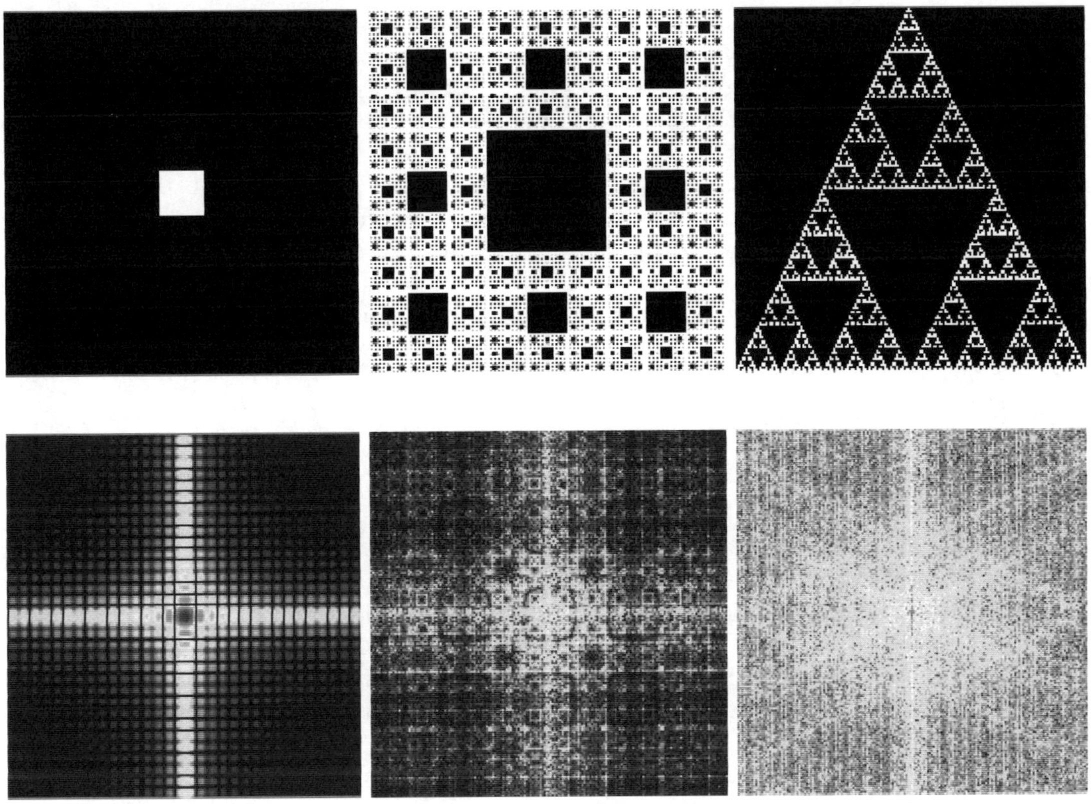

Figure 7. From left to right: Colour coded displays (using the MATLAB 'jet' colour map) of the power spectra (below) plotted on a logarithmic scale for (above) a conventional patch antenna (with a topological dimension of 2), a Sierpinski Carpet ($D = 1.8928$) and a Sierpinski Triangle ($D = 1.5850$).

The electric field is therefore characterized by

$$L^q\widetilde{J}(\mathbf{k}) = L^q\mathcal{F}_2[J(\mathbf{r})] = \mathcal{F}_2[J(L\mathbf{r})]$$

and using the scaling theorem for a two-dimensional Fourier transform, i.e.

$$\mathcal{F}_2[J(L\mathbf{r})] = \frac{1}{L^2}\widetilde{J}\left(\frac{\mathbf{k}}{L}\right)$$

we have

$$L^{2+q}\widetilde{J}(\mathbf{k}) = \widetilde{J}\left(\frac{\mathbf{k}}{L}\right)$$

showing that the electric field is a inverse scaled invariant function, i.e. with $L := 1/L$,

$$\frac{1}{L^{2+q}}\widetilde{J}(\mathbf{k}) = \widetilde{J}\left(L\mathbf{k}\right) \text{ or } L^q\widetilde{J}(\mathbf{k}) = L^{2(1+q)}\widetilde{J}\left(L\mathbf{k}\right)$$

This means that the electric field generated by a fractal patch antenna is itself a re-scaled scale invariant function.

There is an interesting issue in regarding to this result, which is to consider what fractals have the same form in Fourier space. This is similar to the problem associated with conventional functions, namely, classifying those functions that are self-characteristic under Fourier transformation. Although not clearly evident from Figure 7 due to the limited resolution of the image, the Sierpinski triangle is one such fractal whose basic structure appears to be 'reproduce' in Fourier space. Whether this is unique to Sierpinski triangle or otherwise is the subject of further investigation and lies beyond the scope of this publication.

Figure 7 is illustrative of a feature of fractal antennas which is that their side lobe emission (i.e. the power output at wide angles) is significant compared to a square patch antenna. This means that such antennas are not directional which is a potential weakness for long range communications. However, given that in the THz range, such antennas are taken to be operable in the near field, this issue is not so relevant.

4. Introduction to Graphene

Graphene, specifically mono-layer Graphene, consists of a single layer of carbon atoms arranged into a two-dimensional hexagonal-based honeycomb lattice and can be considered as a basic building block for all Graphitic materials of other dimensions [13]. First discovered in 2004 at the University of Manchester [14], it is currently the only known two-dimensional (one-atom-thick) crystalline form of matter [15]. Each cell of the lattice has a delocalised electron cloud above and below the plane. This allows electrons that are introduced to the plane of the lattice to 'skate' freely over the surface with minimal resistance. Thus, electron transport measurements show that Graphene has a remarkably high electron mobility at room temperature, with reported experimental values in excess of $200,000$ cm^{2}V^{-1}s^{-1} at electron densities of 2×10^{11}cm^{-2} [16] with a conductivity of $\sim 10^8$ S/m.

Graphene is a semi-metal (a zero-gap semi-conductor) with a wide range of potential applications including ballistic transistors, nano-ribbons, ultra-sensitive molecular sensing, super capacitance and Spintronics. Its application to constructing fractal Graphene antennas is a direct result of its conductivity in which electrons can flow and oscillate with out the intrinsic resistance that occurs in other materials, thereby providing EM fields with relatively greater energy densities, given the fine structure of the fractal that is required (to produce of multiple frequency channels) and the physical size of a fractal antenna necessary to operate in the THz range.

On a theoretical basis at least, two-dimensional Graphite has been studied since the late 1940s and is widely used for describing properties of various carbon-based materials. An interesting property of Graphene is that the charge carriers are relativistic. Even though they are not actually traveling at relativistic speeds, their structural interaction causes them to act like two-dimensional Dirac Fermions with zero rest mass at the Dirac points on the six comers of the hexagon although this pseudo-relativistic description is restricted to the chiral limit. Fermions are (relativistic) particles that are described by a Spinor field and electrons are taken to have a spin of 1/2. Thus, the charge carriers in Graphene have a unique property in that they mimic relativistic particles as described by the Dirac equation rather than the Schrödinger equation [17]. The relativistic-like description of electron waves on honeycomb lattices has been considered on a theoretical basis for many years and is now being applied in an attempt to understand the electronic properties of Graphene since the experimental discovery of the material [18], [19].

To understand why this is the case let us consider a particle based explanation. As is well known, a Carbon atom has four electrons that are 'free' to pair-bond with the outer most electrons of other molecules, including other carbon-based molecules. This is why Carbon can be used to produce long chain molecules. In (mono-layer) Graphene, each Carbon atom is electron pair bonded to three other Carbon atoms and not four. Thus, each Carbon atom in Graphene has one 'free' or 'delocalised' electron, and, since each hexagonal cell of the lattice consists of six Carbon atoms, we can associate a total is six delocalised electrons with each cell, three of which lie below the plane of the Graphene and another three above the plane. In either plane, we can imagine three electrons (visualised as particles) 'rotating' around the (hexagonal) cell in their 'failed attempt' to pair-bond with another electron. Coupled with the fact that each of the three electrons repels the other, and, that this repulsion includes the electrons in all of the six neighbouring cells, the free electron in each cell continue to 'rotate' at a faster and faster rate obtaining a velocity of the order of 70% the speed of light. This is why the (free) electrons in Graphene are taken to be (near) relativistic particles. However, as always in quantum mechanics, a particle model is not a statement of reality but a liability of fantasy and we are required to visualise the electrons in the plane of a Graphene cell as an electron Cloud as illustrated in Figure 8.

If we consider the combined effect of all the electron clouds for each hexagonal cell in Graphene to be a continuum, then we can consider the planes above and below the Graphene to be composed of an electron fluid, specifically a Dirac or Fermionic fluid (due to the near relativistic speeds of the electrons in the fluid). Like other fluids, this electron fluid can support waves if it is 'pumped' by an external source. If the frequency of this pump is compatible with the physical extent of the Graphene surface (a Graphene ribbon with a finite length and width) then, in principle, it is possible to generate a resonance effect. This is discussed later on in this work in regard to using Graphene to generate THz radiation.

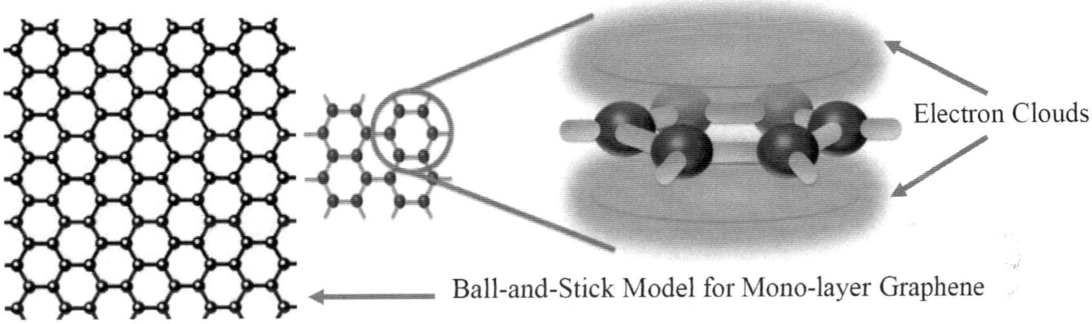

Figure 8. Schematic illustration of the electron clouds (each consisting of three electrons) that exist above a below the plane of mono-layer Graphene.

5. Conductivity of Graphene

While the theoretical 'relativistic properties' of Graphene remains to be fully correlated with experimental evidence, the non-relativistic properties of the material remain a subject of research in terms of (non-relativistic) incident electrons scattering from a two-dimensional electrostatic lattice composed of a regular sequence of localised potentials. These potential are derived from the delocalised electrons associated with the hexagonal nature of the primary structure of Graphene and are replicated throughout the lattice. The delocalised electron cloud represent a repulsive electrostatic potential with respect to any (free) electron propagating over the surface of the material due to the application of an electric field between a source and a drain. These free electrons weakly interact with each delocalised electron cloud scattering from the cloud. This is an effect which ultimately limits the conductivity of Graphene and introduces a frequency dependence.

The electrical conductivity of a material depends on how free the electrons are able to move through the lattice of the material. On the basis of this concept, it is clear that the greater the electron Scattering Cross Section (SCS) is, the lower the conductivity will become. Developing accurate conductivity models for Graphene is a continuing field of research and, to date, there appears to no generalised all frequency model available. However, this is the case with many other materials, especially semi-conductors and meta-materials, for example. Nevertheless, a specific model for the conductivity of the material is crucial in the design and simulation of a Graphene based antenna, fractal or otherwise. This is because, given Equation (4), the source term is frequency dependent as is the current density itself if the conductivity of the material is also frequency dependent when $J = \sigma(\omega)E$. Thus, the purpose of this section if to briefly review the basis for developing models for the conductivity of Graphene in order to assess the frequency dependence of the source term in Equation (4).

5.1. Drude Model

The Drude model is a generic model for the conductivity of a material [20], [21]. For a time independent electric field with angular frequency ω_0, the model is compounded in the result

$$\sigma(\omega) = \frac{\sigma_0}{1 - i\omega\tau}$$

where τ is the mean free time between collisions (the 'relaxation time') with the delocalised electrons of the material and $\sigma_0 = ne^2\tau/m$ is the equivalent conduction of a direct current conforming to Ohm's law, e, n and m being the electron charge, electron number density and electron mass, respectively. For high frequencies, when $\omega\tau >> 1$, $\sigma \sim i\sigma_0/\omega\tau = ine^2/m\omega$. In this case, Equation (4) is reduced to the form

$$(\nabla^2 + k^2)E = -J \tag{14}$$

where $J = \mu_0 ne^2 E/m \sim 3.5 \times 10^{-14} nE$. The electron number density of Graphene varies quadratically with temperature but at room temperature (= 300 °K), $n \simeq 8.5 \times 10^{14}$ m^{-2} [22] and thus, $J \simeq 30.45E$.

Compared with Equation (4), the source term in Equation (14) is frequency independent so that the efficiency of the antenna remains the same for any frequency. This is important for a fractal antenna that, through its geometric configuration, can support many different frequency bands. It means that the limit of the number of bands for which the antenna is designed is limited only by the fine structure of the fractal that can be fabricated in practice and not by a reduction in the conductivity of the material with the frequency of the band. This result is predicated on the conductivity being inversely proportional to the frequency in the high frequency limit; ideally applicable in the THz range.

More advanced models for the (surface) conductivity of mono-layer Graphene have been considered in [23], for example. In this case, a Drude-type approximation for σ is obtained that is valid for mid-infrared wavelengths, where the intra-band conductivity term (due to transformations within electronic states) usually dominates over the inter-band term (describing the transformation between electronic bands, specifically from valence to conduction bands). An approximation for σ in the range 0.01-100 THz is then given by

$$\sigma \simeq \frac{e^2\mu_c\tau}{\pi\hbar^2(1 - i\omega\tau)} \sim \frac{i\sigma_0}{\omega}, \ \omega\tau >> 1; \ \ \sigma_0 = \frac{e^2\mu_c}{\pi\hbar^2}$$

where μ_c is the carrier mobility $\sim 2 \times 10^3$ and $\hbar$ is the Dirac constant. In this case, the relaxation time is given by

$$\tau = \frac{\mu_c\mu_f}{ev_f^2}$$

where μ_f is the Fermi energy (the Chemical Potential) and v_f is the Fermi velocity of Dirac fermions in Graphene ranging from $8.5 \times 10^5 - 3 \times 10^6$ ms^{-1}.

IOP Conf. Series: Materials Science and Engineering 1060 (2021) 012001 doi:10.1088/1757-899X/1060/1/012001

5.2. Quantum Mechanical Models

One important assumption associated with the Drude model is that the electrons move in straight lines between (multiple) collisions. This is known as the free electron approximation and is equivalent to applying a point scattering model for electrons in the context of the quantum theory of conductivity. In this case, the conductivity is taken to be a macroscopic property of the SCS of electrons as they flow through the material. As the population density of scattering sites decreases, one can expect the back-scattering component associated with the SCS to decrease, which is analogous to the mean free time becoming larger, thereby producing a higher conductivity. In this context, if the forward-scattering component of the SCS is large and the back-scattering component is low, then the material will be a good conductor. For electron-electron interactions, the scattering cross-section is determined by the scattered wave function, in particular, the square modulus of the scattered wave function.

For $\mathbf{r} \in \mathbb{R}^n \mapsto V(\mathbf{r})$, the wave function $\psi(\mathbf{r}, k)$ of an electron (for the non-relativistic case) can be taken to conform to Schrödinger's equation, given by

$$(\nabla^2 + k^2)\psi(\mathbf{r}, k) = V(\mathbf{r})\psi(\mathbf{r}, k) \tag{15}$$

where $k^2 = 2m\omega/\hbar = 2mE/\hbar^2$, $E = \hbar\omega$ is the electron energy (for a wave function with angular frequency ω where $\hbar \simeq 1.054571817 \times 10^{34}$ Js is the Dirac constant) and $V(\mathbf{r}) := 2mV(\mathbf{r})/\hbar^2$ is the quantum scattering function; an asymptotic potential where $V(\mathbf{r}) \to 0,\ r \to 0$.

Consider a solution for $\psi(\mathbf{r}, k)$ which is given by the sum of an incident ψ_i and a scattered ψ_s fields, i.e.

$$\psi(\mathbf{r}, k) = \psi_i(\mathbf{r}, k) + \psi_s(\mathbf{r}, k) \text{ where } \frac{\|\psi_s(\mathbf{r}, k)\|}{\|\psi_i(\mathbf{r}, k)\|} << 1,\ \|\psi\| \equiv \left(\int \mid \psi(\mathbf{r}, k) \mid^2 d^n\mathbf{r}\right)^{\frac{1}{2}}$$

If we consider the incident field to be a unit amplitude plane wave given by $\psi_i(\mathbf{r}, k) = \exp(ik\hat{\mathbf{n}}_i \cdot \mathbf{r})$ where $\hat{\mathbf{n}}_i$ is a unit vector which points in the direction of the incident field, then Equation (15) has the approximate form

$$(\nabla^2 + k^2)\psi_s(\mathbf{r}, k) = V(\mathbf{r})\psi_i(\mathbf{r}, k) \tag{16}$$

This is known as the Born or weak scattering approximation and assumes that multiple scattering effects are negligible. This requires that $\|V(\mathbf{r})\| << 1$, i.e. $V(\mathbf{r})$ is a 'weak scattering potential'. The intensity of the wave function $\mid \psi_s \mid^2$ is the SCS.

The general solution (i.e. the Green's function transformation) to Equation (16) for the scattered field is [10]

$$\psi_s(\mathbf{r}, k) = -g(r, k) \otimes [V(\mathbf{r})\psi_i(\mathbf{r}, k)] \tag{17}$$

where $\otimes$ denotes the convolution integral over $\mathbf{r} \in \mathbb{R}^n$ and $g(r, k)$ is the free space Green's function given by the solution to

$$(\nabla^2 + k^2)g(\mathbf{r}, k) = -\delta^n(\mathbf{r})$$

The Probabilistic Current Density (PCD), which is the quantum equivalent to the current density associated with an electric field, is given by

$$\mathbf{J} = \frac{\hbar}{2im}[\psi^*(\mathbf{r}, k)\nabla\psi(\mathbf{r}, k) - \psi(\mathbf{r}, k)\nabla\psi^*(\mathbf{r}, k)]$$

and with $\psi = \phi \exp(iS/\hbar)$, where ϕ and S are real functions of $\mathbf{r}$ and k, we can reduce this expression for the PCD to the form

$$\mathbf{J} = \mid \psi \mid^2 \frac{\nabla S}{m}$$

This result provides a quantum equivalence to Ohm's law in the electromagnetic context where ∇S and $\mid \psi \mid^2$ are equivalent to an electric field and the conductivity, respectively. In this context, $\mid \psi \mid^2$ characterises the Quantum Conductivity (QC) which depends on the scattering potential coupled with the topological dimension over which ψ is evaluated.

Integrating over $\mathbf{r} \in \mathbb{R}^n$, we note that

$$\|\psi\|^2 \leq \|\psi_i\|^2 + \|\psi_s\|^2$$

It is then clear that for the same incident wave (representing an electron introduced into the material from a source, i.e. the value of $\|\psi_i\|^2$) the QC is determined by the SCS alone. We therefore now consider the calculation of of the SCS for $\mathbf{r} \in \mathbb{R}^n$ to derive frequency scaling relationships associated with the QC. The key to this is the functional form of the free space Green's functions that, in addition to the Green's function for $n = 3$ given by Equation (6) as discussed in Section 2.3, for $n = 1$ and $n = 2$ are given by [10]

$$g(x,k) = \frac{i}{2k}\exp(ikx) \text{ and } g(r,k) = \frac{\exp(i\pi/4)\exp(ikr)}{\sqrt{8\pi kr}}, \ kr >> 1$$

respectively. Under the weak scattering approximation, Appendix B derives expressions for the QC using a Coulomb scattering potential. In so doing, we provide scaling relationships between the quantum conductivity and the frequency for $n = 1, 2, 3$.

5.3. Analysis of Results

On the basis of the results derived in Appendix B, it is clear the QC scales with the frequency according to $1/\omega^\alpha$, $\omega > 0$ where, depending on the dimension of the model that is applied $\alpha = 1,\ 1.5,\ 2$ for $n = 1,\ 2,\ 3$, respectively. In this context, the classical Drude model becomes compatible with a one-dimension quantum scattering model, and, on this basis, we can consider Equation (14) to be a model for the electric field generated by a Graphene antenna, whose source term is frequency independent. The application of a Coulomb potential assumes that electron-electron interactions are isolated and further consideration is necessary in regard to what type of (effective) potential an electron encounters when it is incident on a three component electron cloud such as exists in the local vicinity of a Graphene cell.

5.4. Bulk Conductivity

The bulk conductivity is related to the surface conductivity by $\sigma := \sigma/\Delta$, where Δ is the thickness of the Graphene layer which is taken to be much smaller than the excitation wavelength. In this way, Graphene can be modelled as a very thin ($\sim$ nm) three-dimensional material. For a Graphene sheet with relative permittivity ϵ_r, we can write the following equation for the (three-dimensional) electric field

$$(\nabla^2 + k^2\epsilon)E(\mathbf{r},k) = 0, \ \epsilon = \epsilon_r + i\epsilon_i \tag{18}$$

where $\epsilon_i = \sigma/(\epsilon_0\omega\Delta)$ and σ is the surface conductivity (taken to adhere to a Drude model).

The volumetric permittivity, ϵ, is assumed to be that of a Graphene layer with a finite thickness Δ which depends on the surface conductivity. The electric field is then determined primarily by the two-dimensional geometry of the (patch) antenna. In the context of Equation (18), the complex permittivity and the electric field are three dimensional entities. However, for a mono-layer Graphene patch of thickness Δ, only the surface conductivity is relevant, the permittivity over the thickness of the patch being taken to be given by ϵ_r alone. In either case, we can consider the value of ϵ_r to be that of Graphite, when $\epsilon_r \simeq 2.5$. Thus, for three dimensional Graphene models ϵ is anisotropic with out-of-plane and in-plane components ϵ_r and $\epsilon_r + i\epsilon_i$, respectively.

As discussed earlier in this work, Equation (4) is a wave equation for the scalar electric field under the condition $\nabla \cdot \mathbf{E} = 0$. The equivalent equation to Equation (4) for a vector electric field is

$$\nabla \times \nabla \times \mathbf{E}(\mathbf{r}, k) - k^2 \epsilon \mathbf{E}(\mathbf{r}, k) = 0 \tag{19}$$

where we can write ϵ in terms of a complex refractive index with real and imaginary components n and κ, respectively, as $\epsilon = (n + i\kappa)^2$. In the case of Graphene, experimental measurements of n and κ are available in the region 0.15-0.95 μm which depends on the Graphene thickness and the substrate upon which it is mounted [28]. This is the basis for the simulations that are discussed in the following section.

6. Simulations

Simulations are performed for a Sierpinski triangle monopole antenna of nano-metric dimensions. This basic configuration is as follows:

- The Graphene fractal triangle has lateral dimensions of 640nm $\times$ 560nm and a thickness of 3nm;
- the smallest triangle from which the fractal is composed has lateral dimensions that are 1/16 times the 'fundamental triangle';
- the fractal triangle is deposited on top of a 10nm SiO_2 substrate which rests on top of a 10nm thick Si substrate;
- the fractal antenna is fed by a coaxial cable and placed on a ground plane which has a diameter of 800nm and a thickness of 25nm.

Macro-scale Sierpinski triangle monopole antennas of this type are intended to be used for different discrete frequency bands. Nano-scale equivalent antennas are not expected to operate in the same way as the materials have their own properties and cannot be assimilated to be a Perfect Electric Conductor (PEC). At the nano-scale, the volumetric conductivity is described in terms of the real and imaginary components of the relative permittivity, which can be expressed in terms of the components of the (complex) refractive index.

We study a Sierpinski triangle monopole antenna made of Gold, Graphene and Silicon by solving Equation (19) for a complex refractive index in the frequency domain over the range 10-700 THz [29]. A few layers of Graphene are deposited on a nano-metric layer of Silica on a Si substrate. Every material in the nano-structure is modelled through the complex refractive index, the values of n and κ being obtained from both experimental data and theoretical models considered in [30]. The design permits the application of a gate voltage to modify the Graphene surface conductivity over the operating region of 10-300 THz. We compare this design and its operation for different gate voltages using Gold, identifying the differences of using Graphene under the influence of the same gate voltage.

The present dimensions follow the computational constraints associated with the thickness of the fractal that cannot be thicker than 3nm. A better design for the antenna is to scale it up by an order of magnitude or more while retaining the same thickness. Such a design will be able to scan with an order of magnitude smaller frequencies (10 times larger wavelengths). In this work, we model the material properties for wavelengths of 30 - 0.4 μm (10 - 750 THz). This approach differs from similar work in [31], for example, where Graphene is placed underneath a Copper patch, the fractal being on a 100 $\times$ 100 μm scale. Here, we consider a design which extends the frequency range of operation of a fractal antenna by introducing nano-features.

The plasmonic properties of a Graphene nano-antenna provide the ability to tune the antenna in the THz band. This is due to the Graphene surface conductivity that can be modified by applying an electrostatic bias which changes the Fermi level [32]. In [33], Graphene is used as an electrically tuneable load in a nano-scale metal antenna gap. This permits an in-situ control of the antenna frequency. The Graphene patch antenna resonance frequency changes with its width and fractal type which are of growing interest because of their broad-band capabilities for applications using plasmonic metallic materials that generate emissions in the THz frequency (specifically 40 - 750 THz) [34].

6.1. Physical Model

In the model consider here, a fractal Graphene antenna is connected through a coaxial lumped port to a ground plane made of gold. A spherically Perfectly Matched Layer (PML) is then taken to include the Sierpinski triangle, the ground plane and the surrounding volume of air. The bottom of the Sierpinski triangle is flat and connected to a coaxial cable feed. The inner coaxial cable is taken to be gold surrounded by a SiO_2 outer coaxial casing, the ground plane also being taken to be made of gold.

As Graphene is a two-dimensional material, the volumetric conductivity is expressed in terms of the surface conductivity divided by the Graphene layer thickness Δ. The Sierpinski triangle is positioned on a layer of a SiO_2 dielectric on top of a semiconductor Si substrate. Additional matching parts are not required for using the antenna at higher-order resonances with the fractal structure given. The emission and propagation of EM waves is obtained by solving Equation (19) using a Finite Element Analysis (FEA). This spatial frequency domain model is appropriate under the assumption that all material properties are constant with respect to the field strength and the range of frequencies considered. The properties of the materials do not play any role in the macroscopic model, providing the conductivity is high enough to qualify as a PEC. However, at the nano-scale, the material models for n and κ selected in the design change the resonances of the antenna.

For the nano-metric dimensions considered, the second and third-order resonances of the small structures are placed at 320 and 600 THz [29] and we study the case for 10 -1000 THz. In regard to the material properties, for the case of Gold, experimental values for n and κ are provided for different configurations (films of different thickness, evaporated gold, template stripped gold, single-crystal gold) over the range of wavelength 0.3-25 μm [35], [36]. For Silicon, values of n and κ are obtained over the range 0.2-1.45 μm from [37] and [38]. In the case of Graphene, experimental measurements of n and κ are available in the region 0.15-0.95 μm from [28]. These values depend on the Graphene thickness and the substrate.

6.2. Example Results

Example results of the simulations are given in Figure 9 which shows the finite element model for the Sierpinski patch antenna and the Global S parameter for a THz frequency range. In this example, three resonances are found at 35, 110 and 215 THz, the most dominant resonance being for 215 THz. Figure 10 shows the electric field emissions in the far-field at these three resonant frequencies. These distributions are similar to those that would be achieved by a radiator modelled as a PEC except that lower resonance frequencies would be observed due to the properties of the material and structure of the antenna. Depending on the values of n and κ of the material selected for the radiator and the ground plane, these characteristics are changed at the nano-scale.

The thickness of Graphene influences the tuneable characteristics of the antenna and current simulations are dictated by the thickness of the Graphene patch being less than 3 nm. . This determines the minimum size of the mesh for modelling the fractal which must be a fraction of the thickness. To keep the number of mesh elements to a minimum, the dimensions of the antenna have been kept to a minimum. This is the reason for using the THz range in the FEA to generate the example results given in Figures 9 and 10.

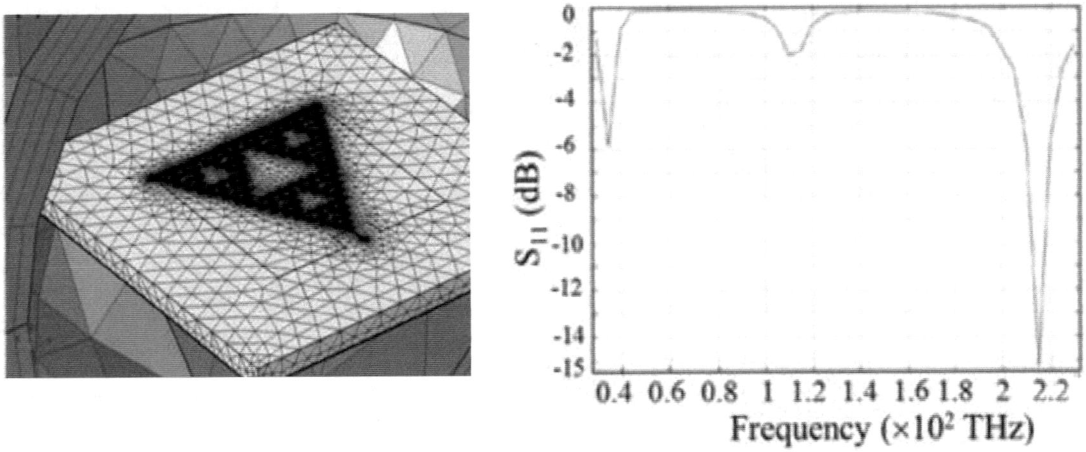

Figure 9. Example simulation for the Sierpinski triangle antenna model (left) showing the Global S-parameter (i.e. S_{11} which is the first element of the S-matrix) for a THz frequency range with three resonances at 35, 110 and 215 THz (right).

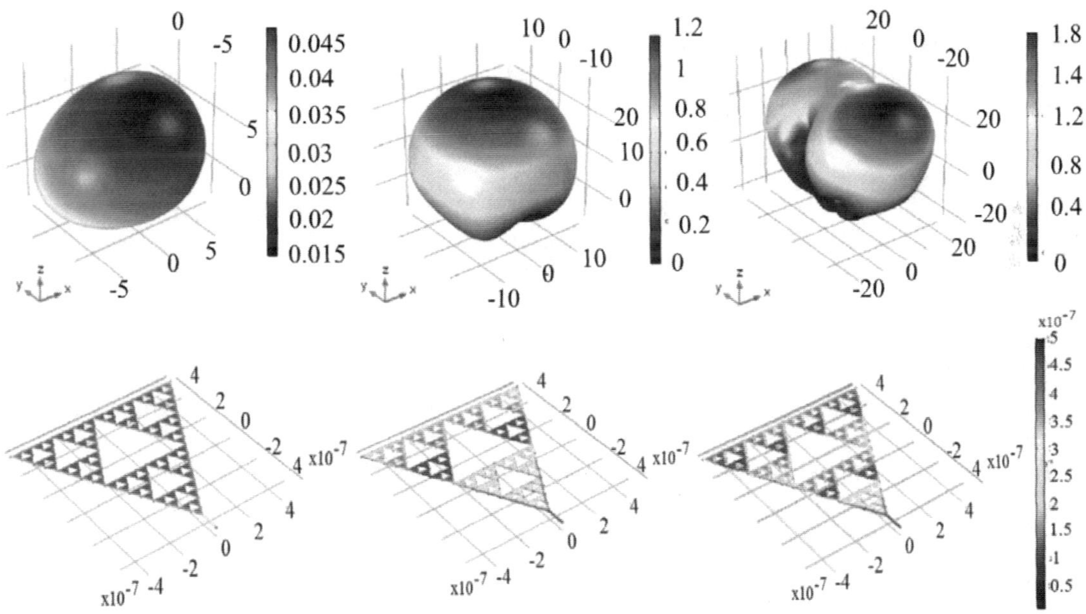

Figure 10. Far electric-field emission patterns (above) and surface electric-field distributions (below) for the Sierpinski fractal patch antenna and the three resonant frequencies, namely, 35, 110 and 215 THz (left to right, respectively).

By further scaling the antenna to operate in the frequency range 10-120 THz, for example, Graphene may have the advantage of being able to tune the refractive index (surface conductivity) if the thickness is approximated to be a single layer (or a few layers) by applying a voltage across the layer and substrates.

The opportunity to model the properties of a Graphene antenna relies on special conditions to be enforced for the current density within the radiator. This can be accomplished by evaluating the complex permittivity of Graphene as the frequency and other parameters are varied and treating the film thickness as an impedance boundary that establishes a relationship between the currents flowing on the film surfaces. This increases the number of degrees of freedom and computational complexity, but, in this way, the total losses through the film can be computed with a more computationally stable discretisation. Moreover, nano-scale effects such as interface roughness can be introduced, evaluated, and correlated with experimental evidence. Hence, back-to-back modelling coupled with experimental activities are necessary to further progress the design of such an antenna.

The example radiation field patterns given in Figures 10 do not exhibit the fine structures associated with the fractal electric fields that will be formed in the far-field as discussed in Section 3.3. This is because of the resolution associated with simulated emissions that are limited within the context of the FEA simulations that have been undertaken.

7. THz Sources using Graphene

Technologies for generating THz sources have been available for many years and include vacuum electronics (the backward wave oscillator) and solid state electronics [39]. For micro- and nano-scale antennas, existing technologies are not suitable as this has a synergy to using a cavity magnetron designed for a microwave oven to feed a patch antenna in a mobile phone, for example. The solution to this problem is to consider Graphene itself as a material to generate a THz source and then integrate that source into a fractal Graphene antenna. To consider how this may be accomplished, the electron fluid model for the delocalised electrons on a Graphene surface as discussed in Section 4 needs to be revisited.

7.1. Electron Fluid Model

Consider the surface of Grphene to be composed of an assembly of electron clouds that collectively constitute an electron fluid. Because the electron clouds are composed of relativistic electrons we refer to this fluid as a Dirac fluid. The fluid represents a negatively charged surface upon which electrons (taken to be introduced by a source) are 'free' to move. For a Graphene patch antenna that is activated by a harmonic source, the electron fluid can then be taken to oscillate at a rate that is compatible with the frequency of the source. The oscillating electrons from which the fluid is composed then generate an EM field at the surface of the Graphene patch. An important feature of this model for a Graphene antenna (fractal or otherwise) is that the energy of the EM emission will not be effected by the frequency of the source if and only if the conductivity of the Graphene varies inversely with the frequency which is the case for $\omega\tau >> 1$ in the THz spectrum assuming a Drude model for the conductivity. In this case, Equation (14) can be taken to be valid for any frequency and thus, for an infinitely thin model, the electric field pattern generated by the patch is determined by the two-dimensional Fourier transform of the patch (as explored in Section 2.5) at any and all frequencies. This is an important reason as to why, on a micro- and nano-scale, Graphene antennas are required for the generation of THz EM fields.

7.2. Graphene THz Resonator

Like any other fluid, an electron fluid can support waves. In the case of Graphene, if an oscillating electric field is present on the surface that is parallel to that surface, then the electron fluid will start to oscillate at the same frequency to produce what is known as plasmon waves, the delocalised electrons in the fluid being, in effect, free negatively charges ions - a cold and strictly negatively charged plasma. So how can we force the electron fluid to oscillate at the

THz range? The answer is to apply an EM field in which the electric field oscillates at the required frequency, namely, in the far infrared (IR) region. This principle is called optical, or more specifically, IR pumping. The far infrared is any radiation with a wavelength of $\sim$ 15 μm to 1 mm (corresponding to a range of about 20 THz to 300 GHz) and is the long wavelength side of the far IR spectrum which overlaps with the THz spectrum.

Suppose we consider a Graphene ribbon mounted on a neutral dielectric substrate and attached to a source (of electrons) and a drain. Taking any plasmon wave motion of the electron fluid between source and drain to be linear, the system will possess a natural frequency that will depends on the length of the ribbon. If the plasmon waves are then optically pumped using an electric field oscillating at a frequency that is the same as this natural frequency, then a resonance will occur. The resonance will produce a standing plasmon wave over the extent of the Graphene strip giving a resonant 'photo-current' which in practice, requires the application of an IR laser. This photo-current will then induce an alternating current in any feed established between the source and drain, a feed that is connected (in a fully integrated sense) to a Graphene (fractal) antenna to generate a THz self-affine EM field. This is the principle by which Graphene can be used to source a fractal Graphene antenna through the stimulated emission of THz radiation with IR-pumping as illustrated schematically in Figure 11.

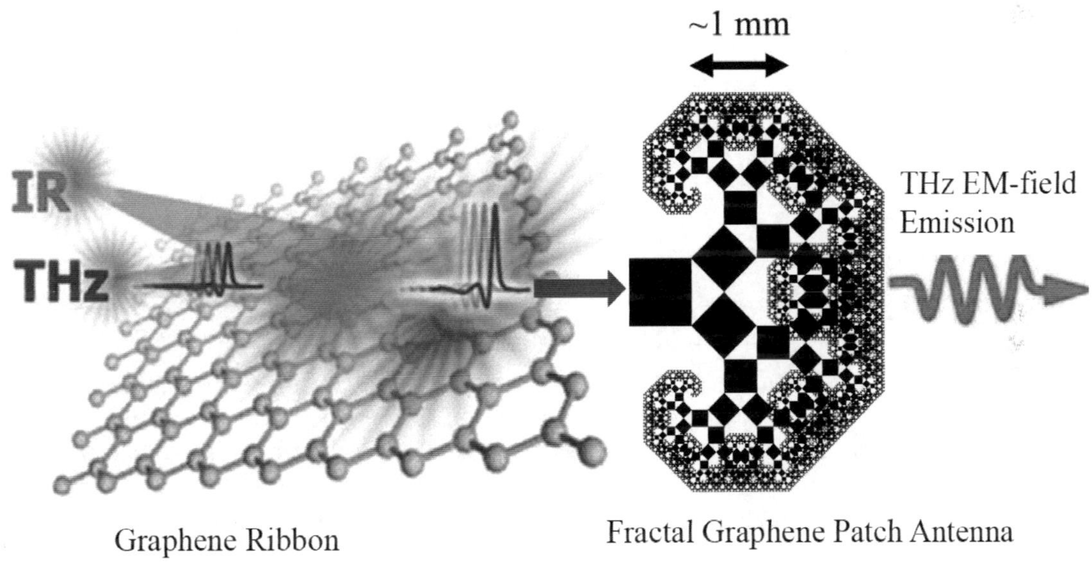

Figure 11. Schematic illustration of a Graphene THz resonator using IR-pumping to source a fractal Graphene antenna.

7.3. Modulation for Optimal Communication

The principle of using a Graphane ribbon to produce a THz resonator discussed in the previous section does not include the issue of how the THz source should be modulated so that information can be communicated in binary form, i.e. as s sequence of pulses, for example. The physics associated with modulating THz radiation from a Graphene source is the subject of current research and lies beyond the scope of this publication. However, the type of modulation that should ideally be applied in order to provide an optimal communications system can be specified. This is because:

(i) efficient micro- and nano-scale methods to generate powerful signals in THz range do not as yet exist thereby limiting THz communication systems to the near field $\sim$ meters;
(ii) the emission and reception of low power THz signals will be subject to distortion by background THz noise, leading to large bit error rates.

Studies on the characteristics of the background noise in the THz range will be undertaken in conjunction with the growth of THz communications systems in the future. However, the key issue is that points (i) and (ii) above necessitate the use of chirplet modulation.

In communications engineering, chirplet modulation has been implemented in a wide range of applications which have their origins in work dating back to the 1950s and 1960s, in particular, the invention and patenting of chirp pulse based communications (e.g. [40], [41]) and forms part of the wireless telecommunications IEEE standards [42]. Chirplet modulation provides an optimal solution to the communication of information through noise. This is why it is used in low power systems where background noise is a major source of bit errors including the noise generated by multiple reflections from the environment in which the system is used (e.g. enclosed spaces). An example of this is 'Bluetooth', a wireless technology standard used for exchanging data between fixed and mobile devices over short distances using short-wavelength radio waves from 2.402 GHz to 2.480 GHz, and, for building personal area networks. It can be shown that, of all the modulation techniques that are currently available or can be conceived, chirplet modulation will always provide the optimal solution for the communication of binary information through noisy channels [43]. Thus, in the development of future THz modulation technologies using micro- and nano-scale Graphene sources, the focus should be on the generation of chirplet modulation/demodulation methods. The best approach to the realisation of THz chirplet modulation is the basis for current and future research.

8. Summary

The principal purpose of this paper has been to investigate the basic physics associated with:

- using a Graphene ribbon to create a THz resonator;
- using a Graphene fractal patch antenna to emit and receive THz radiation.

The physical size of the resonator and radiator is taken to be on the micro- and nano-scales and thereby to yield low power systems that are only suitable for communications in the near field (a few meters at most) but capable of transferring binary data at a rate of Terabits per second using chirplet modulation (assuming the technology for this form a modulation can be realised).

The purpose of developing a Graphene resonator using IR-pumping it to provide a miniature THz source whose physical size and output power is compatible with that of a (patch) antenna radiating a THz EM field (with wavelength on the scale of nm-μm). The purpose of making the antenna from Graphene is because the material is the most (room temperature) conductive material of all and the purpose of using a fractal Graphene antenna is to create a communications system that can operate over many frequency channels.

Under the condition that Graphene adheres to a Drude model for its conductivity and that $\omega\tau >> 1$, the basic model for the emission of a scalar electric field is determined by Equation (14), under the condition that $\nabla \cdot \mathbf{E} = 0$ (and $\epsilon_r = 1$). For an infinitely thin patch, the far-field scalar electric field radiated by the patch is then determined by the (two-dimensional) Fourier transform of the source term which is independent of the frequency and determined solely by the geometrical structure associated with the current density. In this case, for a fractal patch antenna with a fractal dimension of $D \in (1, 2)$, the electric field is given by the Fourier transform of the fractal. In this context, a study on the Fourier transformation of fractals with $D \in (1, 2)$ is required. This is an aspect of fractal geometry that does not appear to have been researched in any significant depth and raises some interesting questions in regard to spectral properties of fractals in general as briefly considered in Section 3.3.

IOP Conf. Series: Materials Science and Engineering 1060 (2021) 012001 doi:10.1088/1757-899X/1060/1/012001

For the case when the Graphene is taken to have a relative permittivity $\epsilon_r > 1$ (which for Graphite is approximately 2.5), Equation (14) must be extended to the form

$$(\nabla^2 + k^2\epsilon_r)E(\mathbf{r}, k) = -J(\mathbf{r}) \tag{20}$$

where, for an applied electric field with amplitude E_0 that is used to generate the (oscillating) current density in the antenna, $J = \mu_0 ne^2 E_0/m \sim 30.45E_0$.

In terms of developing field models that are less conditional, we have considered Equation (19) and presented some example simulations based on a FEA solution to this equation. However, such simulations are currently limited to relatively low resolution models for the fractal. Hence, the far field self-affine radiation field patterns are not resolved in the simulations that have been undertaken to date and presented in this work. The model must be carefully revised to simulate fractal Graphene antennas by using externally performed simulations for the conditions to be enforced in the FEA of Equation (19). It also needs to be evaluated in regard to the relative simplicity of Equation (20) using semi-analytic solutions relative to Equation (19) using FEA.

9. Conclusions

The development of THz communications technology is at the forefront of IT and is an example of a disruptive technology that is set to revolutionise the quantity of digital information in wireless communications and nano-networks. On the basis of the study given in Section 3, we have shown that fractal patch antennas generate fractal EM field patterns in the far field. These field patterns allow for multiple frequency bands to be established and in this context, a fractal antenna may be considered to be a number of antennas in one as illustrated in Figure 6.

We have concluded that a Drude-type model for the conductivity of Graphene is valid. The critical point, is that, if the conductivity is taken to scale inversely with the frequency, then the EM field radiated by a Graphene antenna becomes frequency independent. While further investigations are required on the validity of this scaling relationship using semi-classical and quantum conductivity models of the type considered in Section 5, the general consensus of opinion is that for very high frequencies, the inverse frequency scaling relationship holds reasonably well. Thus, for a Graphene patch antenna, it should be expected that the strength of the field emission is frequency independent.

The electron (Dirac) fluid model described in Section 7.1 provides a useful, if a rather naive physical picture of how Graphene radiates an EM field. The fluid can be considered to oscillate at the same frequency as an applied electric field generated between a source and drain. Similarly, the fluid will oscillate when induced to do so by a incident EM field (the case for a receiving antenna). This is of course the same principle associated with any conductor. However, in a 'normal conductor' the delocalised electrons are significantly more localised than they are in Graphene and are not Fermionic. Because of the special properties of the electrons and their association with a Fermionic fluid, they are often referred to a Graphon's in order to differentiate their properties from plasmon's in other materials. Thus, Graphon's behave like a Dirac fluid and when pumped by an IR laser can be used to establish a resonance in a Graphene ribbon constrained by the physical micro-distance between the source and drain. It is this principle which provides the basis of using Graphene as a THz source.

10. Future Directions

The paper has introduced a number of issues that provide the basis for questions that will drive a range of specific research initiatives. We now consider some examples of such research themes.

10.1. Theoretical Issues

The reader is encouraged to consider the following:

(i) Since the EM far field emission generated by a fractal patch antenna is self-affine, then it is required to investigate what type of fractal antenna should be chosen to optimize the multi-bandwidth properties of a receiver and what type of fractal antenna will produce an EM field which has the same fractal structure. This involves an analysis of fractals in the frequency domain for which remarkably, there appears to be little information [44].

(ii) Given that for a micro-Graphene THz source (with a low power output), the antenna will only be operable in the near field, it is necessary to study the near field emission patterns (in the Fresnel zone) compounded in Equation (9). For the case of an infinitely thin patch antenna model, this will include a study of the field patterns produced by the convolution of $\exp(i\alpha r^2)$, $\mathbf{r} \in \mathbb{R}^2$ with a fractal for $D \in (1, 2)$, an area of interest for which the authors are unaware of any current studies (based on the available open literature).

(iii) The low power rating of the antennas considered necessitates the use of constructing arrays consisting of a regular (or possibly fractal) distribution of individual antennas. Modelling the field patterns of such arrays has not been considered in this work and requires the near field pattern of the array to be analysed through the application of Equation (7). It may then be of value to consider the characteristics of the far field radiation pattern obtained from an array that is itself characterised by a fractal structure, e.g. an antenna of the type shown in Figure 11 where each square component is itself a Graphene fractal.

(iv) A freely suspended Graphene membrane is always partially crumpled because two-dimensional crystals in a three-dimensional space can not be flat due to bending instabilities and thermally induced perturbations. According to the Mermin-Wagner Theorem, 'in one- and two-dimensions, continuous symmetries cannot be spontaneously broken at finite temperatures in systems with sufficiently short-range interactions' [45]. Long-wavelength fluctuations destroy the long-range order of two-dimensional crystals [46] and hence, two-dimensional membranes embedded in a three-dimensional space have a tendency to become crumpled. The spatial characteristics of Graphene therefore include intrinsic defects such as warping, crumpling and rippling. Graphene ripples have various sizes, have a broad distribution and power-law correlation functions of normals together with some typical scale lengths due to the intrinsic tendency of carbon to be bonded. Consequently, there have been a variety of attempts to model the Graphene defects and the defect dynamics of the material [47] and requires further investigations in regard to the efficiency of fractal patch antennas composed of Graphene.

(v) The deformation of Graphene effects the properties of the material and it is therefore important to incorporate this phenomenon in a model for electron transport [48]. This includes the use of fractal surface models on the basis that the deformation of Graphene might be expected to have random self-affine structures [49]. The effect on the conductivity is to change the inverse frequency scaling law associated with a Drude type model to one where the conductivity is proportional to $1/\omega^{\alpha}$, $\alpha > 1$ (for $\omega >> 1$). For the evaluation of the conductivity using quantum electron scattering, a fractal crumpling model can be undertaken by replacing the periodic potential $V(\mathbf{r} + \mathbf{R})$ with $V(\mathbf{r} + \mathbf{R})f(\mathbf{r})$ where $f(\mathbf{r})$ is the (random) fractal, that, unlike the potential, is not periodic. For $\mathbf{r} \in \mathbb{R}^1$, the reflection coefficient (see Appendix B.3) is determined by (using the product theorem)

$$R(k/2) = \frac{i}{k} \int_{-\infty}^{\infty} f(x)V(x + X) \exp(-ikx)dx = \frac{i}{2\pi k}[F(k) \otimes \widetilde{V}(k) \exp(ikX)]$$

where $F(k)$ and $\widetilde{V}(k)$ are the (one-dimensional) Fourier transforms of $f(x)$ and $V(x)$, respectively. Thus, for the scaling relationship associated with Equation (13), we obtain the asymptotic result

$$R(k/2) \sim \frac{2i}{2\pi k}\int\limits_{-\infty}^{\infty}\frac{(\gamma+\ln\mid\xi\mid)\exp(i\xi X)d\xi}{\mid k-\xi\mid^{q}} = \frac{1}{ik\mid k\mid^{q}}\frac{1}{\mid X\mid}, \quad k\rightarrow\infty$$

and we can expect the QC to scale with frequency according to $\sim 1/\omega^{1+q}$, $\omega > 0$ in the high frequency limit.

(vi) The delocalised electrons in Graphene are taken to be relativistic which is the reason why a large population density of Graphon's can be taken to form a Dirac fluid. However, it is arguable that the speed of these electrons ($\sim$ 70% light speed) is not so close to the speed of light as to consider them to be fully relativistic particles. They therefore represent an interesting case when neither a relativistic or a non-relativistic model is ideal and fall into a semi-relativistic regime. A phenomenological (one-dimensional) model for this case has been considered using the equation [50]

$$-i^{1+\alpha}\partial_t^{(1+\alpha)}\Psi(x,t) = -\alpha m^2\Psi(x,t) + \left(\frac{1}{2m}\right)^{1-\alpha}\partial_x^2\Psi(x,t)$$

where, as $\alpha\rightarrow 0$ and as $\alpha\rightarrow 1$, the Schrödinger and the Klein-Gordon equations are recovered, respectively. The wave function Ψ describes spin-less particles and it can be shown that the temporal property of a mass-less Graphon propagator (the time dependent Green's function G) is characterized by $\mid G(0,t)\mid^2\sim 1/t^{1-\alpha}$, $t > 0$. This scaling function is the kernel of the fractional (Riemann-Liouville) integral which is scale invariant. Thus, it may therefore be expected that Graphon propagator's are time fractals.

10.2. Engineering Issues

Issues associated with the engineering of a micro- and nano-scale fractal antennas have not been considered in this paper but, in practice, the antenna would be cut using e-beam lithography after it has been mounted on a neutral substrate. To do this effectively, it is necessary to minimise crumpling when the Graphene is both mounted and then cut with a e-beam. In regard to the latter case, it is required to ascertain the physical scale limits to which a fractal can be cut, subject to maintaining (electrical) connectivity throughout the fractal Graphene patch. Once fabricated, the fractal Graphene patch will need to be inspected. For this purpose, commercially available technology that includes dual E-Beam Lithography (EBL) and Scanning Electron Microscopy (SEM) will be required such as the 'PIONEER Two' nano-fabrication, for example, which combines all maximum performance elements for specialised EBL and SEM into a single comprehensive turnkey system [51].

Irrespective of the technologies that are utilised, to fabricate such an antenna, research on the optimisation of the product is required to provide answers to the following example questions:

- What is the size of mono-layer Graphene that can be used?
- what are the best fractal patch types to use in terms of energy output and the multiple frequency bands expected?
- given a specific fractal type, what is the smallest scale to which the Graphene can be cut without leading to discontinuities and crumpling defects that make the efficiency of the antenna at that scale unacceptable?

In regard to engineering a Graphene resonator by IR laser pumping, research is still at an early stage and the problem of how to provide a resonator that can output chirplet modulated pulses is an area of research that does not appear to have been initiated to date.

IOP Conf. Series: Materials Science and Engineering 1060 (2021) 012001 doi:10.1088/1757-899X/1060/1/012001

10.3. Applications

The quest to fill the 'THz gap' is on-going as it provides one of the last great leaps forward in the field of wireless communications. Those who master the technology first will have short range communications at Tbps in a band that no one else can 'listen in to', until such a time as the 'others' catch up. The applications of the technology is limited only by the near field condition. Given this condition, multi-band Tbps wireless communications will provide a level of data transfer that is unprecedented in the provision of new nano-networks. Such communications systems would not necessarily need to include data encryption and would therefore have significant military applications in regard to localised mobile communications.

Acknowledgments

The authors acknowledge the support of their respective universities and research institutes, to Leonardo Aerospace, Defence and Security, and, in the case of Jonathan Blackledge, the UK Ministry of Defence. The concept of a nano-scale fractal Graphene antenna for THz communications was originally conceived by Jonathan Blackledge and Alberto Boretti through their collaboration at the Military Technological College, Oman.

Appendix A. Intermediate and Far Field Representations of the Electric Field

Given Equation (5), the expressions for the electric field in both the intermediate and far fields are predicated on the binomial expansion of $\mid \mathbf{r}-\mathbf{s}\mid$, namely

$$\mid \mathbf{r}-\mathbf{s}\mid = r\left(1-\frac{2\mathbf{r}\cdot\mathbf{s}}{r^2}+\frac{s^2}{r^2}\right)^{\frac{1}{2}} \simeq r-\frac{\mathbf{r}\cdot\mathbf{s}}{r}+\frac{s^2}{2r}$$

Thus, noting that

$$\frac{r}{2}-\frac{\mathbf{r}\cdot\mathbf{s}}{r}+\frac{s^2}{2r}=\frac{1}{2r}\mid \mathbf{r}-\mathbf{s}\mid^2 \Rightarrow r-\frac{\mathbf{r}\cdot\mathbf{s}}{r}+\frac{s^2}{2r}=\frac{1}{2r}\mid \mathbf{r}-\mathbf{s}\mid^2+\frac{r}{2}$$

we can then write an approximation to Equation (5) given by

$$E(\mathbf{r},k)=\frac{ikz_0}{4\pi r}\exp(ikr/2)\int\limits_{\mathbf{s}\in\mathbb{R}^3}\exp(i\alpha\mid \mathbf{r}-\mathbf{s}\mid^2)J(\mathbf{s})d^3\mathbf{s}=\frac{ikz_0}{4\pi r}\exp(ikr/2)K(\mathbf{r},k)$$

where

$$K(\mathbf{r},k)=\exp(i\alpha r^2)\otimes J(\mathbf{r})$$

and $\alpha = k/2r$. This result provides an expression for the electric field in the Fresnel or intermediate zone. It is an expression for the electric field that occurs at some distance away from the spatial domain in which the current density exists such that $s^2/r^2 << 1$ and is characterised by the convolution with quadratic phase-only function $\exp(i\alpha r^2)$.

If the electric field generated by the current density is taken to be a large distance away from the current density such that $s/r << 1$ then we can consider the case when

$$\mid \mathbf{r}-\mathbf{s}\mid \simeq r\left(1-\frac{2\mathbf{r}\cdot\mathbf{s}}{r^2}\right)^{\frac{1}{2}} \sim r-\frac{\mathbf{r}\cdot\mathbf{s}}{r}$$

from which it follows that a further approximation for the electric field is obtained and Equation (5) is reduced to the asymptotic form

$$E(\mathbf{r},k)=\frac{ikz_0}{4\pi r}\exp(ikr)\widetilde{J}(\mathbf{k}),\ r\to\infty$$

IOP Conf. Series: Materials Science and Engineering 1060 (2021) 012001 doi:10.1088/1757-899X/1060/1/012001

where, with $\mathcal{F}_3$ denoting the three-dimensional Fourier transform,

$$\widetilde{J}(\mathbf{k}) = \mathcal{F}_3[J(\mathbf{s})] = \int\limits_{\mathbf{s}\in\mathbb{R}^3} \exp(-i\mathbf{k}\cdot\mathbf{s})J(\mathbf{s})d^3\mathbf{s},\ \mathbf{k}=k\hat{\mathbf{n}},\ \hat{\mathbf{n}}=\frac{\mathbf{r}}{r}$$

This is an expression for the electric field in the far field of Fourier plane because it is characterised by the Fourier transform of the current density.

Appendix B. Weak Scattering Approximations for a Coulomb Potential for $\mathbf{r}\in\mathbb{R}^n$

It is well known that, under the Born approximation, and, for an incident (unit amplitude) plane wave, Equation (17) reduces to a Fourier transform representation for the scattering amplitude in the far-field [24].

Appendix B.1. Analysis for $\mathbf{r}\in\mathbb{R}^2$

For the two-dimensional case,

$$\psi_s(r,\mathbf{K}) = -\frac{\exp(i\pi/4)\exp(ikr)}{\sqrt{8\pi r}}A(\mathbf{K}),\ r\to\infty$$

where

$$A(\mathbf{K}) = \frac{1}{\sqrt{k}}\int\limits_{-\infty}^{\infty} V(\mathbf{r})\exp(-i\mathbf{K}\cdot\mathbf{r})d^2\mathbf{r},\ \mathbf{K}=k(\hat{\mathbf{n}}_s-\hat{\mathbf{n}}_i)$$

The unit vectors $\hat{\mathbf{n}}_i$ and $\hat{\mathbf{n}}_s$ denote the directions of the incident and scattered electrons (in the Graphene plane), respectively.

The electron scattering potential (i.e. the electron clouds associated with each hexagonal cell in the lattice of Graphene) is regularly spaced in the plane; it is a periodic potential. Thus, the scattering potential is given by $V(\mathbf{r}+\mathbf{R})$ where $\mathbf{R}$ is the lattice vector, $R\equiv\mid\mathbf{R}\mid$ being the radial distance from the centre of one Graphene cell to another. Using the shift theorem, the scattering amplitude is therefore given by

$$A(\mathbf{K}) = \exp(i\mathbf{K}\cdot\mathbf{R})\frac{1}{\sqrt{k}}\int\limits_{-\infty}^{\infty} V(\mathbf{r})\exp(-i\mathbf{K}\cdot\mathbf{r})d^2\mathbf{r}$$

Thus, the effect of electron scattering in the Graphene plane can be investigated by evaluating the scattering cross-section for a single cell alone - the equivalent of Bloch's Theorem [25], [26].

Using polar coordinates (r,ϕ), the scattering amplitude becomes

$$A(\hat{\mathbf{n}}_s,\hat{\mathbf{n}}_i) = \frac{1}{\sqrt{k}}\int\limits_{-\pi}^{\pi}\int\limits_{0}^{\infty} V(\mathbf{r})\exp(-ik\mid\hat{\mathbf{n}}_s-\hat{\mathbf{n}}_i\mid r\cos\phi)rdrd\phi$$

where, for a scattering angle in the plane denoted by θ, $K=k\mid\hat{\mathbf{n}}_s-\hat{\mathbf{n}}_i\mid=2k\sin(\theta/2)$. Thus, using the definition of a zero-order Bessel function J_0, we can write

$$A(K) = \frac{2\pi}{\sqrt{k}}\int\limits_{0}^{\infty} V(\mathbf{r})J_0(Kr)rdr\ \text{ where }\ J_0(Kr)=\frac{1}{2\pi}\int\limits_{-\pi}^{\pi}\exp(-iKr\cos\phi)d\phi$$

For a radially symmetric scattering potential $V(r)$, the scattering amplitude is then determined by the zero-order Hankel transform, i.e.

$$A(K)=\frac{2\pi}{\sqrt{k}}\mathcal{H}_0(K) \text{ where } \mathcal{H}_0(K)=\int\limits_0^\infty V(r)J_0(Kr)rdr$$

Consider the case when the electron-electron scattering over the plane of mono-layer Graphene is characterised by a Coulomb potential for a point charge when $V(r)=e/r$. In this case, the Hankel transform is given by e/K, and the scattering amplitude is

$$A(\theta,k)=\frac{\pi e}{k^{3/2}\sin(\theta/2)}$$

This result illustrates that the scattering cross-section is dominated by forward scattering when $\theta=0$ and that the quantum conductivity is proportional to $1/\omega^{3/2}$, $\omega>0$.

Appendix B.2. Analysis for $\mathbf{r}\in\mathbb{R}^3$

Repeating the analysis provided in the previous section for the case when $\mathbf{r}\in\mathbb{R}^3$ for a spherically symmetric potential (and spherical polar coordinates), the scattering amplitude becomes

$$A(\theta)=\frac{2\pi}{k\sin(\theta/2)}\int\limits_0^\infty \sin[2kr\sin(\theta/2)]V(r)rdr$$

In this case, if we use a Coulomb potential, the integrand does not converge as $r\to\infty$. To solve this problem, we consider the screened Coulomb potential $V(r)=e\exp(-ar)/r$, $a>0$ for which the integral can be evaluated and let $a\to 0$, the scattering amplitude being given by

$$A(\theta)=\frac{\pi e}{k^2\sin^2(\theta/2)}\left(1+\frac{a^2}{[2k\sin(\theta/2)]^2}\right)^{-1}\simeq\frac{\pi e}{k^2\sin^2(\theta/2)},\ a\to 0$$

It is then clear that quantum conductivity is proportional to $1/\omega^2$, the angular dependence of the scattering cross-section being characteristic of Rutherford scattering [27].

Appendix B.3. Analysis for $\mathbf{r}\in\mathbb{R}^1$

In one-dimension, the scattered field (specifically the back-scattered field determined by 'Reflection Coefficient' R) is given by

$$\psi_s(x,k)=-\exp(ikx)R(k) \text{ where } R(k)=\frac{i}{2k}\int\limits_{-\infty}^{\infty}V(x)\exp(-2ikx)dx,\ x\to\infty$$

Thus, for a Coulomb potential, when $V(x)=e/\mid x\mid$, and, noting that

$$\int\limits_{-\infty}^{\infty}\frac{\exp(-ikx)dx}{\mid x\mid}=-2(\gamma+\ln\mid k\mid)$$

where $\gamma\simeq 0.57721$ is the Euler-Mascheroni constant, the Reflection Coefficient is given by

$$R(k)=\frac{ie}{k}(\gamma+\ln\mid 2k\mid)\ \Rightarrow\ \mid\psi_s\mid^2=\frac{\hbar e^2}{2m\omega}\left(6.1489+\frac{1}{2}\ln\omega\right)^2\sim\frac{3\hbar e^2}{m\omega},\ \omega\to\infty$$

This result includes a frequency scaling relationship compatible with the Drude model discussed in Section 5.1. The 'Conductance Quantum' (i.e. the quantized unit of electrical conductance) is given by $e^2/\pi\hbar$ S, and thus, for an electron number density n (i.e. the number of electrons per unit length) we can write the electrical conductivity as

$$\sigma = \frac{n}{\pi\hbar} \mid \psi_s \mid^2 \sim \frac{3ne^2}{\pi m\omega} \text{ S/m}$$

For a Graphene ribbon in which the length of the lattice is significantly larger than its width, we may expect a one-dimensional model to be more significant. This argument also applies to the fine structures associated with a fractal composed of Graphene.

References

[1] Wallace P R 1947 The Band Theory of Graphite, *Physical Review* **71**(9) 622–634 https://journals.aps.org/pr/abstract/10.1103/PhysRev.71.622
[2] Baird D Hughes R I G and Nordmann A (Eds.) 1998 *Heinrich Hertz: Classical Physicist, Modern Philosopher* Springer-Verlag, ISBN 0-7923-4653-X
[3] Maxwell J C 1865 A Dynamical Theory of the Electromagnetic Field, *Philosophical Transactions of the Royal Society of London* **55** 459–512 https://royalsocietypublishing.org/doi/10.1098/rstl.1865.0008
[4] Sizov F and Rogalski A 2010 Review of THz Detectors *Progress in Quantum Electronics* Elsevier **34** 278–347
[5] Novoselov K S, Geim A K, Morozov S V, Jiang D, Katsnelson M I, Grigorieva I V, Dubonos S V and Firsov A A 2005. Two-dimensional Gas of Massless Dirac Fermions in Graphene. *Nature* **438**(7065) 197–200 https://www.nature.com/articles/nature04233
[6] Mikhail M et al Optical Pumping in Graphene-based Terahertz/Far-infrared Superluminescent and Laser Heterostructures with Graded-gap Black-P_xAs_{1-x} Absorbing-cooling Layers 2019 *Optical Engineering* **59**(6) 061606 https://doi.org/10.1117/1.OE.59.6.061606
[7] Ju et al 2011 Graphene Plasmonics for Tunable Terahertz Metamaterials *Nat Nano* **6** 630-634
[8] Llatser et al 2012 Graphene-based Nano-patch Antenna for Terahertz Radiation, *Photonics and Nanostructures - Fundamentals and Applications* **10** 353-358
[9] Zakrajsek et al 2016 Lithographically Defined Plasmonic Graphene Antennas for Terahertz-band Communication, *IEEE Antennas and Wireless Propagation Letters* **15** 1553-1556
[10] Evans G Blackledge J and Yardley P 1999 *Analytic Methods for Partial Differential Equations* Springer Undergraduate Mathematics Series (Springer) ISBN: 978-3-540-76124-2
[11] Khan A Nema R 2012 Analysis of Five Different Dielectric Substrates on Microstrip Patch Antenna *International Journal of Computer Applications* **55**(14) 40-47 https://research.ijcaonline.org/volume55/number14/pxc3882905.pdf
[12] Turner M Blackledge J and Andrews P 1998 *Fractal Geometry in Digital Imaging* (Academic Press), ISBN-13: 978-0127039701
[13] Geim A and Novoselov K 2007 The Rise of Graphene *Nat Mater* **6** 183-191
[14] Graphene (The University of Manchester) https://www.graphene.manchester.ac.uk/about/
[15] Katsnelson K 2012 *Graphene: Carbon in Two Dimensions* (Cambridge University Press) ISBN: 978-1107375802
[16] Sikesb et al 2008 Ultrahigh Electron Mobility in Suspended Graphene *Solid State Communications* (Elsevier) **146**(9–10) 351–355
[17] Brey L and Fertig H 2006 Electronic States of Graphene Nano-ribbons Studied with the Dirac Equation *Phys. Rev. B* **73** 235-411
[18] Bardarson et al 2007, One-Parameter Scaling at the Dirac Point in Graphene *Phys. Rev. Lett.* **99** 106-801
[19] , Gi N 2011 *Relativistic Dynamics and Dirac Particles in Graphene* PhD Thesis Massachusetts Institute of Technology
[20] Drude P 1900 Zur Elektronentheorie der Metalle, *Annalen der Physik* **306**(3) 566–613
[21] Drude P 1900 Zur Elektronentheorie der Metalle II: Teil. Galvanomagnetische und thermomagnetische Effecte, *Annalen der Physik* **308**(11) 369–402
[22] Fang et al 2007 Carrier Statistics and Quantum Capacitance of Graphene Sheets and Ribbons *Appl. Phys. Lett.* **91** 92-109
[23] Hanson G W 2008 Dyadic Green's Functions and Guided Surface waves for a Surface Conductivity Model of Graphene, *Journal of Applied Physics* **103**, 064302 https://doi.org/10.1063/1.2891452
[24] Blackledge J 2005 *Digital Image Processing: Mathematical and Computational Methods* Woodhead

Publishing Series in Electronic and Optical Materials (Woodhead Publishing) ISBN-13: 978-1898563495. https://arrow.tudublin.ie/engschelebk/3/

[25] Bloch F 1928 On the Quantum Mechanics of Electrons in Crystal Lattices *Zeitschrift für Physik* **52**(7–8) 555–600 https://link.springer.com/article/10.1007%2FBF01339455

[26] Kittel C 1996 *Introduction to Solid State Physics* Wiley ISBN 0-471-14286-7

[27] Rutherford E 1911 The Scattering of α and β Rays by Matter and the Structure of the Atom *Philosophical Magazine* **21**(6) 669-688

[28] Ochoa-Martinez et al 2015 Determination of a Refractive Index and an Extinction Coefficient of Standard Production of cvd-Graphene *Nanoscale* **7** 1491-1500

[29] Mishra R Ghatak R and Poddar D R 2008 Design Formula for Sierpinski Gasket Pre-fractal Planar-monopole Antennas [antenna designer's notebook] *IEEE Antennas and Propagation Magazine* **50** 104-107

[30] Boretti et al 2020 *A Preliminary Study of a Graphene Fractal Sierpinski Antenna*, 4th International Conference on Materials Engineering and Nano Sciences (ICMENS2020) Pattaya Thailand https://arrow.tudublin.ie/engscheleart/310/

[31] , Xu Y Hu J and Yin W 2013 *Design of a Novel Reconfigurable Sierpinski Fractal Graphene Antenna Operating at the Thz Band* IEEE Antennas and Propagation Society International Symposium (APSURSI) 574-575

[32] Jornet J and Akyildiz I 2013, Graphene-based Plasmonic Nano-antenna for Terahertz band Communication in Nano-networks, *IEEE Journal on Selected Areas in Communications* **31** 685-694

[33] Yao et al 2013 Broad Electrical Tuning of Graphene-loaded Plasmonic Antennas *Nano Letters* **13** 1257-1264

[34] Rosa L Sun K and Juodkazis S 2011 Sierpinski Fractal Plasmonic Nano-antennas *Physica Status Solidi* Rapid Research Letters **5** 175-177.

[35] Gao L Lemarchand F and Lequime M 2011 Comparison of Different Dispersion Models for Single Layer Optical Thin Film Index Determination *Thin Solid Films* **520** 501-509

[36] Olmon et al 2012 Optical Dielectric Function of Gold *Physical Review B* **86** 235147

[37] Green M 2008 Self-consistent Optical Parameters of Intrinsic Silicon at 300K Including Temperature Coefficients *Solar Energy Materials and Solar Cells* **92** 1305-1310.

[38] Chandler-Horowitz D and Amirtharaj P 2005 High-accuracy, Mid-infrared ($450\text{cm}^{-1} \leq \omega \leq 4000\text{cm}^{-1}$) Refractive Index Values of Silicon *Journal of Applied Physics* **97** 123526

[39] Gilmour A Klystrons 2011 *Traveling Wave Tubes, Magnetrons, Crossed-Field Amplifiers and Gyrotrons* (Artech House) 2ISBN 1608071855

[40] Darlington S 1954 *Chirp Pulse Transmission* US Patent 2678997

[41] Darlington S 1969 *Chirp Pulse Equalization*, US Patent 3618095

[42] IEEE Computer Society 2007 *IEEE Standard 802.15.4a-2007* https://ieeexplore.ieee.org/document/4299496

[43] Blackledge J 2020 On the Chirp Function, the Chirplet Transform and the Optimal Communication of Information, *IAENG International Journal of Applied Mathematics* **50**(2) 285-319 https://arrow.tudublin.ie/engscheleart2/218/

[44] Durable Scope 2020 *Frequency Analysis of Fractals with Fourier Transforms: Exploring the Characteristics of Various Fractals using Frequency Analysis Techniques* https://blog.durablescope.com/post/Analysis_of_Fractals_with_Fourier_Transform/

[45] Werner A 2010 *The Mermin-Wagner Theorem* http://itp.uni-frankfurt.de/~valenti/TALKS_BACHELOR/mermin-wagner.pdf

[46] Hohenberg P 1967 Existence of Long-range Order in One- and Two-Dimensions, *Phys. Rev. Vol.* **158**(2) 383-386

[47] Chen et al 2009 Raman Spectroscopy of Ripple Formation in Suspended Graphene *Nano Letters* **9**(12) 4172–4176

[48] Zwierzycki M 2012 Transport Properties of Rippled Graphene, *Acta Physica Polonica A* **121** 1246-1249

[49] Blackledge J and Rebow M 2014 Two-dimensional Schrodinger Scattering and Electron Transport in Graphene *International Journal for Pure and Applied Mathematics* **4**(8) 913-926. https://arrow.tudublin.ie/engscheleart2/108/

[50] Blackledge J and Babajanov B 2013 The Fractional Schrodinger-Klein-Gordon Equation and Intermediate Relativism, *International Journal for Pure and Applied Mathematics (Mathematica Aeterna)* **3**(8) 601 - 615 https://arrow.tudublin.ie/engscheleart2/76/

[51] Pioneer Two Electron Beam Lithography and Scanning Electron Microscope Imaging 2020 https://www.azonano.com/equipment-details.aspx?EquipID=1752

Express ultra-jet diagnostics of defect resistance for surface layer of a material

A A Barzov[1], A L Galinovskiy[2] and M A Prokhorova[2*]

[1] Lomonosov Moscow State University, Moscow, 119991, Russia
[2] Bauman Moscow State Technical University, 105005, Russia

[*] mary.prokhorova.bmstu@gmail.com

Abstract. In the manufacture of products of modern technology, such technologies of waterjet processing as water jet cutting are successfully used for cutting materials. However, the physical and energy basis of ultrajet technology predetermines a much broader scope of its practical applications. Most of modern technical products are made of composite materials. Moreover, some of them were obtained by innovative methods, for instance, selective laser fusion, which complicates their diagnostics. This article proposes an innovative method for determining a new informative parameter characterizing the defect resistance of various materials. This technique is based on the process of hydro-erosive destruction in the process of ultrajet exposure to a high-energy jet on the surface of the material being diagnosed. Ultra-jet diagnostics (UJD) is based on the analysis of the dependence of the characteristics of local hydro-erosion of the diagnosed area of the surface of the object under analysis (OA) on the parameters of its quality. Surface quality parameters include physical and mechanical properties (PMP) and the degree of operational and technological damage. In this article a new formula has been developed that numerically determines the degree of defect resistance of materials. Experimental testing of this technique on some types of modern materials confirms its reliability. The directions of the future development of the research results are formulated.

1. Introduction

Nowadays the development of new models for assessing the operational properties, including the durability, of modern engineering components is one of the most urgent problems of mechanical engineering [1-3]. The additional informational and physical parameters may be required for estimation of the operational properties, primarily related to surface layer of products made of new materials. The dynamic defect resistance or damage under unsteady loading conditions should be attributed among these parameters. It was shown in [4–8] that by analysing the results of exposure to the surface of the investigated material with high-speed Ultra-Jet (UJ), express determination of not only changes in traditional Physical and Mechanical Properties (PMP) and parameters of the surface layer, but also an estimation of their defect resistance is possible. This estimation is carried out on the basis of comparison of informative signs related to local hydro-erosive destruction: the depth of the formed UJ hydro-cavern, the micro-relief of the surface structure, the mass-geometric parameters of eroded particles, *etc.* [9-11].

Defect resistance is not only about strength, stiffness or fatigue strength, but it is about all of them in complex. The simplicity of the technology for the implementation of method for its determination, as well as high productivity due to the speed of action of the UJ on the surface of the diagnosed part, allows quite justifiably to call this method an express diagnostic method.

2. Ultra-Jet Diagnostics Concept

Ultra-Jet Diagnostics (UJD) is a set of methods and means for creating the implementation of such parameters of a high-energy compact liquid jet, which, when it interacts with a diagnosed medium, for example, during shock-dynamic braking on a solid target, can lead to fixed purposeful changes in the processed material and/or in the liquid itself. In this case, a high-energy compact jet of liquid will be called an *Ultra-Jet*. And the resulting depression in the surface of the material as a result of UJD is *hydro-cavern (*or just *cavern)* [12]. Figure 1 depicts hydro-caverns of two different types of materials: abrasive material (Figure 1 a-b) and polymer material (Figure 1 c-d). Photos were taken with an electronic magnifier at 60x magnification. In other words, the hydro-cavern is like a scratch on the surface of the material made by high speed water jet and is referred to as an "Ultra-Jet".

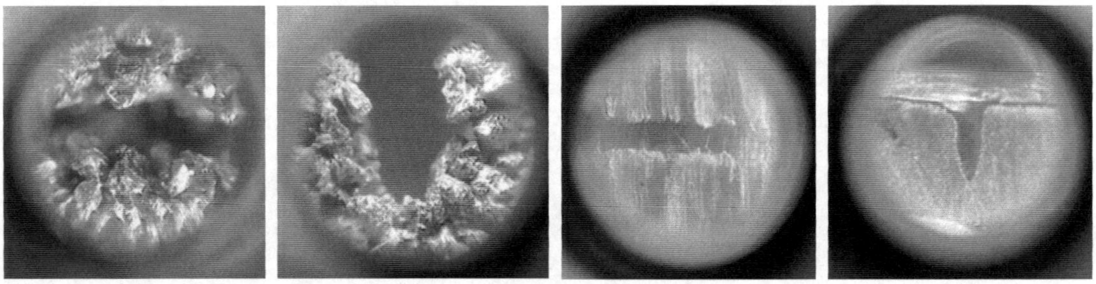

(a) (b) (c) (d)

Figure 1. Illustrations of hydro-caverns of different types of materials:
(a) – top view of a hydro-cavern of abrasive material
(b) – side view of a hydro-cavern of abrasive material
(c) – top view of a hydro-cavern of polymer material
(d) – side view of a hydro-cavern of polymer material.

3. Classic UJD Algorithm

The classic UJD algorithm, according to [12], includes the following steps:

1. The diagnosed surface is exposed to a hydro-jetting effect, which causes hydro-erosion of the surface under certain conditions (working pressure of the jet, its diameter, motion kinematics, *etc.*)
2. Then the parameters of hydro-erosion (particle size of the material, the structure of the material in the place of cut, depth of cut, *etc.*) are investigated and compared with the reference characteristics or with each other on different parts of the surface.
3. The obtained difference in the results of comparisons (absolute and/or relative) is used to judge the quality (resource, physical and mechanical properties) of the controlled area of the surface and the current state as a whole.

4. Suggested diagnostic technique

To increase the identification capabilities of UJ, the following physical and technological technique is proposed. At the first stage, the high-pressured hydro-UJ, which, according to the mechanisms of erosion destruction, carries out softening of its surface layer structure (loss of strength, weakening), is applied to the surface layer of the diagnosed Object of Analysis (AO), *i.e.* forms a sufficiently significant defect damage mainly in the form of a combination of micro- and nano-cracks.

On the second stage, following the previous UJ's action trail, the impact is affected by hydro-abrasive-UJ. Moreover, the effect of its impact will depend not only on hydro-abrasive (micro-fatigue) resistance, according to [13], but also on the degree of previous softening of the surface layer of high-pressure hydro-UJ.

Further, the intact AO's surface layer is exposed to a hydro-abrasive-UJ and the depth of the formed cavern is fixed. It is obvious that the difference between the depths of hydro-abrasive caverns will characterize the defect resistance of the studied material under conditions of energy-extreme hydrodynamic effects on it.

As the first stage of UJD influences the internal and surface structure of the diagnosed material, the new parameter of defect resistance characterizes not only resistance of surface defects but also defects which are close to the surface but hidden in the material structure.

To increase the generality of the obtained comparison results, one can also give additional UJ actions, which follow the hydro-abrasive operation. Schematically, the foregoing is shown in Figure. 2.

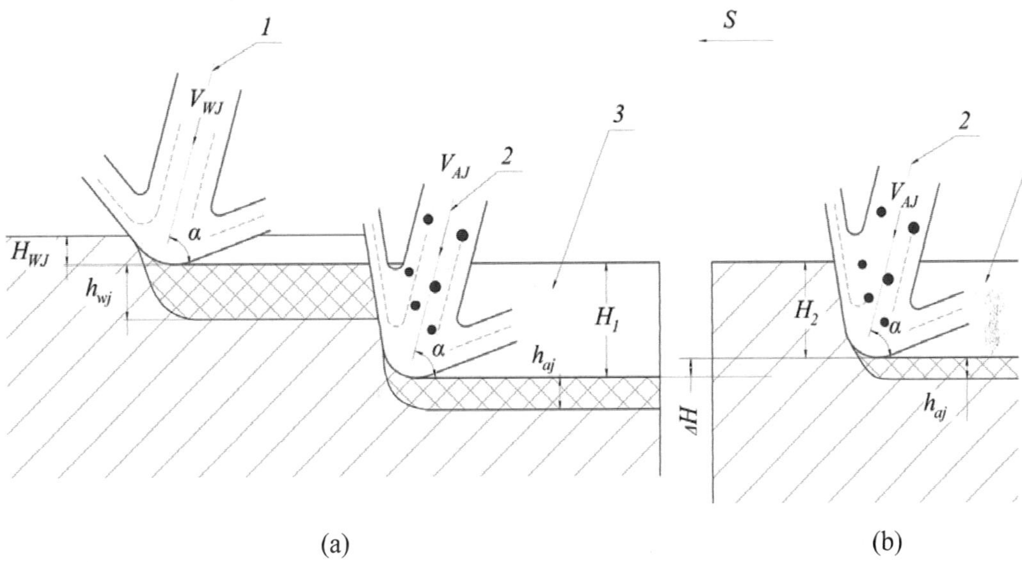

Legend:
1 – diagnostic hydro-UJ
2 – diagnostic hydro-abrasive-UJ
3 – hydro-cavern formed as a result of the first stage of diagnosis
4 – hydro-cavern formed as a result of the second stage of diagnosis
H_1 – cavern depth, formed by the action of hydro-abrasive control system as a result of the 1st stage
H_2 – cavern depth, formed by the action of hydro-abrasive control system as a result of the 2nd stage
H_{WJ} – cavern depth formed by the influence of hydro-UJ at the 1st stage of diagnosis
h_{wj} – surface layer thickness, softened by the action of hydro-UJ
h_{aj} – surface layer thickness, softened by the action of hydro-abrasive-UJ
ΔH – difference between the results of hydro-abrasive actions of the 1st and 2nd stages

Figure 2. UJD implementation scheme: (a) - the 1st stage, (b) - the 2nd stage.

For the mathematical formalization of the foregoing, the following reasoning was carried out. The larger the ΔH value is, the less defect resistance the AO's material has. From a physical point of view, this fact is explained by that after the first operation of hydro-UJ exposure at the 1st stage of diagnosis, the OA's surface layer underwent significant softening, which in turn increased the efficiency of the subsequent hydro-abrasive-UJ exposure. The softening of the surface layer takes place due to the predominance of the influence of the low-cycle fatigue mechanism during the operation of hydro-UJ exposure [14].

At the 2nd stage of diagnosis, the AO's surface was not subjected to preliminary hydro-UJ impact before exposure to hydro-abrasive-UJ, in contrast to the 1st stage, which allows us to consider it as not softened. In order to assess the degree of influence of studied material's softening, we introduce the following formula defining a new property of defect resistance:

$$\tilde{D} = 1 - \frac{H_1 - H_2}{H_2} \quad (1)$$

where: H_1, H_2 – the depths of hydro-cavern after the first and second stages of UJ action onto the surface of the material, mm.

Thus, analysis of the results of successive UJ's impacts makes it possible to single out a new property of materials that are prone to damage of their surface layer under conditions of energy-extreme non-stationary loading, which is commonly to aviation and other similar industries.

It's a remarkable thing that angle alpha (Figure 2) can significantly affect the cavern depth. The most powerful UJ impact is achieved when angle alpha is 45°, which was show in [7, 10]. While the time of UJ's action does not play a significant role. Obviously, the longer the exposure time, the deeper the cavern. However, in our case, the presence of a fast UJ's action is enough.

5. Experimental testing

Table 1 is shown as an example of the implementation of the abovementioned. Table. 1 in generalized form presents materials of UJD for three characteristic representatives of structural materials: high-strength steel RN13-8Mo (HRC=40-50); aluminium lithium alloy V-1469; ceramics based on AL_2O_3 named VO-13; a sample obtained by selective laser fusion (SLF) from brand titanium powder TPP-1 and carbon fibre USAP-AF-LU-V.

Table 1. Defect resistance of various material types

Material	**Identification mark**	$\tilde{D}$**, [%]**
Steel	RN13-8Mo	100
Aluminum	V-1469	94
Ceramics	VO-13	79
SLF of Ti powder	TPP-1	70-76
Carbon fiber	USAP-AF-LU-V	45-60

* The parameters variety for some materials takes place due to the significance of the anisotropy factor of the AO's PMP.

6. Discussion

All in all there were examined for three samples per each type of the diagnosed material. The table 1 shows the average values. The spread of values was less than 5%.

The most defect-resistant material was steel, which confirms its low tendency to softening. Aluminium and ceramics have a lower defect resistance than steel, which is due to the inhomogeneity of their internal structure.

A consequence of the brittleness of ceramics is a significant statistical spread in the values of its strength in comparison with the spread in the strength of metallic structural materials. This is due to the high sensitivity of ceramics to the stress concentration created by defects in the microstructure (pores, inclusions) due to the difficulty of relaxation processes through plastic deformation [15].

TPP-1 and carbon fibre turned out to be the least defect-resistant materials, which is confirmed by the high porosity of these materials, as well as the anisotropy of their properties.

The construction of an empirical dependence of the concentration of defects on their size, as well as the determination of the influence of the concentration of defects of various sizes on the strength of products made of modern materials is an urgent issue [5, 15]. The development of methods for solving these problems could significantly improve the reliability of the proposed method for determining the defect resistance of materials.

7. Conclusion

Based on the results of this study, the following conclusions can be drawn:

1. The fundamental possibility of identifying the tendency of the studied materials to weaken their surface layer under hydrodynamic effects is shown.
2. It was established that, high-strength steel has the maximum defect resistance, and the carbon fiber composite material has the minimum. In this case, a reliable estimation of the surface layer PMP anisotropy is realistic.

The *prospects* for the development of research are the following, in addition to expanding the range of diagnosed materials:

1. To develop a full-scale model that allows to characterize the defect resistance of the material quantitatively by comparing the results of exposure by at least two hydrodynamic effects.
2. Carrying out the UJD operation under changing conditions, in particular during the quality diagnostics of complex-profile parts surface.
3. The estimation of the materials defect resistance or their damage in other types of test and diagnostic impact on them.

Acknowledgments

The studies were carried out under Russian Foundation for Basic Research grants 18-29-18081 and 19-38-90228\19, a grant of the President of the Russian Federation for state support of leading scientific schools of the Russian Federation NSh-3778.2018.8, and grants from the Fund for the Promotion of Innovations under the UMNIK-18 program (v) following the contract No. 14727GU/2019 and No. 14549GU/2019.

References

[1] Jones R, Singh Raman RK, McMillan A J 2018 *Eng Fract Mech.* **187**: 190-210.
[2] McMillan A J and Jones R 2020 *Engineering Reports, In Press*
[3] Yadollahi A, Mhatabi M J, Khalili A, Doude H R, Newman J C Jr 2018 *Fatigue Fract Eng Mater Struct.* **41**: 1602- 1614
[4] Lee S, Hoa V D 2019 *In the book: Gagarin Readings - 2019 Abstracts of the XLV International Youth Scientific Conference.* 854-855
[5] Barzov A A, Galinovsky A L, Provatorov A S 2019 *Innovative mechanical engineering* 51-57.
[6] Galinovsky A L, Mulyar S G, Khafizov M V 2012 *Izvestiya Vysshikh Uchebnykh Zavedenii* Engineering. **No. 9.** 65-69
[7] Abashin M I, Barzov A A, Galinovsky A L, Shuteyev V A 2011 *Scientific and technical statements of St. Petersburg State Polytechnic University.* **No. 2**(123) 141-147
[8] Sudnik L V, Galinovsky A L, Kolpakov V I, Mulyar S G, Abashin M I, Provatorov A S 2014 *Science and education: a scientific publication of MSTU im. N.E. Bauman* **No. 3** 15-23
[9] Abashin M I, Galinovsky A L, Moiseev V A, Osipkov A S, Provatorov A S 2014 *Nanoengineering* **No. 5**(35) 23-27
[10] Galinovsky A L, Abashin M I, Khafizov M V 2013 *Moscow: Bauman Moscow State Technical University Press*
[11] Barzov A A, Galinovsky A L, Abashin M I 2012 *Bulletin of Moscow State Technical University. N.E. Bauman* **No. 11**(11) 14
[12] Barzov A A, Galinovskiy A L, Kuznetsov I E 2009 *Bulletin of the Ufa State Aviation Technical University* **12**(4) 121-126
[13] Abashin M I 2013 *Moscow: Bauman Moscow State Technical University Press* 16
[14] Cherepanov G P 1974 *The mechanics of brittle fracture* (Moscow: Nauka) 640 pp.
[15] Barzov A A, Galinovskiy A L, Khafizov M V 2013 *Proceedings of higher educational institutions. Mechanical engineering* **(8)** 40-46

IOP Conf. Series: Materials Science and Engineering 1060 (2021) 012003 doi:10.1088/1757-899X/1060/1/012003

Comparative analysis results of the homogenization of nanosuspensions using ultra-jet processing method

Kyaw Myo Htet[1*], A L Galinovskiy[1], A A Barzov[2] and J Soehnlein[3]

[1] Bauman Moscow State Technical University, 105005, Russia
[2] Lomonosov Moscow State University, Moscow, 119991, Russia
[3] Technical University of Munich, 82024, Munich, Germany

E-mail: * htet2066@gmail.com

Abstract. This work discusses the urgent problem of finding new highly effective dispersion methods for suspensions prepared using nanomaterials such as (boehmite, graphene, carbon nanotubes). In this research, two experiments are considered. The first experiment uses a solid target from a single crystal of diamond. The second experiment is conducted in the absence of a target. This means that ultrasound interacts with the surface of a tool for collecting the suspension at the initial time, and then with the processed suspension in the tank. Based on comparative experimental data, it has been shown that the ultra-jet processing method has several advantages.

1. Introduction

Currently, dispersion and suspension technologies have founded wide application in the national economy and underlie the production of such products as; cosmetics and medicines, paints, varnishes, inks, various technological lubricants, *etc.* The suspension technology can obtained in a number of technological methods: abrasive-liquid jet processing method, mechanical processing method, electro-hydraulic processing method, electron-beam processing method, "Milling" processing method, Ultra-blasting method, rotary-blasting method, disc processing method and ultrasonic processing method. The dispersion technology can obtained in a number of technological methods: ultrasonic processing method, electrical dispersion method, homogenization, emulsification, "Milling" processing, ultra-blasting method, rotary blasting method, disc processing method [1, 17].

2. Literature Review

Analysis of literature sources, in particular the works of Barzov, Tarasov, Galinovskiy, Abashin, Khafizov, Kobernik, Provatorov *et al.* showed that ultra-jet micro suspension of liquids in hydro-technologies environments [1, 4, 5, 13, 12], ultra-jet technology [6, 8, 11], and nanomaterials [7, 9] had been described. The study of the suspension of liquid using ultra-jet technology [17], under the leadership of Galinovskiy and Barzov, considered two experiments and the theoretical basis of this developed technology [16]. The first experiment used a solid target from a single crystal of diamond and has been conducted earlier in the works of the author's team [2, 3]. In the previous experiment are studied the comparative analysis results of the distribution of the particles size of nanomaterials using ultrasonic technology and ultra-jet technology. Its advantages were noted when dispersing suspensions with carbon nanotubes, boehmite and graphene, where the minimum particle size was an order of magnitude smaller, in comparison with ultrasonic treatment. In the second work the results of particle size with suspensions using ultra-jet technology with a target and without target are considered and discussed with the results of previous works.

3. Experiment description

This experiment has been conducted in the absence of a target, *i.e.* ultrasound interacted with the surface of a tool to collect the suspension at the initial time, and then with the processed suspension in the tank (tool). The purpose of this experiment is to evaluate the possibility of dispersing suspensions directly in the hydraulic installation path (figure 1).

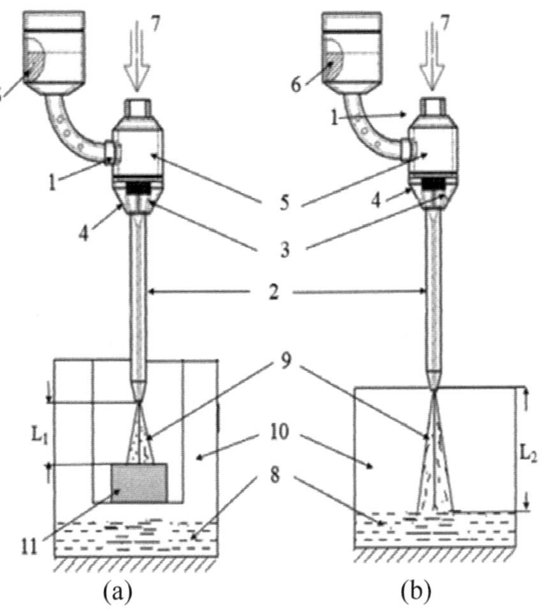

Legend:
L1 – The distance between the hydraulic nozzle and the target surface of the tank in mm
L2 – The distance between the hydraulic nozzle and the surface of the tank-trap in mm
1 – Suction channel
2 – Jet forming tube
3 – Clamping nut
4 – Clamping sleeve
5 – Body of the integrated diamond inductor
6 – Feeding the suspension with powders (boehmite, graphene, nanotubes) in the initial state
7 – Water supply
8 – Water with processed powder (boehmite, graphene, nanotubes)
9 – High-speed cavitating jet
10 – Capacity trap
11 – Carbide target.

Figure 1. Schematic diagram of ultra-jet dispersion of nano-containing suspensions: (a) – with the target; (b) – without the target.

Suspensions based on distilled water with boehmite (produced by the Scientific Research Institute of Impulse Processes with Pilot Production "OKP NII IP with OP", Minsk, Republic of Belarus), with graphene (OOO "Nanotech Centre" Tambov city) and with carbon nanotubes ("Arkema", France) were processed using ultra-jet machining. In order to prepare a mixture for the initial state suspension, 3 grams of nano-powders mixed with 1 litre of distilled water in a glass beaker have been used [8].

In the next stage of this research, the possibilities of ultra-jet suspension processing method have been studied. The studies are based on the methodological base which was formed in the framework of the scientific school "Ultra-jet processing and diagnostics of materials and liquids" (NSh-3778.2018.8) at the Department of SM-12 of BMSTU [2, 3, 9].

Figure 1 shows the Flow Waterjet hydraulic system used in the experiment with jet diameter of 0.01–0.2 mm, high-pressure (pressure 100–700 MPa) and high-speed (600–700 m/s) fluid flow of the multiplier type, as well as a schematic diagram of the ultra-jet dispersion of suspensions [4, 5]. Characteristics of the studied samples of nanomaterials are presented in table 1.

Suspensions based on distilled water with boehmite, graphene and carbon nanotubes are fed into the mixing chamber of the ultra-jet unit using a special metering device. The suspension acceleration is 800 m/s, which corresponds to a maximum working pressure in the hydraulic system of 400 MPa. It has been previously found that the ultra-jet speed determines the liquid treatment efficiency [13].

As a capacity (10), specially developed technological equipment has been used (figure 2). The target is a synthetic diamond fixed in a mandrel, which, in turn, is fixed in the nozzle of the container cover using screws (4) (figure 2).

IOP Conf. Series: Materials Science and Engineering 1060 (2021) 012003 doi:10.1088/1757-899X/1060/1/012003

Table 1 Characteristics of the original nanomaterials samples used for preparing suspensions and their further dispersion

Manufacturer	Composition	Technological and operational features of the powder
Research Institute of Pulse Processes with Experimental Production "OCP NII IP with OP", Minsk, Belarus	**Al (OOH) "Boehmite"**	Mass fraction of the main substance, not less than 99[%]. A specific surface according to the BET method, up to 400 [m^2/g], density, 3.06 [g/cm^3], size of individual particles, 0.1-0.8 [nm].
OOO "Nanotech Center" Tambov city	**Graphene**	Number of graphene layers is 15-25. Thickness of the nano-plates is 6-8 nm. Size of the nano-plates in the plane is 2-10 microns. The content of nano-plates is 4-7% of the mass. The oxygen content in relation to carbon is 9-13% of the mass. Sulfur content mass is $\leq$0.7%. Specific absorption coefficient is 30-33 lm / (g·cm).
Arkema (France)	**Carbon nanotubes**	External diameter is 10-20 [nm]. Tube length is more than 2 [microns]. Specific surface according to the BET method is >300 [m^2/g]. Content of impurities is not more than 1[%].

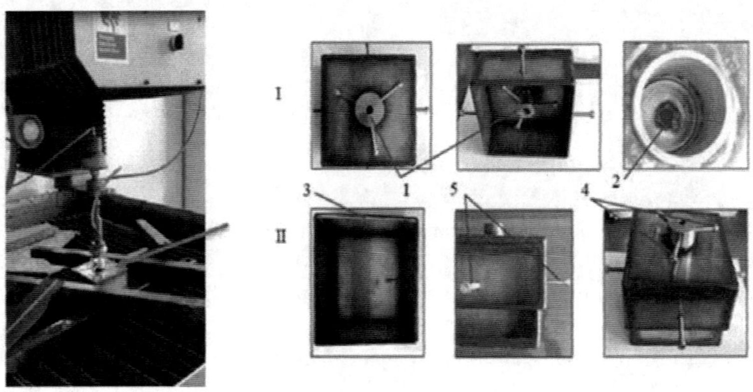

Legend:
1 – Nozzle for installing a focusing tube in it
2 – Cylindrical target with a diamond single crystal fixed in it
3 – Container for collecting the suspension (bottom of the tank)
4 – Screws – Fixing elements for the top cover to the bottom of the container for collecting the suspension
5 – Screws – Target fixation elements.

Figure 2. Tooling for dispersing micro and nano-suspensions:
I – Upper lid of the tank; II – Bottom of the tank and tank assembly.

4. The results of the experiment

After conducting the experiment, the analysis has been performed using a Microtrac Bluewave laser particle size analyzer (Microtrac S3500). It uses the technology of three lasers technology (Tri-laser) shown in figure 3. The laser particle size analyzer (the diagnosable size range is from 0.01 to 2800 μm) allows particle size distribution in suspensions, emulsions, and powders using laser granulomere [14, 15]. Results of the analysis are average particle sizes by quantitative and volume distributions, as well as the minimum recorded particle size in the samples (see table 1).

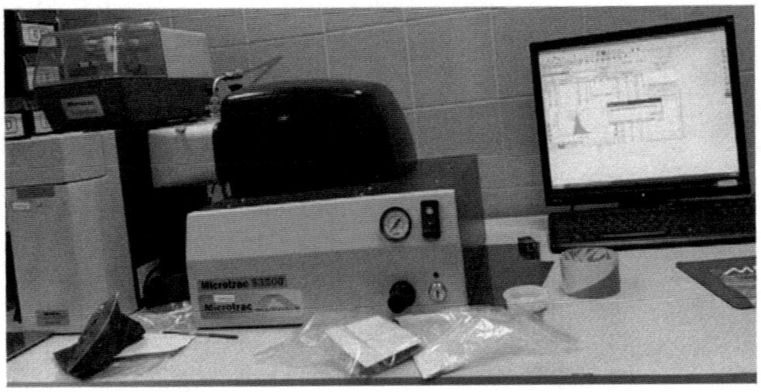

Figure 3. Microtrac S3500 Laser Particle Analyzer.

Results of the comparative analysis of particle sizes in the composition of suspensions are presented in figures 4-6 in the initial state, after ultra-jet treatment with a target, and without a target respectively.

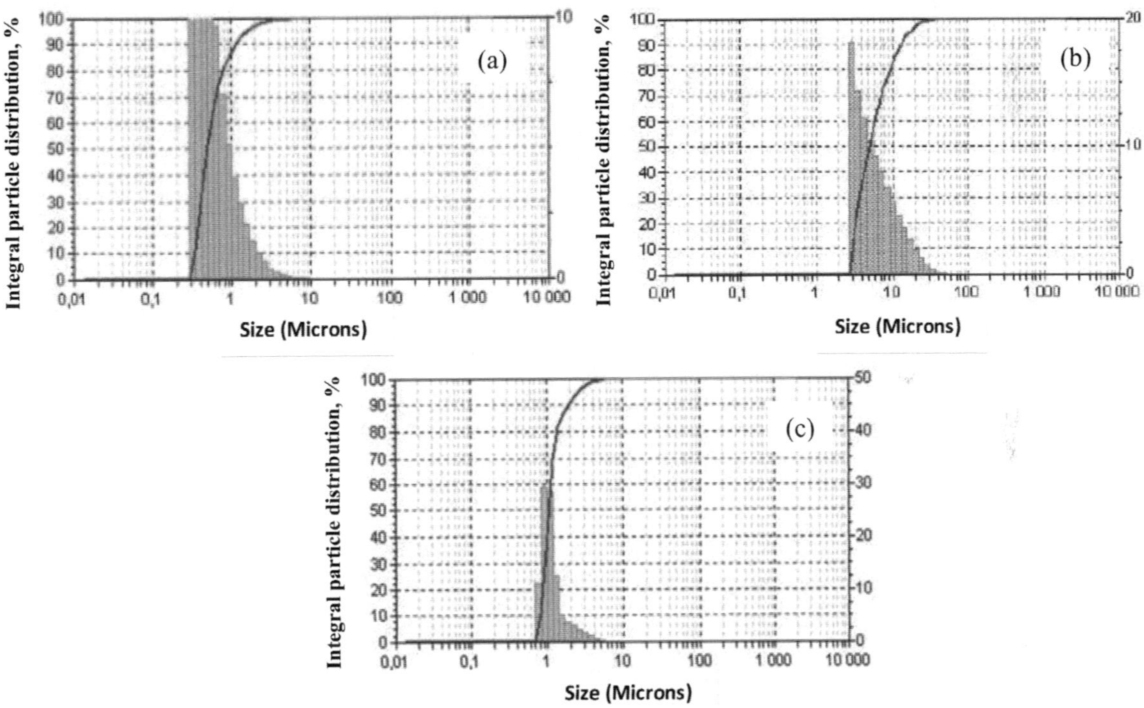

Figure 4. Graphs of average particle sizes values by quantitative distribution for initial samples; (a) – Boehmite, (b) – graphene, (c) – carbon nanotubes.

5. Discussion

Comparative experiments on the application of the ultra-jet method with a target with the ultrasonic dispersion method show that the proposed method allows to reduce the particle size from 2 to 10 times depending on the nanomaterial under consideration in previous articles [2, 3]. The result of ultra-jet technology without a target of this experiment are close to the obtained results of previous work using the ultrasonic treatment. Therefore, the prospect of the development of the research is the establishment of time limits for maintaining a given level of dispersion of boehmite, graphene and carbon nanotubes, which is extremely important for issuing practical recommendations on their use for the preparation of binders, the introduction of additives, paints and other materials. The comparative analysis of the obtained particle size distributions of the studied suspension samples (figure 5 and figure 6) with their initial distributions (figure 4) show that the ultra-jet treatment method can be used practically and effectively.

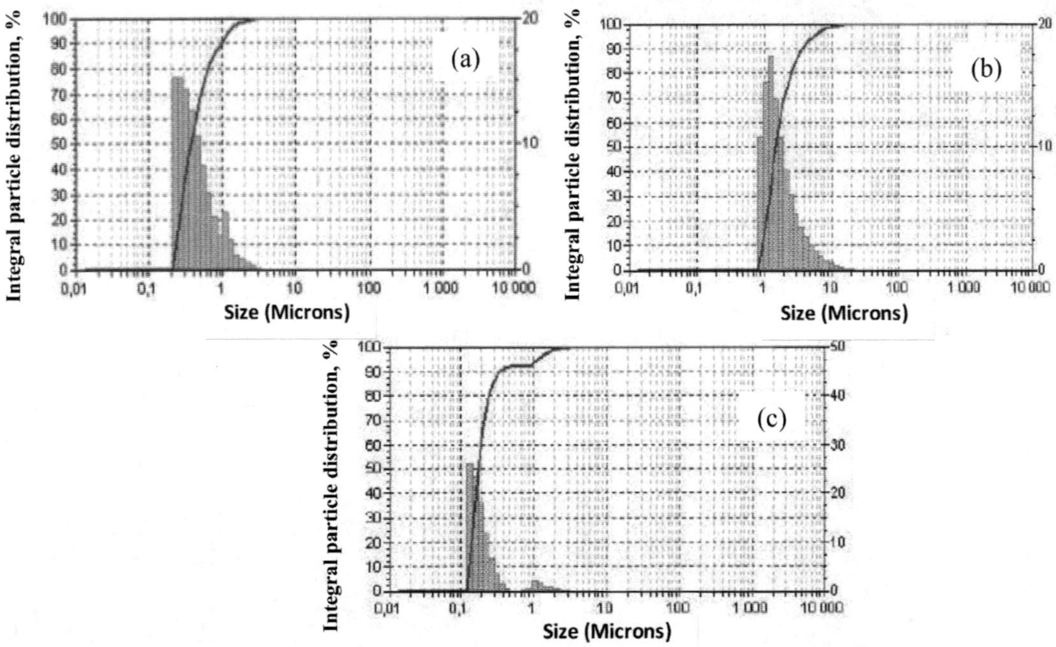

Figure 5. Graphs of average particle sizes values by quantitative distribution for samples after ultra-jet impact without a target; (a) – boehmite, (b) – graphene, (c) – carbon nanotubes.

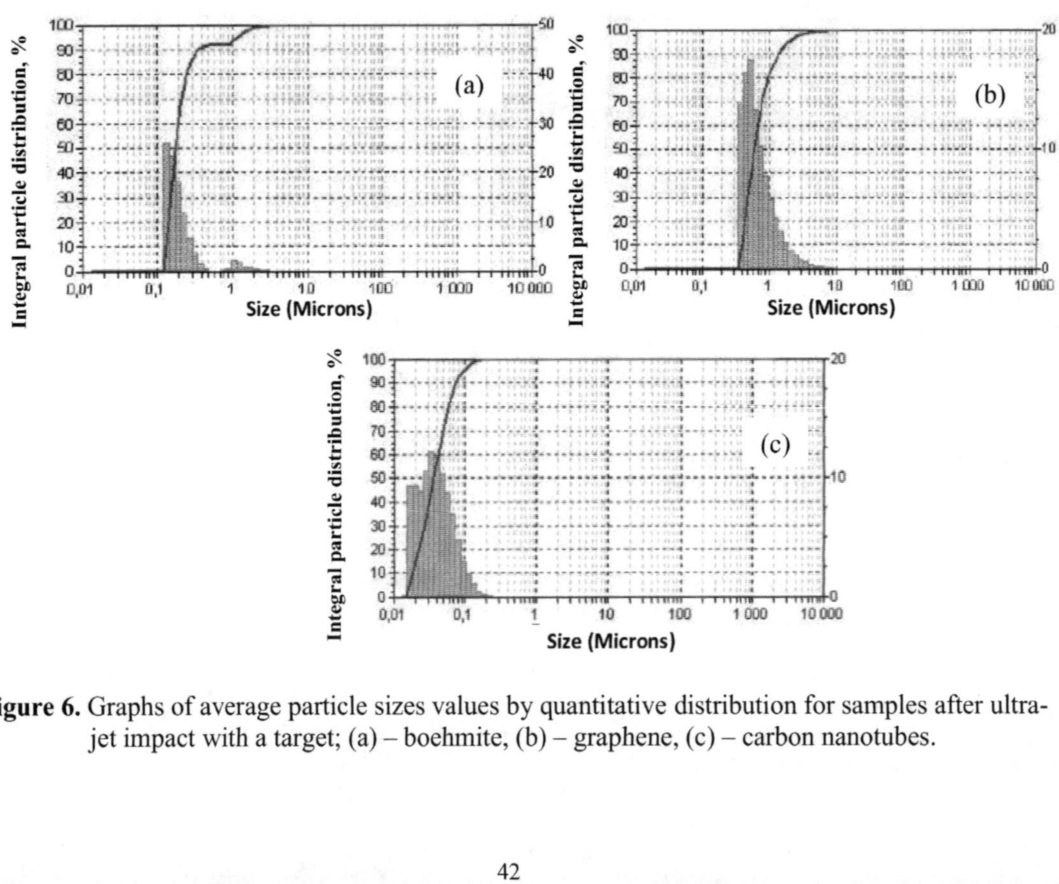

Figure 6. Graphs of average particle sizes values by quantitative distribution for samples after ultra-jet impact with a target; (a) – boehmite, (b) – graphene, (c) – carbon nanotubes.

6. Conclusion

It has been found that dispersion is carried out not only upon the impact on a solid target, but also during the passage of the hydraulic installation path, i.e. mixing chamber and focusing tube (Figure 1). Moreover; the results of this process are close to those obtained using ultrasonic treatment. It can be assumed that ensuring the circulation of the suspension according to the scheme proposed in Figure 1 b may improve the controlled indicator - particle size. The presence of a positive effect in this scheme implementation requires additional study and modernization in order to increase the efficiency and quality of suspension.

Acknowledgments

The studies were carried out within the framework of grants of the Russian Foundation for Basic Research 18-29-18081 and 19-38-90228\19, a grant of the President of the Russian Federation for state support of leading scientific schools of the Russian Federation NSh-3778.2018.8 and grants from the Fund for the Promotion of Innovations under the UMNIK-18 program in accordance with the contracts No. 14727GU/2019 and No. 14549GU/2019.

References

[1] Galinovsky A L, Khafizov M V, Barzov A A 2012 *(Bulletin of Bauman Moscow State Technical University)* 27-34
[2] Kyaw Myo Htet et al 2020 *IOP Conf. Ser.: Mater. Sci. Eng.* **709** 044092
[3] Kyaw Myo Htet, Galinovsky A L 2019 *All materials: Encyclopedic Handbook* **11** 2-7
[4] Barzov A A, Galinovskiy A L, Litvin N K, Khafizov M V, Abashin M I 2012 *Fundamental and applied problems and technologies* **2-3**(292) 3-8
[5] Stepanishchev N A, Tarasov V A, Boyarskaya R V 2013 *Engineering journal: science and innovations* **6**(18) 47
[6] Abashin M I, Barzov A A, Bochkarev S V, Galinovsky A L, Maslov B G 2015 *Welding International* **29**(9) 730-733
[7] Gedanken A 2004 *Ultrason Sonochem.* **11**(2) 47-55
[8] Barzov A A, Galinovsky A L, Golubev E V, Ilyukhina A A, Kobernik N K, Zarubina O V 2018 *Fundamental and applied problems of engineering and technology* **3**(329) 73-81
[9] Barzov A A, Veltischev V V, Galinovskiy A L, Mugla D R 2018 *Proceedings of higher educational institutions. Mechanical engineering* **1**(694) 55-62
[10] Barzov A A, Galinovsky A L, Golubev E S, Sysoev N N, Fedyanin A A, Filimonov A S 2018 *Engineering journal: science and innovation* **6**(78) 4
[11] Ilyukhina A A, Kolpakov V I, Galinovsky A L, Khakhalin A V 2018 *Moscow University Bulletin. Series 3. Physics. Astronomy* **4** 101-107
[12] Bochkarev S V, Tsaplin A I, Galinovsky A L, Abashin M I, Barzov A A 2017 *Metallovedenie i termoobrabotka metallov — Physical Metallurgy and Metal Heat Treatment* **6**(744) 58–63
[13] Kyaw Myo Htet, Provatorov A S 2019 In the book: Gagarin Readings - 2019 *Abstracts Collection of the XLV International Youth Scientific Conference* 830-831
[14] Kyaw Myo Htet., Shevchenko Yu V 2018 In the collection: *New approaches and technologies for the design, production, testing and industrial design of rocket and space technology products collection of works of the II International Youth Conference* 269-272
[15] Kyaw Myo Htet, Ivanushkin D V 2018 In the collection: *New approaches and technologies for the design, production, testing and industrial design of rocket and space technology products collection of works of the II International Youth Conference* 260-265.
[16] Barzov A A, Galinovsky A L, Puzakov V S 2009 *Bulletin of the Moscow State Technical University. N.E. Bauman. Series Mechanical Engineering* **2**(75) 72-83
[17] Abashin M I, Galinovsky A L, Moiseev V A, Osipkov A S, Provatorov A S 2014 *Nanoengineering* **5**(35) 23-27

IOP Conf. Series: Materials Science and Engineering 1060 (2021) 012004 doi:10.1088/1757-899X/1060/1/012004

New design and technological solutions for creating a composite materials recycling tool and methods of their ultra-jet diagnostics

Li X, A L Galinovsky*, M I Abashin, I N Kravchenko

Bauman Moscow State Technical University, 105005, Russia

E-mail: * galcomputer@mail.ru

Abstract. The article presents the results of two interrelated studies. The first one is related to assessing the possibility of using a water-polymer ultra-jet as a diagnostic tool for cutting devices that are used for plastic recycling. The second one is related to the development of a new tool design. As a result of the research, it is shown that both the method of water-polymer diagnostics and the proposed design of the knife have prospects to practical application. Rapid wear of the tools for recycling polymer materials is the primary problem that is advisable to eliminate in first order because the idle equipment leads to financial losses. Therefore it is necessary to develop tools with high values of wear resistance and which at the same time have a low cost. A model of a prefabricated tool for recycling composites is proposed, which potentially has high performance properties. The main feature of new tool is its geometric configuration that allows to increase the tool resource and significantly saves the material for its manufacturing. The increase of resource occurs due to the following: if the cutting plate of the tool is worn, it can be removed off the tool, flip it 180 degrees and reinstall. Economy of the material is also due to a decrease in size of the plate, which is fixed to the tool body with countersunk screws head. The research results are confirmed by comparative analysis of various diagnostic methods and strength analysis of the tool.

1. Introduction

Ultra-jet diagnostic technology (UJD) was developed in Bauman Moscow State Technical University and has been widely used to assess the quality and resource of functional coatings, residual life of structures, the performance properties of the surface layer, study of abrasive-liquid material resistance, *etc.* [1-3]. The advantage of UJD is quick access to the necessary information (for determining the quality of the diagnosed material for instance) about the operational and technological parameters characterizing the surface layer state of the material of various technical objects and the ability to create simulated local effects on the material [2]. One of the most attractive benefits of UJD is that it can be used already on finished parts, real final components, so there is no need to manufacture special samples for testing.

The result of the interaction of ahigh-speed liquid jet with the surface of the diagnosed material is the creation of a hydro-cavern. The *hydro-cavern* is a pit on the surface of the material made by the action of the UJ on it. UJD is based on the dependence analysis between the local characteristics of the hydro-cavern in diagnosed area of the surface and its quality parameters.

Wide prospects for UJD are opened in case of using *suspensions* of different composition and structure as a special diagnostic tool. An ordinary suspension is a combination of liquid and something else that does not dissolve in that liquid.

When an ultra-jet reaches a solid target made of some kind of the material, this material produces very small particles. The water that contains that particles is called suspension in our case too. Suspensions can consist of more than two components, depending on the characteristics that are to create: ultra-jet of water, cavitating jet, abrasive-liquid suspension, multiphase suspensions. For instance, suspension with ice crystals can be used as such tool in special machinery industry [2, 3]. It can be used as a tool that simulates the aggressive environment of ice impact. Their use suggests that the object of diagnostics is affected by an impact similar to the actual operating conditions, for example, water-jet effects for pumps that are used for pumping petroleum products, coal, *etc.* It should be noted that in contrast to some well-known methods, such as evaluating the wear resistance of materials with abrasive-liquid jets, the proposed UJD technology uses ultra-jets with a power flow density of about 0.5 MW/mm^2, which corresponds to the power density of industrial lasers [4].

2. The problem and proposed solution

Wide technological and diagnostic capabilities of UJD allow us to find new areas of application. Thus, the analysis showed that the existing cutting tool (knives), on the market, for the recycling of polymer materials had a number of drawbacks: low resource, the need for frequent overloading and dismantling of the tool, the complexity of installing and balancing rotors – the main advantage of installations for recycling, *etc.* [5, 6, 7]. One of the obvious ways to modernize the tool is changing the material from which it is made and in this case, there may be a large number of possible applications not only of various materials, but also coatings, hard-facing methods, *etc.* At the stage of technology development, that is, at the preliminary stages, it is necessary to have diagnostic methods that will allow fast and reliable access to information about the operational properties of the tool/knife material. Field tests are the most reliable and provide unambiguous information about the tool's useful operation time and quality, but they take very long time, about 160 hours on average.

Taking into account the experience of using the UJD, an attempt was made to apply a combined ultra-jet, which would include a material – polymer powder – that interacts with the knife. In other words, a suspension consisting of water and polymer powder particles was used. For experiments, polymer powders identical or similar in properties to the processed material were used, that is, the technologically closest and affordable physical and chemical analog [8].

The modification of the ultra-jet can increase the functional and information capabilities of the UJD method. At the same time, the results of ultra-jet interactions, first of all, the nature and geometric parameters of the formed hydro-caverns, products of hydro-erosive destruction, mass-geometric parameters of the object of research contain objective control and diagnostic information about its operational and technological characteristics [9].

3. Ultra-jet flexibility

It should also be noted that the UJD method, among other things, has the ability to vary a number of technological parameters. In particular, it is possible to change the concentration of the powder (that is added to the ultra-jet of liquid), the jet velocity, the speed or vary the distance from the jet-forming focusing tube to the surface of the sample. This, firstly, provides the possibility of expanding the range of changes in the power flow density of the ultra-jet. Secondly, it provides a variation of the energy disequilibrium of physical processes that are similar to the actual operating conditions of this cutting tool. This provision determines the relevance of the research topic and allows us to formulate a goal consisting in a scientific and methodological justification of the possibility of using an ultra-jet as a diagnostic tool.

4. Description of the experiment

A schematic diagram of the UJD using a modified polymer powder ultra-jet is shown in Figure 1.

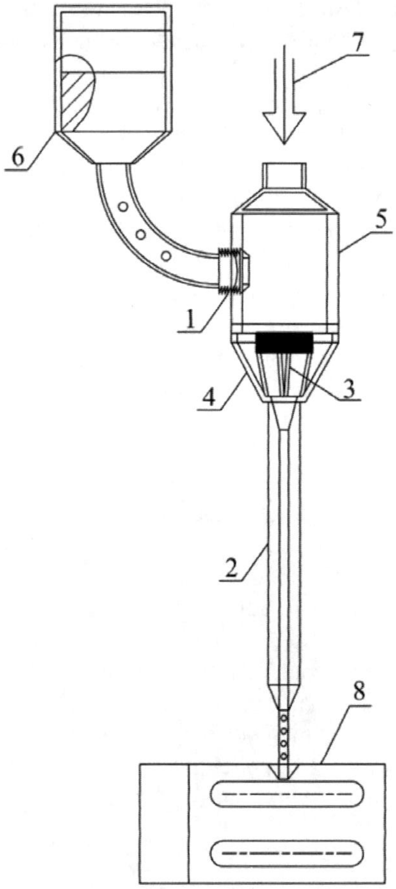

Legend:
1 – Feed channel for modifying powder
2 –Jet forming tube
3 – Clamping nut
4 – Collet
5 – The mixing chamber;
6 – Measuring container for feeding powder (polyamide-12)
7 – The flow of the working fluid (water)
8 – The object of study (the knife)

Figure 1. UJD implementation scheme.

The Polyamide-12 powder that is usually used in selective laser melting technology in our case was used as a modifier for the ultra-jet of water. So, we managed to achieve a water-polymer ultra-jet just by adding polymer powder into the ultra-jet of water.

The UJD experiment using a modified ultra-jet (water-polymer ultra-jet) was performed using a water-jet cutting machine made by Flow System (USA) – "Flow Mach 3" model. The pressure in the system was P=400 MPa, the distance from the cross section of the focusing tube to the surface of the knife was h=4 mm, the feed rate of the nozzle head was S=0.41 mm/s, the concentration of polymer powder was C=0.2 kg/min (table 1).

Table 1. Experimental parameters

Parameter	Value
P (MPa)	400
h (mm)	4
S (mm/s)	0.41
C (kg/min)	0.2

As samples for the research and testing of the UJD method, we selected knives for the PM250 rotary crusher model, manufactured of bimetal (Steel 3 (base) – Steel 9KHF) [10]. A total of four knives were considered with the following modes of heat treatment in accordance with table 2.

It should be noted that the use of bimetal blanks for the manufacture of knives is a novelty and has been tested for the first time.

Table 2. Information about the knives and their modes of heat treatment

Marking	Mode of heat treatment
№1 (Z + V)	quenching 840°C, cooling in a flow of water
№2 (Z + V + O)	quenching 840°C, cooling in water flow, tempering 160°C, 2 hours
№3 (Z + M)	quenching 880°C, cooling in oil
№4 (Z + M + O)	quenching 880°C, cooling in oil, tempering 160°C, 2 hours

5. Results

For a comparative analysis of the results of UJD, the knives were tested using traditional methods: friction tests, full-scale tests and micro-hardness measurement [12, 13].

Four sets of bimetallic knives made from *9KHF* – Steel 30 with marking №1-4 were transferred to the limited liability company "Astrelati" for industrial (full-scale) tests. The knives were installed in a rotary crusher model PM250 together with standard knives made of steel 6KHV2S [14]. Figure 2 presents the photos of PM250 and installed knives. The working time of the cutting tool was 160 hours [15, 16]. The recycled (shredded) material was defective polypropylene structures.

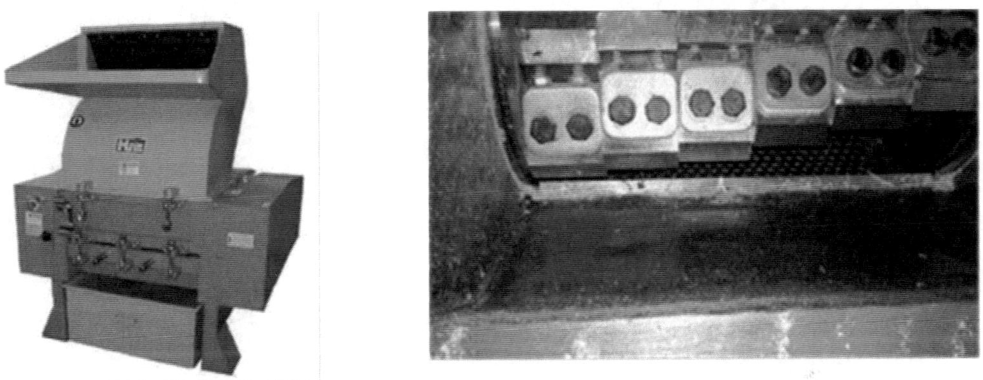

Figure 2. (a) Rotary crusher, (b) PM250 modules and a cutting tool fixed in the rotor of the crusher.

After the experiment and dismantling of the tool, the following parameters were measured: changing (increasing) the radius of rounding and the number of chips on the cutting edge of the knife, changing the front rake angle etc. [17, 18]. The obtained data were compared with the value of the depth of the hydro-cavern after UJD [19], which was determined using the profilometer-profilograph BV-7669 (produced by the research institute "Measurements", Russian Federation). The example of obtained data by this equipment is presented in figure 3. It depicts the contour of the hydro-cavern and its depth (the profile of the surface).

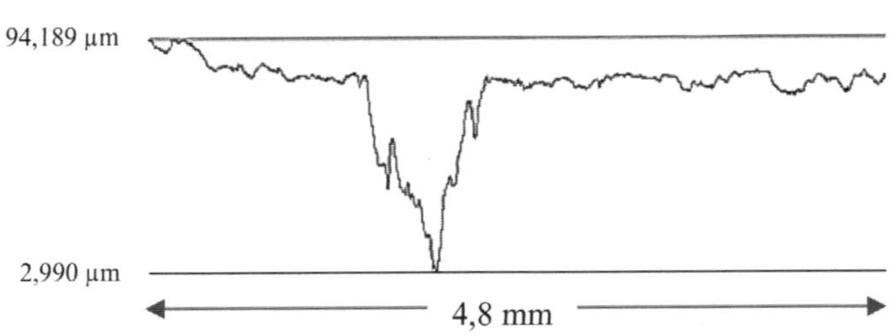

Figure 3. Example of data produced by BV-7669.

The friction tests were carried out using a classic friction test rig (figure 4). The contact load was 0.2 kg and was determined by the weight of the rod [9]. Micro-hardness was measured using a

Micromet-II micro-hardness tester with a load of 500 g. according to GOST 9450-76 (Measurement Of Microhardness by Pressing Diamond Tips) [11].

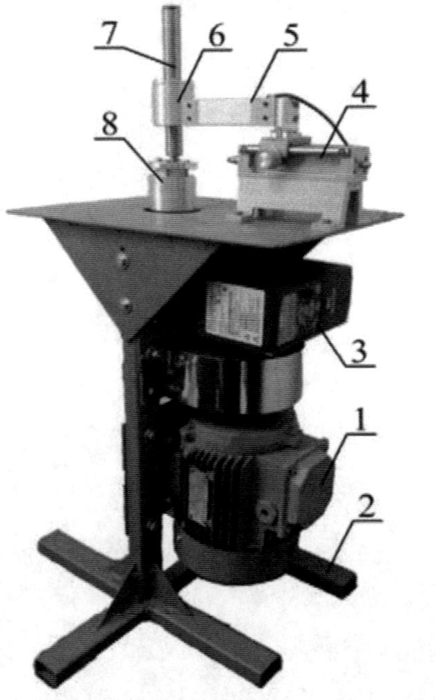

Legend:
1 – electric motor
2 – foundation slab
3 – engine control unit
4 – linear drive with flywheel
5 – load cell
6 – linear bearing
7 – cylindrical shaft of linear guide
8 – rotating table with sample
9 – experimental sample (knife)

Figure 4. The device for friction testing.

Based on the results of measurements, the values of the pairwise correlation of each of the informative parameters were calculated (table 3).

Table 3. Data of comparative correlation analysis of cutting tool diagnostics methods

	a	b	c	d	e	f	g
a	-	0.93	0.88	0.87	0.91	0.97	-0.74
b	0.93	-	0.91	0.94	0.92	0.81	-0.79
c	0.88	0.91	-	0.99	0.99	0.81	-0.95
d	0.87	0.94	0.99	-	0.99	0.77	-0.95
e	0.91	0.92	0.99	0.99	-	0.85	-0.95
f	0.97	0.81	0.81	0.77	0.85	-	-0.68
g	-0.74	-0.79	-0.95	-0.95	-0.95	-0.68	-

[a] Rounded cutting edge radius (mm).
[b] Changing the front rake angle (degrees).
[c] The number of chips per unit length of the cutting edge.
[d] The ratio of the cutting edge's ablation area to the length (mm).
[e] Hydro-cavern depth (mm).
[f] Depth of the friction test cavity (μm).
[g] Micro-hardnes (MPa).

6. The idea of a new tool

In this regard, it was suggested to analyze other possible ways to improve the tool that combines a combination of materials, in particular, the use of a tool with a removable working plate (figure 5). As a result, the design of the knife was proposed, which uses a replaceable double-sided plate made of high-speed steel P6M5, which has shown itself well as a result of water-polymer UJD (figure 6). A replacement plate with a diamond-shaped cross-section that allows, firstly, to make its turn to change the cutting edge, and secondly, to redistribute the loads acting on the working edge perpendicular to the plane in the scheme, from the mounting screws to the knife body.

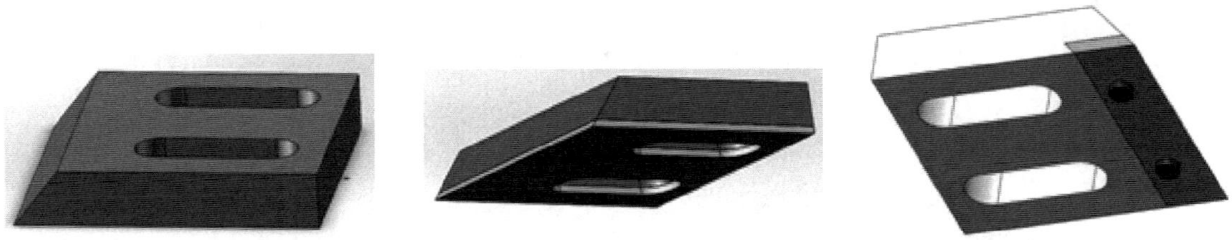

(a) (b) (c)

Figure 5. Traditional and promising tool for recycling PCM:
(a) – Traditional knife, (b) – Bimetallic knife, (c) – Knife with replaceable plates.

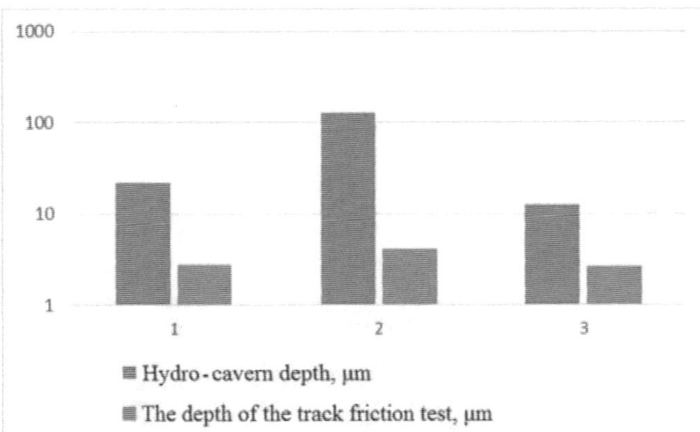

Legend:
1 – 6KHV2C
2 – 9KHS (as a part of bimetal)
3 – R6M5

Figure 6. Comparison of hydro-cavern depths and friction tracks for various tool materials.

The potential increase in the performance of the assembly tool is as follows:

- The use of high-wear-resistant alloy replacement plates as a material reduces the time associated with knife service (reloading, rotor balancing, installation and adjustment of the gap between the knives), reduces the amount of metal impurities in the processing of composite materials components, which increases the cost of raw materials
- Reduce the time to change knives by allowing the cutting plate to be rotated 180 degrees after using one side. At the same time, these types of work can be carried out without removing the tool from the rotor, which avoids balancing and setting a gap between the knives.

The plate is attached to the knife body using titanium screws with a countersunk head (figure 5 b). In contrast to the traditional design of ordinary and bimetal knives (figure 5 a, b), the new knife saves material by reducing the size of the plate and the consumption of tool material.

The proposed design was calculated for strength in the SolidWorks program by simulating the load on the product. The calculation results are shown in figure 7. According to the scheme (figure 5 c), a sample of the knife was made for the purpose of conducting full-scale tests (figure 8).

7. Discussion

Table 3 shows that the depth of the hydro-cavern corresponds to the results of other tests and the method may be equally valid for practical application.

The ideas for improving the recycling tool are based on the use of combination of several materials and the use of exchangeable working plate; however, further consideration revealed the disadvantages of bimetallic knives, due to the fact that, like an all-alloy knife, the bimetallic one also needs to be removed from the rotor for recasting, which requires a lot of time and subsequent balancing of the crusher rotor [20]. In addition, a significant amount of tool material is simply not used in this design, so, that's why the new design of the cutting tool was proposed. The use of an additional diamond-shaped section of the plate increases the reliability of fastening and reduces the number and cross-section of fixing screws, which has a positive effect on the cyclic strength of this product.

Figure 7. Stresses calculated for a promising cutting tool.

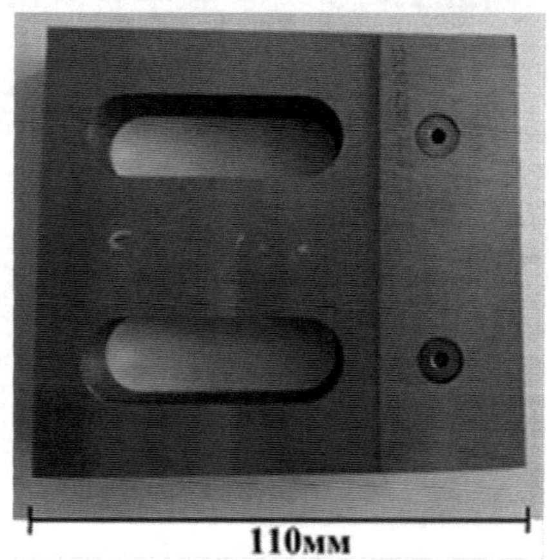

Figure 8. Photos of manufactured knives with a removable plate.

8. Conclusion

Based on the data obtained (table 3), it can be argued that the UJD method is highly informative, not inferior to the traditional method of friction tests. In addition, which is especially important in practice, there is a high correlation between the results of field tests and informative signs of UJD.

Furthermore, in comparison with the method of friction testing, water-polymer UJD has a number of advantages, such as: the rapid diagnostic process; the formation of a powerful wave perturbation in the zone of influence of water-polymer ultra-jet; simulating the working conditions of the tool for a long time of operation; the creation of shock variable loads when polymer particles collide with the surface of the knife. Additionally, a promising design of the knife with the ability to reverse the plate has been developed, which has significant advantages that will soon be tested during field tests. Most important features of this knife include reduction of service time for it and increased cost of produced materials.

Acknowledgments

The studies were carried out within the framework of grants of the Russian Foundation for Basic Research 18-29-18081 and 19-38-90228\19, a grant of the President of the Russian Federation for state support of leading scientific schools of the Russian Federation NSh-3778.2018.8 and grants from the Fund for the Promotion of Innovations under the UMNIK-18 program in accordance with the contracts No. 14727GU/2019 and No. 14549GU/2019.

References

[1] Sidorenko P D and Karpes'o I Ya 2018 *Teknologiya metallov* [Metal technology] **2**(77) 54-8
[2] Li Sjuejan', Semashko B C, Galinovskij A L, Belov V A and Abashin M I 2019 *Teknologiya metallov* [Metal technology] **10** 41-7
[3] Lahtela V, Hyvärinen M and Kärki T 2019 *Polymers* **11**(1) 69
[4] Barejan A G 2004 *New promising materials and technologies for their production* **2** 56-7
[5] Mohammadi M, Jämsä-Jounela S-L and Harjunkoski I 2019 *Comput. Chem. Eng.* **123** 155-169
[6] Ferreira G V B, Barletta M and Lima A R A 2019 *Sci. Total Environ.* **655** 292-304
[7] Okan M, Aydin H M and Barsbay M 2019 *J. Chem. Technol. Biotechnol.* **94**(1) 8-21
[8] Abashin M I, Barzov A A and Galinovskij A L 2012 *Vestnik MGTU* [Herald of the BMSTU], *Ser. Natural Sciences* **6** 7
[9] Huysveld S, Hubo S, Ragaert K and Dewulf J 2019 *J. Clean. Prod.* **211** 1-13
[10] Alifanov A V, Miljukova A M and Curan V V 2015 *Trudy BGTU* [Proceedings of BSTU] **2**(175) 280-4
[11] Palanikumar K, Karunamoorthy L and Karthikeyan R 2004 *Int. J. Manuf. Sci. Prod* **6**(3) 119–128
[12] Bochkarev S V, Tsaplin A I , Galinovskii A L, Abashin M I and Barzov A A 2017 *Met. Sci. Heat Treat.* **59**(5-6) 384-8
[13] Abashin M I, Barzov A A, Bochkarev S V, Galinovsky A L and Maslov B G 2015 *Weld. Int.* **29**(9) 730-3
[14] Savinov Ju P, Chernov N M and Medvedev K A 2005 *Teknologiya metallov* [Metal technology] **10** 23-6
[15] Barannikova S Li and Zuev L Y 2018 *Metalurgija* **57**(4) 275-8
[16] Hainsworth S V, Delaney R J and Rutty G N 2008 *Int. J. Legal Med.* **122**(4) 281-91
[17] Konig W and Grab P 1989 *Annal. CIRP* **38**(1) 119–24
[18] Sakuma K and Seto M 1983 *Bull. JSME* **26**(218) 1420–27
[19] Bhatnagar N, Ramakrishnan N, Naik N K and Komanduri R 1995 *Int. J. Machine Tool Manuf.* **35**(5) 701–16
[20] Reutova Ja I 2016 *Vestnik BGTU imeni V G Shukhova* [Bulletin of BSTU named after V.G. Shukhova] **4** 136-9

IOP Conf. Series: Materials Science and Engineering 1060 (2021) 012005 doi:10.1088/1757-899X/1060/1/012005

Ways of abrasive blasting technology development

V A Sychuk and O V Zabolotnyi

Lutsk National Technical University, Lutsk, Volyn region, 43018, Ukraine

svamator@gmail.com

Abstract. Having analysed problems in abrasive blasting processing, we present solutions. To reduce the probability of failure of the abrasive blasting nozzle, it is necessary to reduce or eliminate the fact of its wear using the air layer between the inner working wall of this tool with the flow of abrasive particles. It is also preferable to remake the layout of the nozzle from solid to assembled unit in order to replace with a new part of the nozzle, which can still become wear-off because of some reasons. As a result of implementation of such abrasive blasting nozzle, there was a need to develop the method of its use. Regarding the quality control of the processed surface, it is usually carried out in such ways: visually during processing; after processing; due to certain processing modes, which are predicted to provide a given surface quality. Using the methods presenting above it is impossible to quantify the quality of the surface, especially during machining. Accordingly, the processed part in some cases returns to machining again. Therefore, it is suggested that active control should be applied, namely to measure, for example, the roughness of the resulting surface during processing. If the degree of roughness is not reached, a signal is sent to the worker or the automated system about the need to finish the surface, or change the processing modes, which will provide the desired result.

1. Introduction

Abrasive blasting of surfaces of parts is a productive and effective type of processing [1, 3, 5, 6]. It is known that surfaces can be of different shapes which is important for processing complex shaped or hard-to-reach surfaces of parts. Also, it is it is possible to process a large surface area of parts rather quickly. Other important advantages of abrasive blasting are relative simplicity of abrasive blasting processing equipment and no need for high professionalism of workers.

But at the same time, this type of processing has a number of problems [4, 7]. Abrasive blasting nozzle, through which the working mixture of abrasive and fluid, for example air passes, is subjected to rapid wear-off and replacement of worn nozzle is expensive. It is also problematic to control the quality of treated surface during processing to be more concrete it is difficulty in determine during processing if the surface has acquired the necessary quality parameters (for example roughness). Besides this type of processing is harmful to the worker because of much dust and a silicosis disease may appear.

Some of these problems can be solved in ways presented in works [2, 8, 9, 10, 11]. For example, the abrasive blasting nozzle can be made according to the design shown in figure 1. The principal difference from abrasive blasting nozzles of other designs is that the wear resistance of the internal working surface of the nozzle is achieved not by the use of super-hard materials but by the formation of air layer on the working surface of the nozzle, which prevents the contact of abrasive particles with the inner working surface during the operation of the abrasive blasting processing. Accordingly, the proposed special tool must be used taking into account certain rules of its use. The scheme of assembled abrasive blasting nozzle is shown in figure 2.

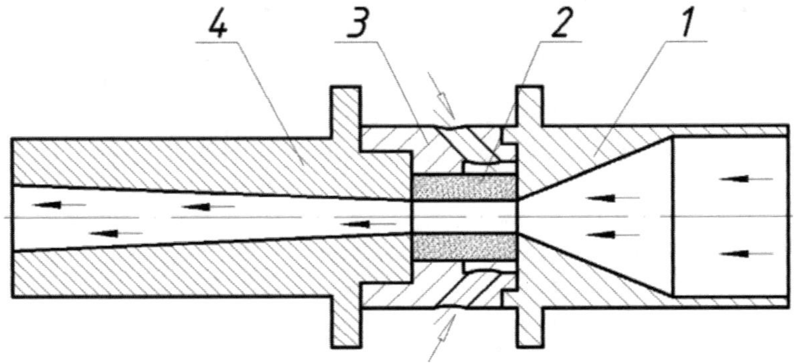

Figure 1. Developed assembled design of the abrasive blasting machine nozzle:

1. Inflow conical wide neck
2. Porous permeable insert (acceleration zone)
3. Housing for placing porous permeable insert
4. Outflow conical narrow neck

– Air flow with abrasive particles – Additional air flow.

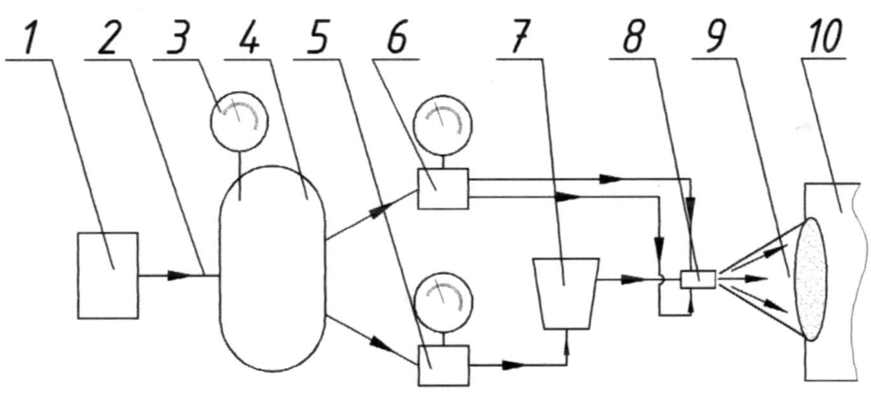

Figure 2. A scheme of abrasive jet processing of parts using assembled nozzle:

1. Compressor
2. Hose
3. Pressure gauge
4. Receiver
5. Regulator of air pressure of the main flow
6. Air pressure regulator of the additional flow
7. A mixer of abrasive and air
8. Assembled nozzle of new design
9. Mixture of air flow with abrasive, a working flow
10. The surface of the workpiece.

The key point in the abrasive blasting treatment according to this scheme is the primary supply of clean air to the assembled nozzle in the radial direction, and only then the supply of the main flow of a mixture of air and abrasive which is passing along the nozzle. Also an important point in processing the surfaces of parts is the control of the processing itself. A control over the condition of the surface of the part subjected to abrasive blasting is usually done visually during the processing or determining of roughness by the classical contact method after processing. It is also possible to predict future required surface conditions due to certain processing modes (type and size of the fraction, flow rate, distance of the working nozzle from the processing surface, *etc.*)

The disadvantage of the first method is the almost subjective assessment of the result of processing. The disadvantage of the second method is the possible reprocessing of the product if the condition of a surface was not achieved according to demands. The third method may not achieve the required surface quality at all because this method only predicts the result and does not guarantee it. Therefore, in order to automate the control of the obtaining surface of the part and the passage of abrasive blasting process, authors of this article proposed a special system of active control. The essence of the system is a constant analysis of the machining surface in real time with the ability to complete the surface treatment or even adjust the modes of processing to the process itself. Figure 3 shows a scheme of active control of abrasive blasting processing.

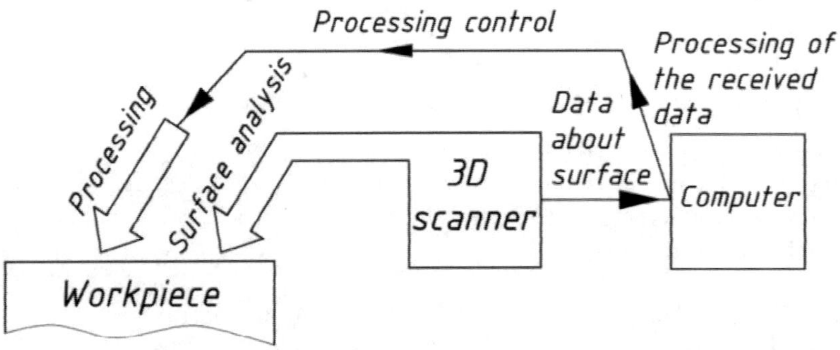

Figure 3. A scheme of active control of abrasive blasting processing.

According to the scheme shown above, the 3D scanner is needed to obtain information about the condition of the newly treated area of the surface of the part. Then the information is transmitted to a computer that analyses it to determine the current surface roughness of the part. In case of disparity between the current roughness and the necessary one, the system adjusts processing modes and, if necessary, finishes an area of a surface which does not correspond to the set roughness.

The machining process itself can be performed by the employee or the automatic equipment which is controlled by a CNC system. In the first case, the employee will be given special light signals about the process and, in the second case, special commands will affect the CNC system of the equipment by adjusting its operation. It should be noted that the obtaining information about the quality of the surface is gained in a non-contact way, which does not affect the actual surface relief of the part, and therefore is reliable.

2. Methodology

To test the proposed idea, an experimental work was carried out, which consisted of the following stages:

1. Obtaining information on 3D scanning of surfaces of parts of different roughness
2. Analysis of the received information for estimation of roughness
3. Determination of the surface roughness of the parts that have been scanned using the known classical method
4. Comparison of obtained results for compliance
5. Analysing the work which was done

Therefore, sandpaper of different fractional composition was chosen as experimental surfaces of different roughness. Figure 4 and table 1 show the brand and some characteristics of the selected sandpaper. The next step was to scan (figure 5) areas of sandpaper of different fractional composition using a 3D scanner DAVID SLS-1.

Figure 6 shows a digital model of the scanned area of the sandpaper surface. It is worth noting that the maximum scanning resolution we chose was 12 μm. Then from digital models of scanned surfaces, crossed by 3 planes, 3 classical common profiles were obtained (figure 7). Next, to calculate the roughness of the Ra parameter, the profiles were divided (figure 8) into a certain number of parts in order to determine the coordinates of the points relative to the midline.

Figure 4. Klingspor sandpaper.

Table 1. Characteristics of 8 experimental surfaces of the chosen sandpaper.

Type of sandpaper (ISO-6344)	Abrasive fraction size, μm	Type of sandpaper (ISO-6344)	Abrasive fraction size, μm
P36	500…630	P100	125…160
P40	400…500	P150	80…100
P60	250…315	P180	63…80
P80	200…250	P240	50…63

Figure 5. The process of 3D scanning of areas of sandpaper surfaces of different fractional composition.

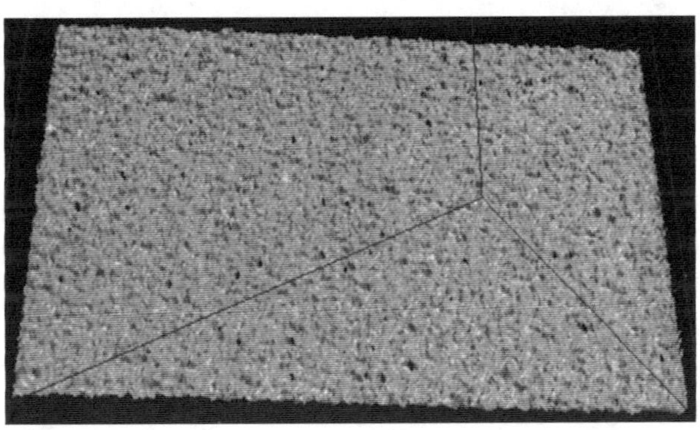

Figure 6. The obtained digital model of the surface of sandpaper with indicated scanning resolution.

After getting positive and negative height values (40 in total), according to middle line, the average value of roughness was calculated in Excel. All scanned areas (3 profiles in each area) of different types of sandpaper were analysed using this methodology. Obtained results are presented in Result section. Also, to compare the roughness parameters of the sections of sandpaper estimated by this method, they were also investigated on a special automated profilograph-profilometer (figure 9) to obtain the appropriate profiles (figure 10). The roughness of all types of researched sandpapers (3 profiles in each type of sandpaper) were estimated using methodology shown above.

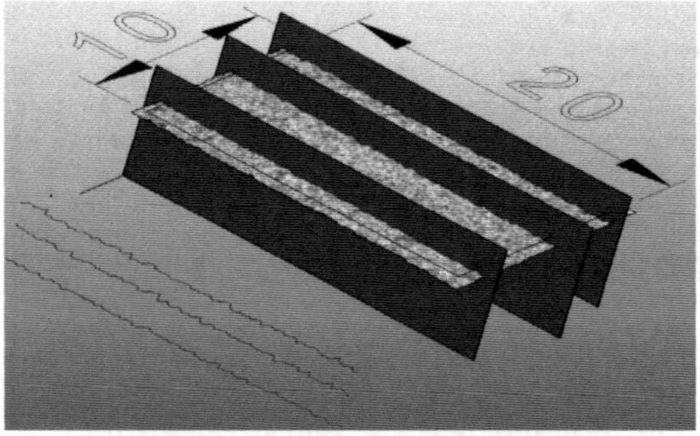

Figure 7. Obtaining classic profiles to determine the surface roughness.

Figure 9. The modernized system of profilographer-profilometer model 202 of "Caliber" plant.

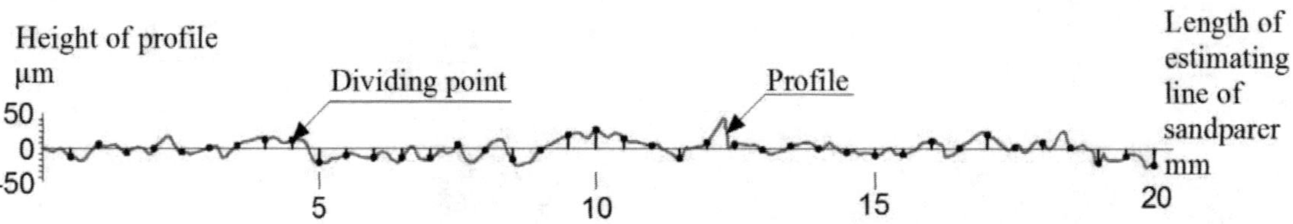

Figure 8. Preparation of the profile to estimate the Ra roughness parameter.

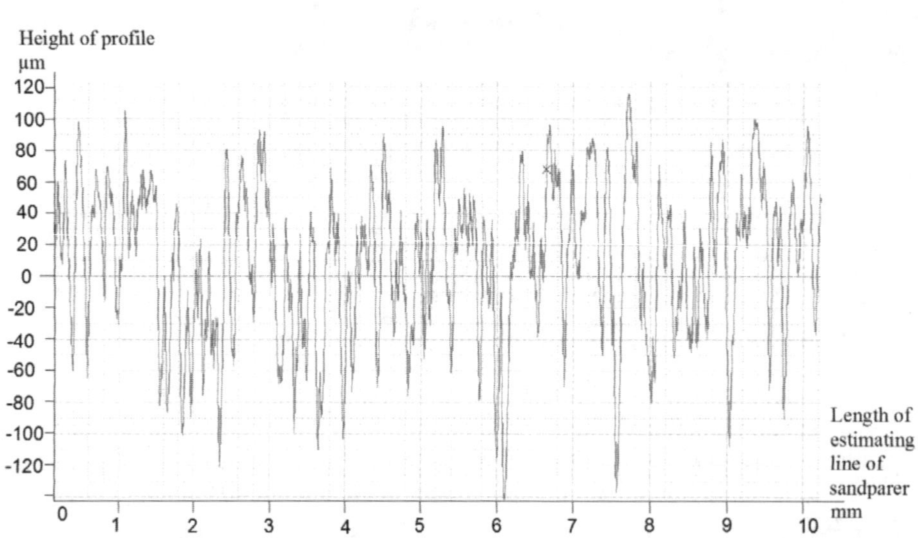

Figure 10. A sample of resulting profile which was generated in PowerGraph software.

3. Results

After performing all the necessary research and estimations, results were summarized in the graph shown in figure 11. It is also necessary to present a graph which shows the dependence of grain size to the ISO-6344 mark of sandpaper (figure 12).

Having received important data there is a need to make conclusions and tasks for future research.

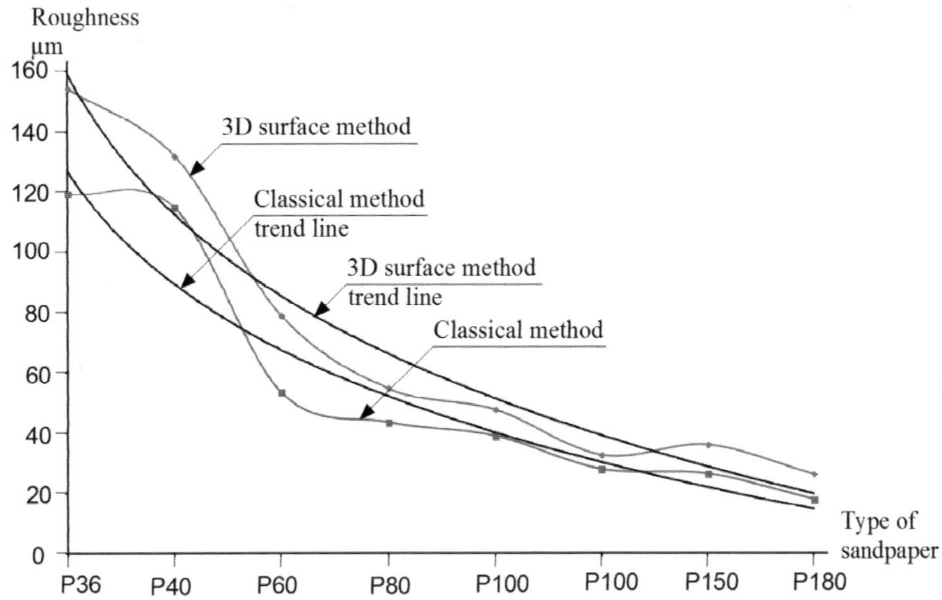

Figure 11. Estimated dependence of surface roughness of abrasive paper on different grain size according to ISO P marking.

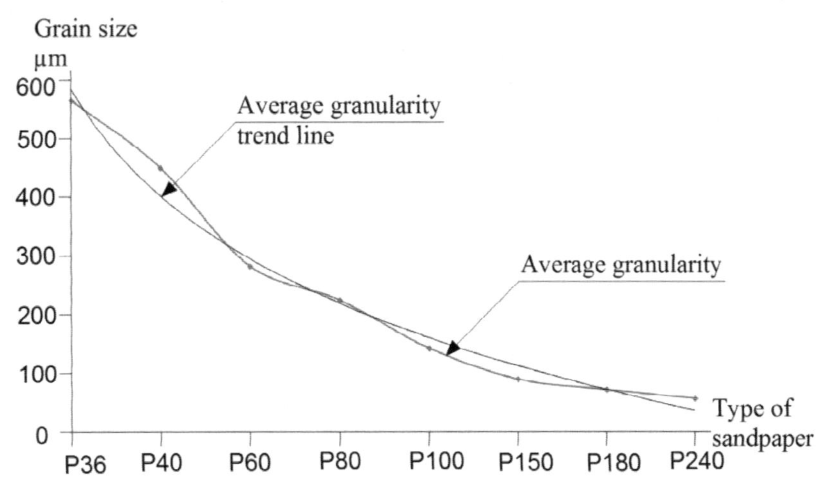

Figure 12. Average granularity of abrasive paper depending on ISO P marking.

4. Conclusion

According to the graphs shown in figure 11 and figure 12, the following conclusions can be drawn:

- Graphs of surface roughness obtained by the classical contact method, 3D scanning method and the dependence of grain size on the marking of sandpaper are almost of the same shape, respectively, and their graphs of average values are also of the same shape;

- A small offset of the roughness graph obtained by the 3D method relative to the roughness graph obtained by the classical method can be explained by the fact that the abrasive grains are attached to the sandpaper with glue, which in contact determination of roughness peels off, so roughness is "more contrasting".

In our future work, the estimation of roughness in real time will take place without the use of a number of computer programs, and using a special mathematical algorithm which will be saved into the analyser itself.

Another important step is to establish the link between the estimated current roughness and the change of processing parameters, which will be carried out by automated control devices which are provided by the CNC system.

References

[1] Hansel D 2005 Abrasive blasting systems *Metal Finishing* **102** pp 21-37

[2] Herbert D 2008 Improving air-blasting operations with off-the-shelf option *Metal Finishing* **106** pp 22-25

[3] Hower H 1991 New in the surface preparation by abrasive blasting blast plating *J. Prot. Coat. And tinings* **8**, pp 38-45

[4] Settles G and Garg S 1996 Scientific view of the productivity of abrasive blasting nozzles *Journal of Thermal Spray Technology* **5** 35-41

[5] Spencer L 1975 Abrasive blasting *Metal Finish. Gvideb. Direct., Hackensack* **7** pp 67-75

[6] Provolotskij A 1989 *Abrasive jet processing of machine-building parts* (Kyiv: Tehnika) p 171

[7] Andylahaj A 2013 Scientific foundations of effective finishing abrasive treatment of parts with submerged jets *Dissertation of Doctor of Technical Sciences. Specialty - 05.02.08. (Mariupol)* p 533

[8] Sychuk V, Zabolotnyi O and McMillan A 2015 Developing new design and investigating porous nozzles for abrasive jet machine *Powder Metallurgy and Metal Ceramics* **53** pp 600-605

[9] Sychuk V, Zabolotnyi O and Somov D 2019 Technology of effective abrasive jet machining of parts surfaces *Advances in Design, Simulation and Manufacturing*

[10] Sychuk V, Zabolotnyi O, Golodjuk R and Koljadunskyj M 2016 Assembled nozzle for abrasive jet machines *Patent of Ukraine* **104492**

[11] Sychuk V, Zabolotnyi O, Somov D, Golodyuk G and Slabkyi A 2018 Surface roughness measurement of casted parts after abrasive blasting processing *Int. Conf. MMS*

IOP Conf. Series: Materials Science and Engineering 1060 (2021) 012006 doi:10.1088/1757-899X/1060/1/012006

Application of interface guided waves for structural health monitoring of hybrid bonded joints

Mark Jahanbin

The Boeing Company

E-mail: mark.jahanbin@gmail.com

Abstract. The application of ultrasonic interface guided waves for the inspection of adhesively bonded joints is studied. A thin adhesive layer forms a bond line between the surfaces of distinct structural elements. This work demonstrates the formation of specific ultrasonic guided waves which are propagating on the boundaries of bonded joints and can be used for health monitoring of adhesive bonds. Interface waves are extremely sensitive to changes in elastic and plastic characteristics such as density and viscosity of the adhesive layer. Using finite element simulations, the changes in propagation of wave form, attenuation of leaky interface wave and Time of Flight (TOF) were observed and recorded as a baseline. Then the pervasive effect of bonding failure on these parameters are used for damage detection in adhesive bonded joints. The cohesive damages such as dis-bond and delamination are slowing down interface wave speed depending on size and location, and the adhesive properties also change the Time of Flight of propagating wave. The results of this study show that interface waves can be used to inspect adhesively bonded joints and, possibly, to determine the strength of the bond line and predict the failure mechanisms of bonded structures.

1. Introduction

Metallic–composite hybrid structures in aircraft are often joined using mechanical fasteners. The fasteners add weight, increase localized stress and damage the structure due to drilling, and also add labour and part inventory costs. Total elimination or reduction in the number of fasteners in hybrid joints can reduce cost and weight. Adhesive bonding is emerging as an alternative to mechanical fasteners for joining hybrid structures. Adhesive bonds are already inherent to composite materials at the laminate level where plies are co-cured together. The usage of adhesively bonded hybrid structures with customized mechanical properties and adaptive functional performance is gradually changing the traditional design concept, especially in aerospace structures.

While adhesive joining techniques offer numerous design advantages over conventional joining methods, there is a great reluctance to use adhesive bonding, because it is sensitive to process variables (surface treatment, cure cycle, etc.) and it is not feasible to reliably inspect once the bonding process is completed. The development of a non-destructive technique to assess the quality of a bond has long been recognized to be critical to the acceptance of adhesive bonding as a substitute for conventional fastening techniques in high-strength applications. Regulatory agencies demand the development of repeatable and reliable non-destructive inspection techniques to ensure the strength of the bonded joint [23, 24].

Historically, specific modes of ultrasonic waves, so called interface waves, were introduced by Rayleigh, Stoneley, and Scholte and primarily used in geology for geophysics phenomena such as seismic waves for earthquake measurements. These ultrasonic waves propagate in all directions away from the excitation source or transducer but are primarily confined to the interface between two distinct media. One possible wave propagating along the interface between two solids is a Stoneley wave [1]. Interface Stoneley waves exist for certain combinations of isotropic media [2]. Claus and Kline first introduced the application of Stoneley-like guided ultrasonic waves for bond inspection [3-5]. Others used the surface and interface wave approach to study the bonding in solid media and adhesive joints [6-9]. Rose introduced a feature mapping approach for bonded joint with the concept of surface waves by usage of angle beam transducers introduced for non-destructive inspection [10] and later Rokhlin used the critical angle beam for assessment of interfacial properties [11]. Rose and Pilarski also introduced the mode conversion of interface wave to surface wave and bonded Lamb wave in weak joint [12]. Since the 1990s many researchers have used interface waves to characterize the quality of bonded joints but the problem of measurements for strength of the bond line remained unsolved [13-23]. Presently, there are no viable damage growth inspection methods in industry to evaluate the bond integrity of adhesively bonded structures and, moreover, certification requires assembly and component proof testing in which the results depend heavily on the coupon configuration.

The method used in this work is versatile and can be used to monitor the structural integrity of composite laminates and adhesively bonded joints in an in-situ manner and throughout the service life of the structure. It utilizes a guided wave mode to inspect the adhesive bond, regardless of topography and through-the-thickness structure of bonded sections.

The fundamentals of ultrasonic guided interface wave are briefly presented in the methodology section. Typical cohesive failure common to bonded joints such as dis-bond crack at free edge and delamination at first plies of the hybrid bonded joint has been explored by the authors [18-19]. In this study the numerical simulation of an adhesively bonded joint with aluminium and composite section is performed using finite element method. Finally, this model is used for different failure scenarios testing, associated with possible adhesion and cohesion damages in bonded joints

2. Methodology

Guided waves are acoustic waves that are guided by boundaries. Depending on the structural geometry, guided waves can either propagate between boundaries, which are known as plate waves (Lamb waves) or propagate on the surface of the object (surface-bonded waves or acoustic surface waves).

2.1. Modelling interface wave at hybrid joints

This section describes the modelling and finite element simulation of hybrid joints and interface wave at the bond line of composite–aluminium section. The benefit of FEM is that it provides the opportunity to understand the physics and possible implications of the results, before constructing the test coupons and experimental studies. The schematic of a hybrid bonded joint is shown in figure 1. The upper section of the model is metallic (aluminium) and the lower section is a laminated CFRP composite with the ply orientation in principal [0, 45, -45, 90, -45, 45, 0] degree orientations stacking sequence.

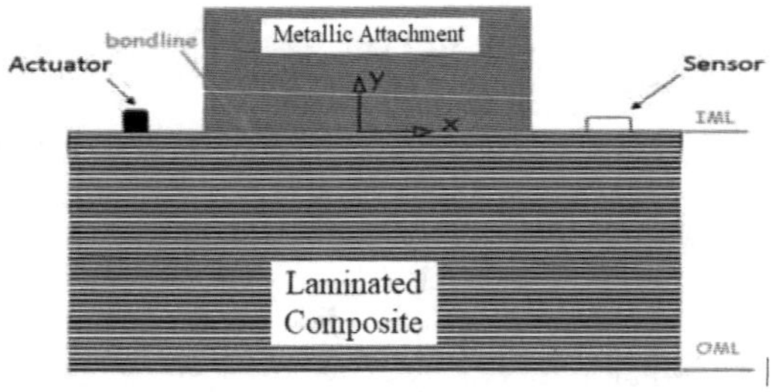

Figure1 Schematic of hybrid structure [17]

The interfacing bonded ply is called Inner Mould Line or IML and the farthest ply from the interface bond line is Outer Mould Line or OML. The orientation of IML is critical for determination of interface wave speed at the bond line. This configuration replicates situations in which metallic fittings are attached to the composite structure in wing fabrication or in repair circumstances where a composite patch is used for metallic skin or repairing composite surface with metallic patch. This model also can be used to investigate the bonding condition of two composite sections by changing the metallic section to another laminated or honeycomb composite. In some applications to avoid corrosion between metals and CFRP, layers of different materials might be used at the interfaces, and for composite–composite hybrid interfaces, a layer of adhesive might be used; all can be simulated by introducing a third material or viscous properties of adhesives.

Each lamina in the composite laminate is a single ply with typical ply thickness of 0.2 mm. The interface waves decay away from the bond line so the orientation and thickness of other plies through the thickness of the composite section is not of interest. However, typically the 2nd and 3rd plies play an important role in the interlaminar buckling and delamination damage of the composite section.

The metallic section has dimensions 76.2 x 101.6 mm, while the composite section has dimensions 254 x 101.6 mm. The length of the bond-line is 101.6 mm at the interface of composite and metallic sections. Aluminium is the material used for the metallic section with elastic constants, E= 69 GPa, and Poisson's ratio, ν = 0.3, each ply of the unidirectional composite and has four elastic constants of E1 = 137.9 GPa, E2 =10.3 GPa, G12 = 6.2 GPa, and ν = 0.34.

ABAQUS finite element code was used for wave propagation simulation and analysis. Plane strain continuum shell elements were utilized. These elements include the effects of transverse shear deformation, which is critical for studying interface waves. The anisotropic section is built up by composite layup tools to form a balanced, symmetric laminate and all the unidirectional plies are meshed accordingly. A typical mesh had around 520,000 nodes and 520,000 linear quadrilateral two dimensional plane strain (PS4) elements. The model is constrained for translation and rotation at the two upper corners of the metallic section and the two lower corners of the composite section. An excitation source or actuator is placed on the composite section on one side, and a receiving sensor is placed on the other side of bond line. A Hanning-windowed, five-cycle burst of a sinusoidal signal was used as the pulse excitation signal. The initial forcing frequency was set at 1 MHz to model the interface wave. A 10 MHz excitation has also been used to investigate the influence of frequency on the damage identification process.

2.2. Bonded Joint mechanism

The high frequency of interface waves and independence from the through-thickness topology of the structures offers a promising wave mode for the detection of very small Interfacial defects. Dis-bonds and delaminations of hybrid interfaces and localized matrix defects or fibre fracture in laminated composites are among the critical failure modes of these structures. In prior work [19, 20] the application of interface waves for the detection of interlaminar and intralaminar defects in hybrid joints was demonstrated. In reality there is an adhesive layer in between the substrates and coatings that is referred to as the bond line, as shown in figure 2.

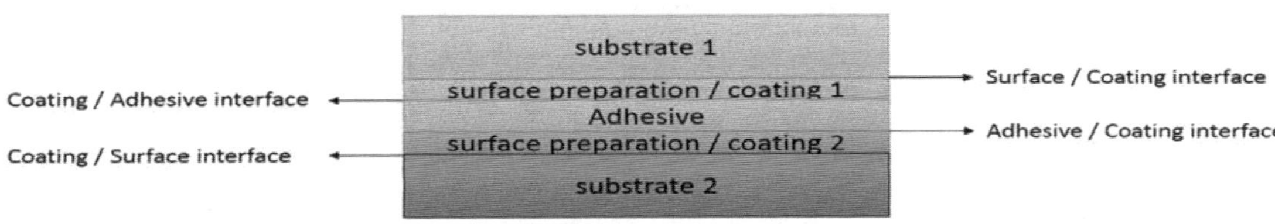

Figure 2. Idealized adhesive bonded joint.

The bonded load path is a chain of material and interfaces that form the adhesive bonded joint. The strength of the bond is determined by the weakest link in this chain. The interface wave frequency of this application can be optimized to the efficient wavelength as explained in reference [20] to cover the entire chain of the bond line such that it can capture any adhesive or cohesive degradation of the elements shown in figure 4. The interface wave travels on the bond line and damage can cause distortion in the path of the traveling waves. If the damage is due to failure of materials, coatings and adhesive, then the bonding process is probably reliable, but if the failure is due to an anomaly in any of the four interfaces between substrates and coatings or between the coatings and the adhesive the bonding process is not reliable. The latter is the primary focus of the study presented in the following sections.

2.3. Numerical Simulation of Bonded Joint

The physics of interface wave propagation in hybrid structures is considered. The analytical Stoneley wave solution is used to determine the frequency of excitation for the interface. Stoneley-like waves form at the bond line of isotropic and anisotropic media and the mode conversion due to interfacial damages can be used for detection purposes. Numerical simulations have been performed using the Explicit Dynamic solver of the ABAQUS Finite Element code to predict the existence of interface waves and to define their propagation characteristics based on the mechanical properties of the bonded joint and adhesive layer. Interface wave propagation is simulated in the interfaces between laminated Carbon Fibre Reinforced Polymer (CFRP) and aluminium alloy 6061, with HYSOL EA 9394 adhesive in between. Table 1 lists the materials used in this simulation.

Table 1. Material Properties of hybrid bonded joint.

Sections	Modulus: E, G (GPa)	Density, ρ (g/cm^3)	Poisson's ratio, ν
Composite ply (CFRP)	E1 = 137.90 E2 = 10.34 G12 = 6.89	1.55	0.34
Adhesive HYSOL	E = 3.72 Viscosity of adhesive: 1 × 10^{-6} kPa*s (1 centipoise)	1.38	0.40

The Adhesive simulated in this modelling matches the low viscosity adhesive we are using for experimental validation. The adhesive Dexter Corporation's Hysol EA-9394 is an amine-cured epoxy paste adhesive which can be cured at room temperature. The adhesive has density of 1.38 g/cc, a porosity of about 6% which makes it compatible for post peel ply application to CFRP composite. Hysol has glass transition temperature of 82oC and a coefficient of thermal expansion of approximately 60 x 10-6 oC-1 (between -30o C and 70o C)

Theoretical studies suggest that the interface waves propagate with speeds lower than the lowest speeds of bulk waves in the denser media. The finite element simulation shows that the velocities of the guided waves depend on the frequency of excitation, material orientation, and specific material properties at the interface boundary.

The properties and quality of the adhesive layer can change the velocity of the interface wave. For a hybrid bonded joint with the similar configuration, the approximate velocity of interface wave is about 3 mm/μs, with the assumption of a perfect bond without adhesive layer. The velocity of the interface wave increases due to the presence of the adhesive layer. The wave velocity is the function of surface coating interfaces and the density, viscosity and elastic–plastic deformation of the adhesive layer.

In this work, the reaction of the interface waves to the changes in the bonding interfaces at the bond line of an adhesively bonded joint is investigated. Previously it was shown that the velocity of interface waves decreases as the size of dis-bond cracks and delaminations at the bond line increase. In this study, the focus will be on interface degradation that depends on the bonding process. Therefore, the changes in all elements of the interface such as surface preparations, coating types and adhesive itself have been analysed to evaluate the detection characteristics of interface waves. This work demonstrates the cases in which interface waves are sensitive to bond defects and wave form distortion can be used to detect damage. The density and viscosity changes of interfaces also interfere with the normal wave behaviour. This study is in the most part focused on the interfacial damages to characterize the adhesion defects and predict the possibility of bond failure. Figure 3 (a) shows the wave form attenuation in the vicinity of an adhesive layer with pristine density and modulus properties, and figure 3 (b) shows the attenuation of the same wave form when it encounters a low viscosity adhesive with reduced stiffness.

The detection technique described in the next section, requires further assessment of the interface wave parameters such as reflection, attenuation, and energy dissipation.

(a)

(b)

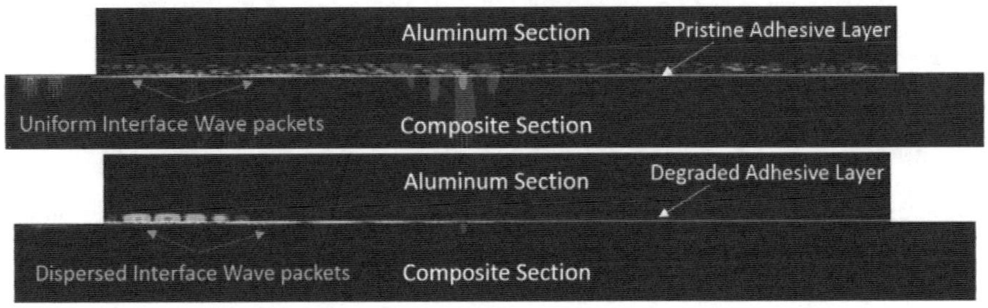

Figure 3. (a) Pristine adhesive property – normal wave behaviour
(b) Degraded adhesive property – anomalous wave behaviour

3. Damage modelling in adhesively bonded structure

The damage and failure of adhesively bonded joints is fuzzy and complex. It is important to identify and understand the nature and location of failure in a bonded joint and composites. The bonding process and compatibility of material selection with appropriate surface preparation and coatings are primary factors for constructing a good bond. Some other factors such as environmental effects, corrosion, heat damage, fluid ingression and contamination could be considered as secondary causes of failure of adhesive bonds.

In industry, these requirements for the bonded joint are critical to the adhesion failure because they affect bonding process reliability. Any other form of mode I, II, III and mixed fracture modes, which causes cohesive material failure are irrelevant to the bonding process. Hence damage modelling of the bonded joint is focused on the damage nucleation at surface–coating - adhesive interfaces and on the degradation of the adhesive layer itself. The bond region is resin-rich and the reduced stiffness and density of resin-rich area causes the wave velocity to decrease. If the changes in physical and chemical properties of adhesive cause thickness change, then the damage can be detected easily; because it will be visible in the form of skin pull-up, pull-down or wrinkles at the adhesively bonded joint.

The other factor contributing to the defects of adhesive bonds is cohesive failure which might happen separately, in conjunction with or in addition to adhesive failure. The failure at substrates, surface preparations or coating and adhesive itself, which are shown in figure 4, are all cohesive failures. These types of damages are not as much critical as the adhesive failure which are shown as interfaces between bonding chain in figure 4.

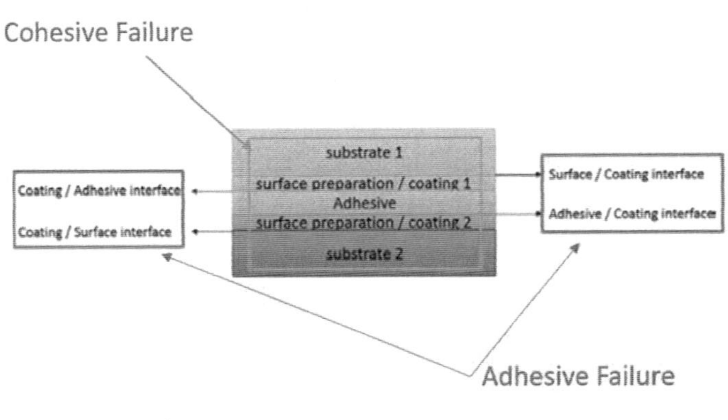

Figure 4. Cohesive Failure *vs.* Adhesive Failure

4. Results

In this section the results of numerical simulations for both adhesive and cohesive type of damages are presented. Damage was simulated in form of drack, delamination and also through adhesive layer density reduction. Dis-bond of hybrid interfaces, delaminations, and localized matrix defects or fibre fracture in laminated composites are among the critical failure modes of hybrid structures.

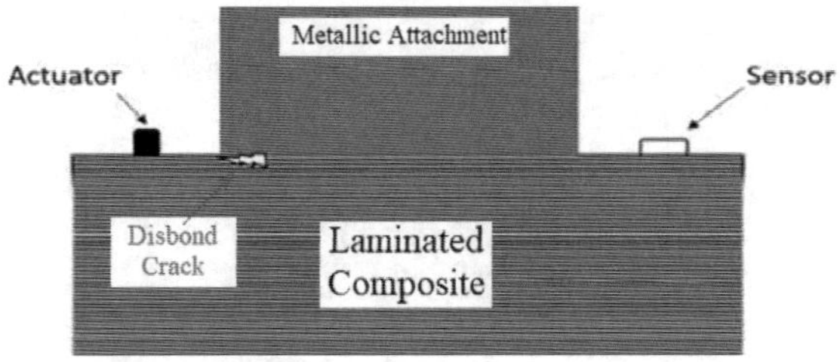

Figure 5. Dis-bond nucleation / initiation locations

For damage simulation, we introduce a dis-bond crack at one free edge of the bonded section (see figure 5) and monitor the disruption in the continuity of interface wave propagation to calculate the delay in time of flight (travel time of interface wave from actuator to receiver). Figure 6 shows the undamaged or perfectly bonded interface as well as some of the different lengths of dis-bond cracks that are considered for this simulation.

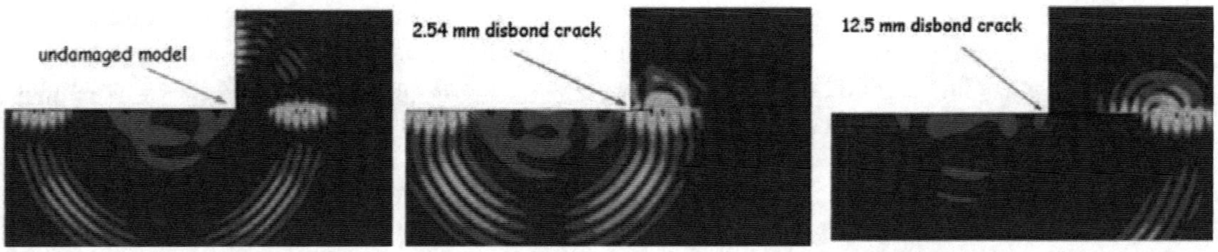

Figure 6. Dis-bond crack size effect on interface wave propagation

Both actuator and sensor are placed on the exposed surface of the composite section. Several different crack sizes are considered ranging from 2.54 mm to 25.4 mm for a 101 mm bond. Simulations were also performed for a pristine undamaged (UD) structure to provide the baseline. The travel time of the propagating wave from the actuator to the receiver is recorded. Wave motion in the x coordinate direction is perceived at a slightly different time than wave motion in the y-direction and hence these two travel times are differentiated. The nominal delay time of recorded displacement in y direction is about 1.5%. The results are presented in figure 7 for four crack simulations. As the size of damage increases the time of travel of the wave pulse increases. The interface wave is faster when the direction of wave propagation aligns with the composite ply direction (i.e. 0°) whereas 90° ply orientation causes the slowing of the interface wave. We observed nominal 5% delay in time of flight for recorded signals at sensor when the fibre orientation is perpendicular to the interface wave travel path on bond line. For both cases, the time of travel appears to be linear with the crack size.

Table 2 and figure 8 show the interface wave velocity change with decrease in density of the adhesive layer. Some factors that contribute to the attenuation of interface wave in degraded adhesive model are absorption and scattering of the wave. In reality, the pervasive effects of chemical and viscous effects are not linear, even though the numerical simulation represents a linear trend on interface wave propagation velocity.

For a typical SHM setup, the baseline is pristine or undamaged condition. Then, the most appropriate signal processing methods are used to detect damage by comparing signal voltage, energy dissipation, velocity, and time of flight, among others. The study reported in this paper did not elaborate on all of these methods and focused mostly on the time of flight, which corresponds to the velocity of wave and mode conversion in case of any damage at the bond line. The low-viscosity adhesive EA 9394 used for simulation and experiments is available with different densities and moduli of elasticity.

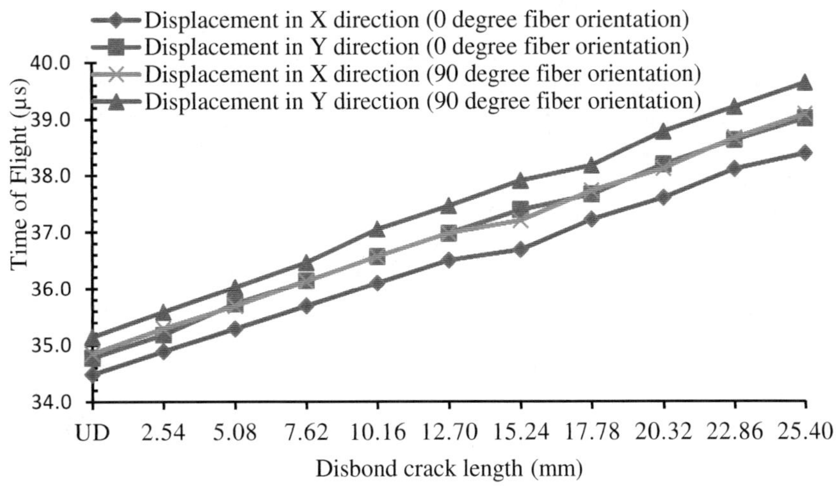

Figure 7. Travel time for interface wave signal versus the dis-bond crack size

Table 2. Interface waves Time of Flight (TOF) between points with 101.6 mm distance

Points	Density Change (g/cm^3)	TOF (µs)
UD	$\rho = 1.38$	14.89
D1	$\rho = 0.70$	14.85
D2	$\rho = 0.35$	14.79
D3	$\rho = 0.17$	14.77
D4	$\rho = 0.09$	14.65

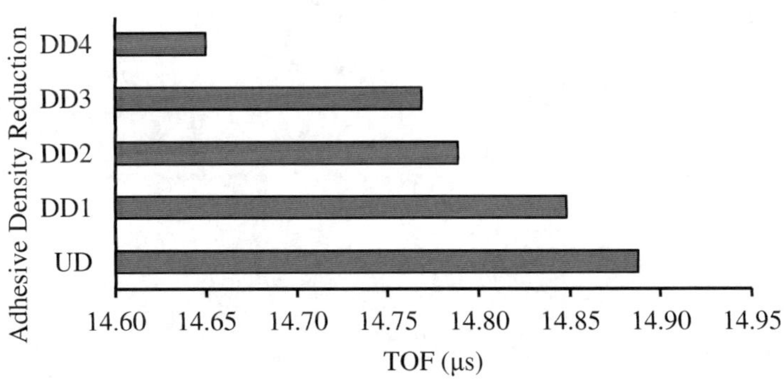

Figure 8. Change in time of flight of interface wave with adhesive density reduction

Usually, the reasons for adhesive characteristics deterioration are not purely physical, they may be caused by chemical or other factors, such as environmental, that can contribute to degradation of the adhesive bond. The debonding failure occurs in the plastic regime of the adhesive layer. The detection technique requires further assessment of the interface wave parameters such as reflection, attenuation, and energy dissipation. In this study, the extreme change of density magnitude has been used as the most efficient way to build up the model and to verify the effect of lowering the density or viscosity of the adhesive. There was no attempt to replicate effects of moisture/humidity, chemicals, temperature variation, vacuum, radiation, and many of the pre-post boding effects, such as surface preparation. They may be studied in the future. This work was focused on demonstrating typical and common adhesive and cohesive failures in bonded joints.

This phenomenon is in good agreement with the general equation of wave velocity in solid media where $V \propto (E/\rho)1/2$ which confirms the reduction in density is a contributing factor to time of flight

IOP Conf. Series: Materials Science and Engineering 1060 (2021) 012006 doi:10.1088/1757-899X/1060/1/012006

decrease or velocity increase. Another factor to characterize adhesive bond quality is the energy method and the measurement of the energy dissipated per unit volume by viscous effects, *i.e.*, the EVDDEN factor, which is an output in finite element analysis. The interface energy is confined to the region near the boundary. The energy method is more reliable compared to the time-of-flight measurement and attenuation. For instance, TOF monitoring is tedious and needs very accurate instrumentation for small size test coupons. The measurement of attenuation might also not be preferred because it is difficult to consistently identify the attenuation due to leakage when there are other causes for loss of energy in measurement.

Previous studies have found that the distribution of the energy of the interface waves above and below the interface changes repeatedly with propagation distance due to interference between the two modes which have slightly different phase velocities. The pervasive effect of cohesive failure on interface wave is very similar to that demonstrated by the studies on dis-bonds and delaminations. The high frequency interface wave is typically effective within the range of bonding construction which is less than one wavelength. Hence, if there is any discontinuity in the pathway of the traveling wave, the wave behaviour, such as the time of flight, changes and the effects change linearly with the size and location of the discontinuity.

Figures 9 (a) and (b) shows the scattering of interface waves due to a cohesive damage crack at the free edge on the left. This dis-bond and delamination at the interface of the coating and the adhesive causes a Stoneley to Rayleigh wave mode conversion plate like Lamb wave formation at upper or lower bond of adhesive layer. Either case will change the slowness profile of the interface wave and reduction in wave form velocity.

The results of these modelling techniques and damage simulations might be of particular interest for the development of experimental test setups and the basic idea of actuator – sensor placement for a novel pitch–catch ultrasonic structural health monitoring system, designed for laminated and hybrid structures.

(a)

(b)

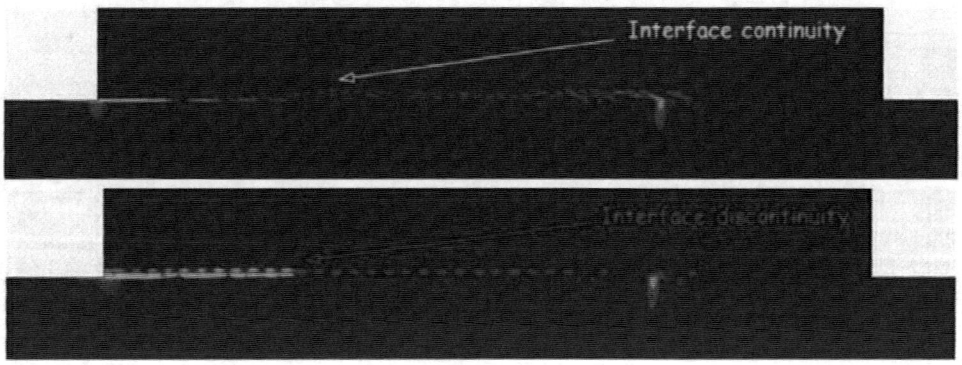

Figure 9. (a) Undamaged adhesive bond, (b) Damaged cohesive bond

5. Conclusion

This research is devoted to the use of a Structural Health Monitoring (SHM) system to assess the quality of adhesive bonds between two materials, by measuring the propagation of ultrasonic interface waves through the adhesive, guided by the physical interface between the materials and the adhesive. Anomalies in interface waves generated by an ultrasonic signal applied to the bonded materials are a mixture of wave effects resulting from differences in velocity, phase, and amplitude, originating from differences in material viscosity, density, thickness, continuity, and specifically differences at the physical boundaries between materials.

This research summarizes the results of investigating the following measurable aspects to differentiate good bonds from bad bonds. First, wave velocity (wave-front time-of-flight over a given distance) is measured. Second, wave attenuation (diminishing amplitude over a known distance due to absorption and scattering of the interface wave) and third, the wave energy dissipation (diminishing

IOP Conf. Series: Materials Science and Engineering 1060 (2021) 012006 doi:10.1088/1757-899X/1060/1/012006

amplitude integral over a known distance due to absorption and scattering of the interface wave). The results of this study may lead to the development of a repeatable and reliable inspection method for the assessment of adhesively bonded joints.

References

[1] Stoneley R 1924 Elastic waves at the surface of separation of two solids *Proc. Roy. Soc. London A* **106** 416–28
[2] Scholte G 1947 The range of existence of Rayleigh and Stoneley waves *Geophys. Suppl. MNRAS* **5**(5) 120–6
[3] Claus R O and Kline R A 1979 Adhesive bondline interrogation using Stoneley wave methods *J Appl. Phys.* **50**(12) 8066–9
[4] Claus R O 1980 Optical Measurements of Ultrasonic Waves on Interfaces Between Bonded Solids *IEEE Trans. Sonics Ultrasonics* **27**(3) 97–102.
[5] Claus R O 1980 Surface and near-surface defects in glass-to-glass bonds" *Proc. SPIE Int. Soc. Opt. Eng.* **250** 38–44
[6] Kumar V and Murty G S 1982 Influence of External Pressure on Interfacial Wave Propagation at a Loosely Bonded Solid-Solid Interface" *IEEE Trans. Sonics Ultrasonics* **29**(3) 138–42
[7] Rokhlin S I 1983 An ultrasonic bridge for the study of viscoelastic properties of thin interface films *J. Acoust. Soc. Am.* **73**(5) 1619–23
[8] Kumar V 1983 Attenuation and velocity of waves propagating along a steel-steel interface *J. Appl. Phys.* **54**(2) 1141–3.
[9] Yew C H 1984 Using ultrasonic SH waves to estimate the quality of adhesive bonds: A preliminary study *J. Acoust. Soc. Am.* **76**(2) 525–31
[10] Rose J L, Nestleroth J B and Balasubramaniam K 1988 Utility of feature mapping in ultrasonic NDE *Ultrasonics* **26**(3) 124–31.
[11] Wang W and Rokhlin S I 1991 Evaluation of interfacial properties in adhesive joints of aluminum alloys using angle-beam ultrasonic spectroscopy *J. Adhes. Sci. Techol.* **5**(8) 647–66.
[12] Pilarski A and Rose J L 1992 Lamb wave mode selection concepts for interfacial weakness analysis *J. Nondestr. Eval.* **11** 237–49
[13] Hsieh T M and Rosen M 1993 Ultrasonic leaky waves for non-destructive interface characterization *Ultrasonics* **31**(1) 45–52
[14] Singher L, Segal Y, Segal E and Shamir J 1994 Considerations in bond strength evaluation by ultrasonic guided waves *J. Acoust. Soc. Am.* **96**(4) 2497–505
[15] Lowe M J S and Cawley P 1994 The applicability of plate wave techniques for the inspection of adhesive and diffusion bonded joints *J. Nondestr. Eval.* **13**(4) 185–200
[16] Singher L 1997 Bond strength measurement by ultrasonic guided waves *Ultrasonics* **35**(4) 305–15
[17] Vine K, Cawley P and Kinloch A J 2001 The correlation.of NDT measurements and toughness changes in adhesive joints during environmental attack *J. Adhes.* **77**(2) 125–61
[18] Jahanbin M, Santhanam S and Ihn J-B 2016 Interface Waves in Hybrid Metal-Composite Structures *ASME 2016 SMASIS2016-9007* **1** V001T05A002
[19] Jahanbin M, Santhanam S, Ihn J-B, Bossi R and Cox A 2017 Application of Interface Waves for near surface damage detection in hybrid structures *10169-96_ProcSPIE_SS17-SSN08-35*
[20] Wang H, Han Q-B and Qian M-L 2012 Leaky interface wave measurement at a solid-solid interface with laser ultrasonics *Chinese Physics Letters* **29**.10.104304
[21] Qingbang H and Hao W 2014 Laser-induced interface waves at solid/solid interface *AIP Conf. Proc.* **1581**(33) 437–43
[22] Cho H and Rokhlin S I 2015 Interface wave propagation and edge conversion at a low stiffness interphase layer between two solids: A numerical study *Ultrasonics* **62** 213–22

IOP Conf. Series: Materials Science and Engineering 1060 (2021) 012007 doi:10.1088/1757-899X/1060/1/012007

Radio-shielding metamaterials transparent in the visible spectrum: approaches to creation

A Osipkov*, M Makeev, E Garsiya, A Filyaev, K Sinyagaeva, D Kirillov, D Ryzhenko, G Yurkov

Bauman Moscow State Technical University, Moscow 105005, Russia

*E-mail – osipkov@bmstu.ru

Abstract. The approaches to creation of the materials providing simultaneously high indicators of transparency in the visible spectrum and shielding in a wide radio frequency band are considered in this paper. The analysis and comparison of the main designs of such materials, as well as approaches to their creation, including multilayer and conductive mesh structures, is carried out. The results of our own theoretical studies of the disordered mesh structure are presented, which allow one to obtain a light transmission coefficient from 90 to 98 % in combination with an electromagnetic interference shielding efficiency from 50 to 65 dB. The best results practically achieved to date (shielding efficiency equal 45 dB in the range from 10 kHz to 20 GHz with a light transparency of more than 80 %) were obtained on mesh structures by photolithography, which is a significant limiting factor of this approach. The created multilayer structures show, in general, lower characteristics. However, the technology for their production is better scaled, and the optimization of the thicknesses and chemical composition of multilayer structures can significantly increase them. In this regard, technological aspects may come to the fore when taking into account the possibility of subsequent scaling of the technology and economic indicators when choosing an approach for the implementation of the materials with the required characteristics.

1. Introduction

The rapid development of information technology in the modern world leads to the widespread use of various electronic devices, which increasingly makes the problem of their electromagnetic compatibility more relevant. The operation of many electronic systems and micro-electromechanical devices requires their isolation from exposure to extraneous electromagnetic waves and noise, or vice versa, the need arises to protect against the radiation of the devices themselves, which today is becoming a serious problem due to its adverse effects on human health [1, 2].

To date, the most common shielding method is the creation of protective metal cases. However, under modern conditions there is an increasing demand for materials that should not only have high shielding properties, but also have a different combination of other required operational properties, such as flexibility, transparency in different spectral ranges, weight and size characteristics, air permeability, etc. This makes it impossible to use traditional metallic materials and requires the development of new approaches and materials to protect electronic devices from electromagnetic waves.

This work focuses on the creation of electrically conductive materials that are transparent in the visible spectrum range, which, in addition to the task of electromagnetic radiation shielding, are widely used as transparent electrodes in portable electronic devices. At the same time, solutions used in industry allow to provide either high transparency (more than 90 %) at low electromagnetic interference (EMI) shielding efficiency (SE) values (up to 20 dB), or high SE with a significant reduction in light transparency. In this regard, it is necessary to search such structures and technologies that which allow to achieve high transparency and EMI shielding efficiency simultaneously.

Currently, two main approaches are used to obtain such structures: continuous multilayer structures and conductive mesh structures, as well as their combination. Each approach has its advantages and disadvantages, allowing to achieve various combinations of transparency and shielding efficiency in conjunction with other operational characteristics. This article is devoted to a review of promising design solutions and analysis of approaches to creating materials with high transparency and EMI shielding efficiency.

2. Multilayer structures

Such structures can be realized on the basis of doped wide-gap semiconductor materials from metal oxides and nitrides (ZnO [3], TiO_2, In_2O_3, SnO_2, Zn_3N_2, TiN, etc.), thin metal layers, and conductive organic compounds [4-6]. The most commonly used transparent conductive layers are Indium tin oxide (ITO coatings) systems, which currently have the best ratio of electrical conductivity and light transparency in the visible range (at least 70 %, with a sheet resistance of 15 - 20 Ω/sq, which corresponds to SE of about 22 dB), good adhesive and satisfactory corrosion properties [7-8]. Moreover, their optical properties can be improved (light transparency of more than 90 % at a sheet resistance of 20 Ω/sq) by applying antireflection layers [9]. Figure 1 shows the results of measurements of such coatings obtained by the authors of the current work.

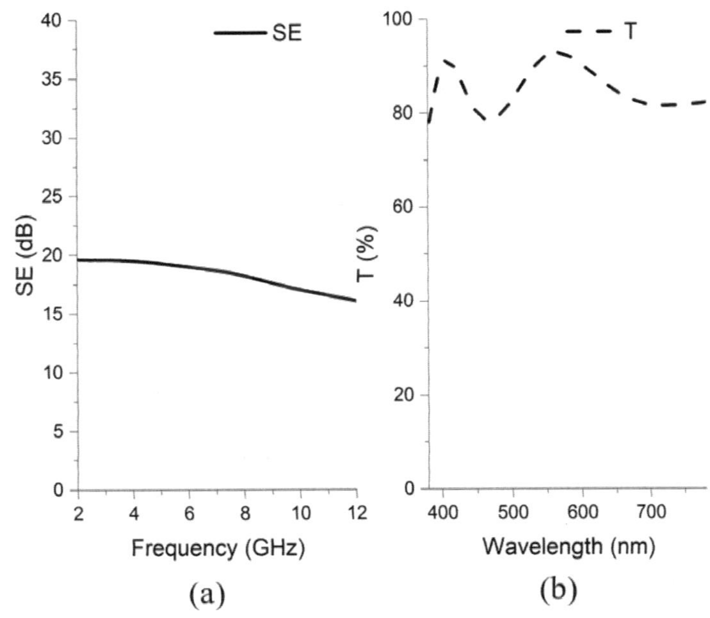

Figure 1. EMI SE (a) and transparency (b) of ITO-coating.

One of the disadvantages of ITO coatings is the high temperature of their growth and annealing, which complicates the deposition of such films on flexible polymer substrates and leads to their somewhat worse characteristics compared to ITO layers on glasses. In view of the recent intensive development of the technology of flexible LCD touch screens, an urgent area of research is improving the technical characteristics of flexible conductive layers [10-11].

In [12], the authors compare the characteristics of an ITO coating (95 % In_2O_3 + 5 % SnO_2) with the structure of ZrO_2 (90 nm) – Ni (8 nm) – ZrO_2 (90 nm) – Cu (22) – Ni (5 nm) – ZrO_2 (35 nm) based on thin metal films, which are made in the form of thin alternating layers of metal with high electrical conductivity and magnetic permeability, separated from antireflective oxide layers, in order to increase light transparency. Both coatings were applied to glass substrates. The ITO coating after annealing in vacuum at a temperature of 350 – 400 °C with a transparency of 95 % has a sheet resistance of 10 Ω/sq (SE 15 - 20 dB in the range of 6 - 17 GHz and 26 – 27 dB in the range of 0.7 – 5.0 GHz). A structure based on thin metal films with a transparency of about 60 % has a sheet resistance of 2 – 3 Ω/sq (SE 26 – 33 dB); however, its deposition technology does not require high-temperature annealing, which allows it to be applied to non-heat-resistant flexible polymeric materials. Similar characteristics were obtained in [7] on the structure (ZrO_2, HfO_2, TiO_2) + Ni + (ZrO_2, HfO_2, TiO_2) + Au + Ni + (ZrO_2, HfO_2, TiO_2) deposited on glass. The shielding efficiency of such samples was not less than 25 dB in the range of 100 kHz – 10 GHz, the light transparency at a wavelength of 580 nm was not less than 55 %, and the sheet resistance of the screen was 2.5 – 3.0 Ω/sq.

In [13], a structure with a total thickness of 312 nm from seven alternating layers of silver and zinc oxide (Ag (11 nm) / ZnO (81 nm) / Ag (22 nm) / ZnO (81 nm) / Ag (22 nm) / ZnO (81 nm) / Ag (11 nm)) was deposited on a polycarbonate substrate, which had an SE in the range from 30 kHz to 1 GHz of more than 40 dB with a transparency of about 60 %. In [14], a dielectric-metal-dielectric shielding structure based on a ultrathin (8 nm) copper alloyed silver film (the electromagnetic Ag shielding, EMAGS) was proposed, with a transparency of 96 % and a SE of about 26 dB in a wide frequency band 32 GHz, covering all the X, Ku, Ka, and K. bands. Coating of two such structures made it possible to achieve SE more than 30 dB in the range of 8 – 40 GHz with a transparency of 93 %, and their separation by a layer corresponding to a quarter of the wavelength was up to 50 dB. It is worth noting that the proposed EMAGS structure can be applied on a large surface area using vacuum deposition with roll-to-roll system.

In order to further improving the characteristics of transparent conductive multilayer coatings, a number of structures based on a combination of thin metal layers with wide-gap oxide semiconductors are proposed. In [15], optical materials consisting of alternating ITO and Ag layers were developed and produced. There shielding efficiency at layer thicknesses of 40 nm and 20 nm, respectively, exceeded 40 dB in the range from 0.1 GHz to 18.0 GHz, and even reached 70 dB at a frequency of 0.1 GHz. Experimental study of the coatings' transparency with different ratios of layer thicknesses showed that the highest transparency about 90 % in the visible range was achieved for ITO and Ag thicknesses of 60 nm and 20 nm, respectively. The transparent layers based on conductive organic compounds have not yet allowed to achieve high functional characteristics. In [4], a thin film of polyaniline (PANI) synthesized by the method of self-stabilized dispersion polymerization (SSDP) [5-6] at a sheet resistance of 500 Ω/sq, which corresponds to SE 2.7 dB, showed a transparency of 55 - 60 % at wavelengths of 500 - 600 nm. Thus, the ITO coatings that are widely used today do not allow to obtain a high SE without a significant decrease in light transparency, which, as a rule, does not exceed 25 dB with a transparency of 90 %, while on flexible substrates they have shown a significant decrease. Higher values of these parameters can be obtained using multilayer structures based on thin metal layers (Ag and Ni, less often Au) in combination with antireflective oxide dielectric or semiconductor materials (ZrO_2, ZnO, HfO_2, TiO_2, ITO, etc.).

3. Mesh structures

A constructive feature of such structures is the creation of the conductive mesh on or inside the surface of a transparent substrate with certain parameters, such as: orderliness of conductors; cell geometry; geometry of conductors forming conductive tracks; electrical conductivity of the tracks.

Disordered conductive mesh are most often formed on the basis of silver nanofibers (AgNWs) randomly oriented on the substrate surface [16-21], while ordered grids are obtained by microelectronics methods [22-36]: photolithography or direct laser etching, etching through masks, and etc. At the same time, some authors have attempts to create ordered grids using methods of self-organization [37-42]. Different combinations of transparency, shielding, or sheet resistance can be achieved by adjusting the parameters of the mesh structures [22].

3.1. Disordered meshes

A common method of forming such meshes is application of a suspension containing metal nanoparticles (nanofibers) to a transparent substrate [16-21]. The metal particles form a disordered conductive network on the surface (Figure 2a) after removing the dispersing medium during the drying process.

In [16] a polyimide coating with AgNWs was obtained (average diameter 30 nm, length 30 μm), which made it possible to obtain transparency of about 90 % in the range 450 – 750 nm and SE about 25 dB in the frequency range 100 – 1500 GHz. In order to increase the electrical conductivity of the layer, galvanic deposition of copper was carried out, which made it possible to reduce the sheet resistance from 500 to 450 Ω/sq with a decrease in transparency to 80 % during the deposition of Cu for 6 min. An increase in the deposition time of up to 8 min reduced the resistance to 28 Ω/sq, which provided a shielding efficiency of 55 dB with a decrease in transparency up to 58 %.

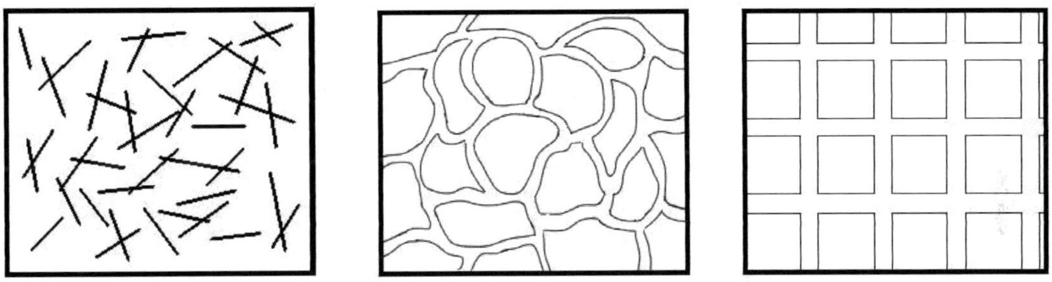

Figure 2. An example of structures based on metal meshes:
(a) Network of randomly oriented AgNWs [16],
(b) Structured mesh obtained by self-organization during drying of emulsion with Ag nanoparticles [38],
(c) Ordered copper grid produced by photolithography [26].

In [17] a shielding film based on a disordered network of AgNWs, calcium alginate and polyurethane were presented. This structure has a high transparency (T = 92 %) and a shielding efficiency (SE = 20.7 dB). With an increase in the surface density of silver nanofibers from 58 mg/m^2 to 174 mg/m^2, the shielding coefficient can be increased to 31.3 dB while maintaining the transparency of 81 %. In addition, these transparent films have high mechanical stability under difficult operating conditions with 98 % and 96 % shielding preserved even after 30 minutes of ultrasonic treatment and 5000 bending cycles (radius 1.5 mm), respectively.

In order to improve the ratio of transparency and shielding (electrical conductivity of the layer), a number of studies have proposed various methods for ordering the location of Me nanoparticles on the surface of a substrate [37] or for creating grooves of a certain geometry with their subsequent filling with metal. In [43-45], such grooves are obtained using self-organization mechanisms associated with the destruction (cracking) of the sacrificial layer. The parameters of the resulting grooves depend on the chemical composition (for example, on the mass ratio between methyl methacrylate (MMA) and butyl acrylate (BA) in the emulsion) and the thickness of the applied sacrificial layer.

In [43], a method for producing a Ag mesh on an quartz substrate with a line width of 0.5 μm and a grid pitch of 30 – 80 μm by using the destruction of a sacrificial layer of 3 μm thickness, deposition into the obtained grooves Ag with a thickness of 200 nm, and subsequent etching of the sacrificial layer was described. In comparison with the ordered Ag grid (line width 5 μm, grid pitch 250 μm, layer thickness 200 nm), which gives a transparency of 95 % in the range 300 – 800 nm and shielding efficiency from 17 to 20 dB in the range 12 – 18 GHz, the obtained sample (the mass ratio between MMA and BA in the emulsion is equal to 2) has very similar transparency characteristics; however, the SE is only about 10 dB. By increasing the mass ratio between MMA and BA in the emulsion to 3, it is possible to increase

SE to 28 dB with a transparency of 92 %. An important feature of this technology is the possibility of applying conductive coatings on surfaces of complex shape and large area, including domed glass.

From our point of view, another promising approach that can be used to form conductive meshes of a given geometry is based on the technology of self-organizing "breathing figures" [37-39], which allows one to produce porous polymer films with a controlled pore size (Figure 1b)]. The concept of "breathing figures" describes the condensation of many drops of moist air upon contact of water vapor with a cold surface. At the same time, not all the processes and results of this method obtained in practice have been explained and modeled theoretically. Therefore, we will look towards the development of this method in our further works.

An analysis of the works on the creation of radio-shielding (conducting) transparent materials based on disordered meshes showed that some ordering of the conductors improves their transparency and shielding efficiency; however, the characteristics of such materials are still far from materials based on ordered grids obtained by lithography methods. At the same time, the development of technologies and methods of self-organization has great potential for improving the performance of the resulting structures with a slight increase in the cost of the technological process and the possibility of its scaling.

3.2. Ordered Grids

The most common methods for ordered grids producing are based on photolithography processes [22-30], which allow one to obtain a given geometry in a controlled manner (Figure 2c), while achieving maximum values of the shielding and light transparency of such materials. In [27], a Ta / Al / Ta multilayer grid (the width of conductive tracks is 15 μm and the grid pitch is 400 μm) was created by combining the magnetron sputtering process and photolithography. With an Al layer thickness of 150 nm this grid has a transparency of 86 % in the range 500 – 2000 nm, sheet resistance 10.2 Ω/sq and SE 46 dB at 6 GHz. It is worth noting that the deposition of an ITO layer on such a structure with a slight decrease in light transparency (from 84 % for ITO on glass to 80 % for ITO over Ta / Al / Ta for 500 nm) allows one to achieve sheet resistance of the ITO + Ta / Al / Ta structure 7.8 Ω/sq which improves the shielding efficiency of the created coating with a slight loss of transparency.

In [28], by UV lithography followed by galvanic deposition of Cu with annealing in air, an oxidized copper grid of 2 μm thickness was obtained, which shows a transparency of about 80 % in the range 400 - 800 nm and SE from 30 to 20 dB in the range 12 - 18 GHz. Similarly, metal grids with a grid pitch of 50 μm from Au and Cu were produced on a polymer substrate in [26]. These structures showed a transparency of 72 % and 87 % (at a wavelength of 550 nm) with a sheet resistance of 0.2 Ω/sq and 0.7 Ω/sq for Au and Cu, respectively, which is significantly lower than the resistance, for example, of ITO films with the same transparency.

It is worth noting that today most microelectronic factories and laboratories are equipped with equipment designed for substrates with a diameter of 300 mm, and the limiting factor is precisely the photolithography plants, which makes it difficult to obtain such conductive grids on large samples. In order to solve this problem, the process of forming the polymer "stamp" by the lithographic method was described in [46], which is used then to form a network of grooves of a given topology over a large area using the nanoimprinting method. In order to metallize the obtained topology, a paste based on silver nanoparticles was introduced into the formed system of grooves, followed by galvanic interrogation of Ni. The resulting materials with a light transparency of 80 – 85 % (without antireflection layers) have SE 45 dB at a frequency of 8 GHz.

The lithography process can be used, not only to create a protective mask with subsequent etching of the structure through it, but also to create a conductive structure that is networked in the process of photochemical or photothermal synthesis from solutions of photosensitive heterometallic complexes. Such an approach was proposed in [29], where exposure to an emulsion layer of silver halide grains dispersed in gelatin on a PET substrate makes it possible to create a network of such grains with a line width of about 15 μm and a grid pitch of 300 μm. At the same time, the authors can achieve a distance of SE 40 dB in the range 1 – 800 MHz with a decrease in light transparency by 5 % compared with the original substrate.

As an alternative to the lithography process, laser etching can be used. A similar approach, for example, was applied in [31], where the authors create a two-layer Au-Ni grid with a line width of 10.5 μm and a grid pitch of 200 μm by laser etching using a femtosecond laser. With a quartz substrate thickness of 16 mm on a two-layer sample, a 76 % transparency was obtained in the range of 400 – 900 nm at SE 60 – 70 dB in the range 1 – 5 GHz, while a single-layer sample showed 84 % transparency at SE 45 dB.

In a number of works [32, 33], 3D printing was used to create grid metal structures. In [32], using this method, grid and line structures with different parameters were obtained. For a silver grid with a line width of 15 μm and a grid pitch of 1000 μm, a transparency of about 93 % at 550 nm was achieved with a sheet resistance of 0.2 Ω/square. In the work, it is shown that line structures make it possible to achieve higher transparency with an increase in the layer resistance compared to grids.

A structure with high transparency and shielding indices can be constructed on the basis of clusters of metal rings that form a mesh similar to a conducting substrate [34-36]. In [34], such a structure, created by multi-step UV lithography followed by plasma-chemical etching of an Al layer 300 nm thick deposited on a quartz substrate (Figure 3), has a transparency of 95 % in the range 400 - 2400 nm at SE 26 dB at 12 GHz.

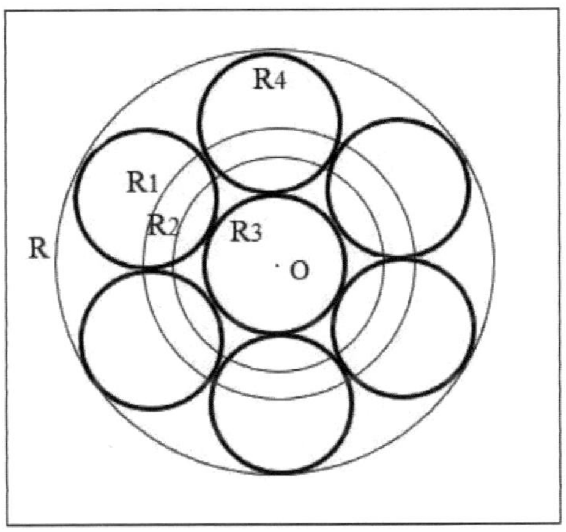

Figure 3. The structure of metal rings in the form of a floral pattern [34].

It is worth noting that the creation on the surface of an ordered grid of metal conductors often leads to the appearance of Moiré waves, which sharply reduce the contrast of the image. For this effect, we proposed the use of an artificially disordered mesh [47, 48]. Moreover, periodic (with toroidal topology) Voronoi diagrams of the set of points randomly distributed inside the unit cell with a uniform probability density were taken as the basis. In order to increase the optical homogeneity of the meshes, the Voronoi diagrams were additionally subjected to Lloyd relaxation [49]. The design of the proposed mesh is shown in Figure 4.

Based on the theoretical studies, the authors of this work proposed options for the geometric parameters of the mesh, which allow one to obtain a light transmission coefficient from 90 to 98 % in combination with a shielding coefficient in the radio wavelength range from 50 to 65 dB (Figure 5), which is typical for existing opaque materials. The technology of producing such meshes is currently under development. Thus, the analysis of practical results on the creation of materials based on a grid of ordered conductors showed the possibility of achieving high shielding efficiencies of up to 45 or more dB in the range from 10 kHz to 20 GHz with a light transparency of more than 80 %. Theoretical calculations show great potential for improving these indicators; however, the limiting factor of this approach is the complexity of the process and the issue of its scalability. At the same time, it is worth noting the recent work on obtaining such grids by the methods of additive technologies, which, with their further development, can solve these limitations.

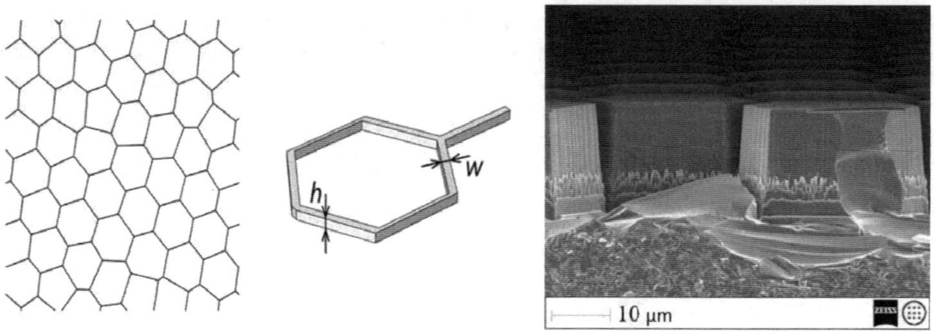

(a) (b) (c)

Figure 4. (a) Transparent mesh formed by opaque conductor wires
(b) single unit cell of the simplified model used in this study for numerical simulations
(c) SEM image of the obtained structure

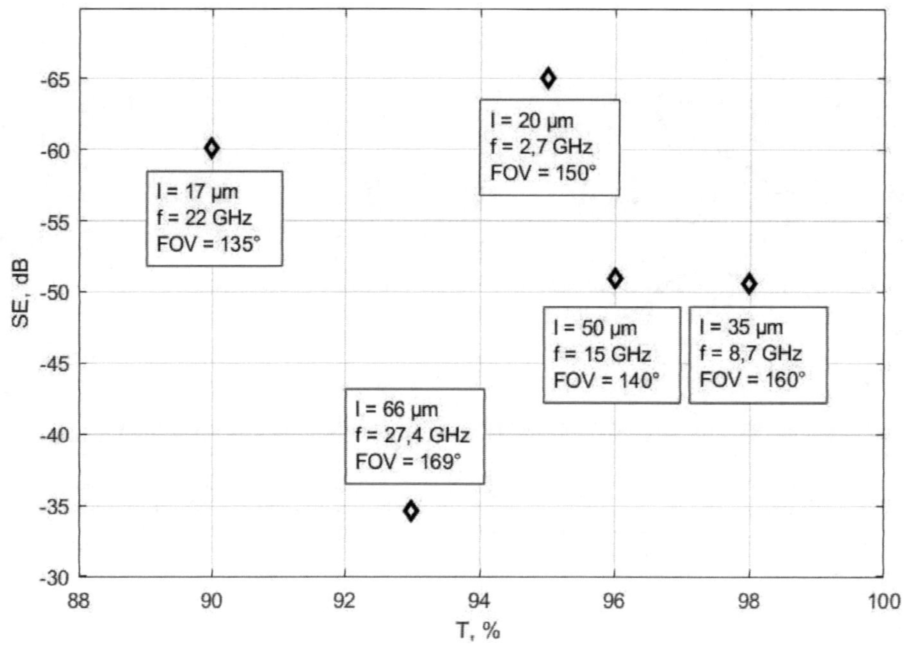

Figure 5. Theoretically calculated values of shielding (SE), transparency (T), and field of view (FOV) coefficients for various mesh sizes (l)

4. Hybrid structures

The methods for producing such materials combine two or more approaches to the production of electrically conductive transparent coatings, discussed earlier, in order to use the advantages of each method. Such structures are often a multilayer material containing a metal mesh in combination with other layers that also have electrically conductive properties [50-54]. In [50, 51], the processes of fabricating hybrid structures based on either randomly oriented AgNWs on polymer substrates in combination with a layer of reduced graphene oxide (RGO) [50] or a high-conductivity layer of two-dimensional graphene nanoplates (GNS) [51], respectively (structures A / RGO / AgNWs and GNS / AgNWs, where A is a protective layer of acrylic polymer). The shielding coefficient SE of the A / RGO

/ AgNWs structure was more than 24 dB, which is approximately two times higher than that of an ITO film on glass (about 10 dB) with a similar relative transparency (84 - 85 %).

The addition of a graphene oxide layer to a layer of randomly oriented AgNWs nanofibers [18] gives a slight increase in SE (the GNS / AgNWs film shows SE up to 26 dB in the Ku and K frequency ranges with a transparency higher than 78 %). It is worth noting that the examined films have good flexibility and high structural stability under various operating conditions. In [52, 53], hybrid ones were built on the basis of an Al metal grid with square cells formed on a quartz substrate by photolithography and graphene layers. Moreover, in [53] a different combination of metal and graphene layers is considered. With an optical transparency of 94 %, a sheet resistance of 20.7 Ω/sq was obtained in [52], and the maximum EMI SE value was 21 dB at 12 GHz, which is comparable with the best samples of ITO coatings on glass substrates [15].

The GMTD hybrid structure [53] (Figure 1)] with two graphene layers and two metal grids with a period of 160 μm provides an SE exceeding 48 dB in the Ku band and SE exceeding 32 dB in the Ka band with a visible transparency of ~ 85 % at 700 nm. In [54], a hybrid structure was obtained by hot pressing two PET films, on one of which an ordered silver grid was formed by photolithography (with a period of 1, 2 or 3 mm), and the other was coated with a network of randomly arranged AgNWs with the addition of RGO, addition of which leads to an increase in the electrical conductivity of the layer. Such a structure with an Ag network with a step of 1 mm has the highest SE value of 43 dB at a frequency of 1.8 GHz with a transmission coefficient of 74 %, while the SE of the AgNWs / RGO layer does not exceed 10 dB.

Thus, hybrid structures can slightly improve the shielding and transparency of the resulting layers compared to, for example, metal meshes or ITO coatings, while providing other operational characteristics of the formed layer, such as structure flexibility, mechanical stability, and resistance to environmental factors environment, *etc.* At the same time, this approach leads to some complication in the technological process of obtaining such structures, therefore, the feasibility of its application will depend on the particular problem being solved.

5. Discussion

Based on the results of the analysis of scientific and technical literature, a consolidated graph has been constructed (Figure 6), which displays the practically obtained shielding efficiency (SE) and light transparency (T) for the various types of structures considered above. When constructing it, the frequency dependences of SE and T, methods and measurement technics, which can lead to significant errors, have not been taken into account. However, the obtained data still allows to identify promising approaches for obtaining materials with a certain combination of transparency and radio shielding efficiency.

If the task is to obtain materials with SE less than 30 dB and maintain light transparency at 80 %, all the considered approaches (multilayer coatings, including ITO, ordered and disordered mesh structures) can be used to obtain specified characteristics. However, the use of ITO coatings is currently limited to application to inflexible substrates (such as glass), since their manufacture with SE up to 25 dB and a transparency of about 90 % requires the operation of high-temperature annealing. An increase in the SE requirements to 40 dB or light transparency up to 90 % or more leads to the need to complicate the structure and the technological process of obtaining such a layer, including the increase of the layers number in a multilayer structure or adding continuous conductive layers based on wide-gap semiconductors or graphene to disordered conducting networks, thus obtaining hybrid structures. In this case, replacing disordered fibers with an ordered conducting grid in such a hybrid structure with graphene oxide layers can increase SE from 26 to 48 dB with the same light transparency (up to ~ 85 %).

Broadly speaking, the highest characteristics (SE more than 45 dB at a light transparency of about 85 %) were obtained experimentally on ordered mesh structures. Theoretical calculations show that, by optimizing the parameters of the conductive grid, it is possible to achieve 60 dB or more on such materials with a transparency of about 90 %, but the imperfection of the resulting structures can significantly reduce the calculated parameters.

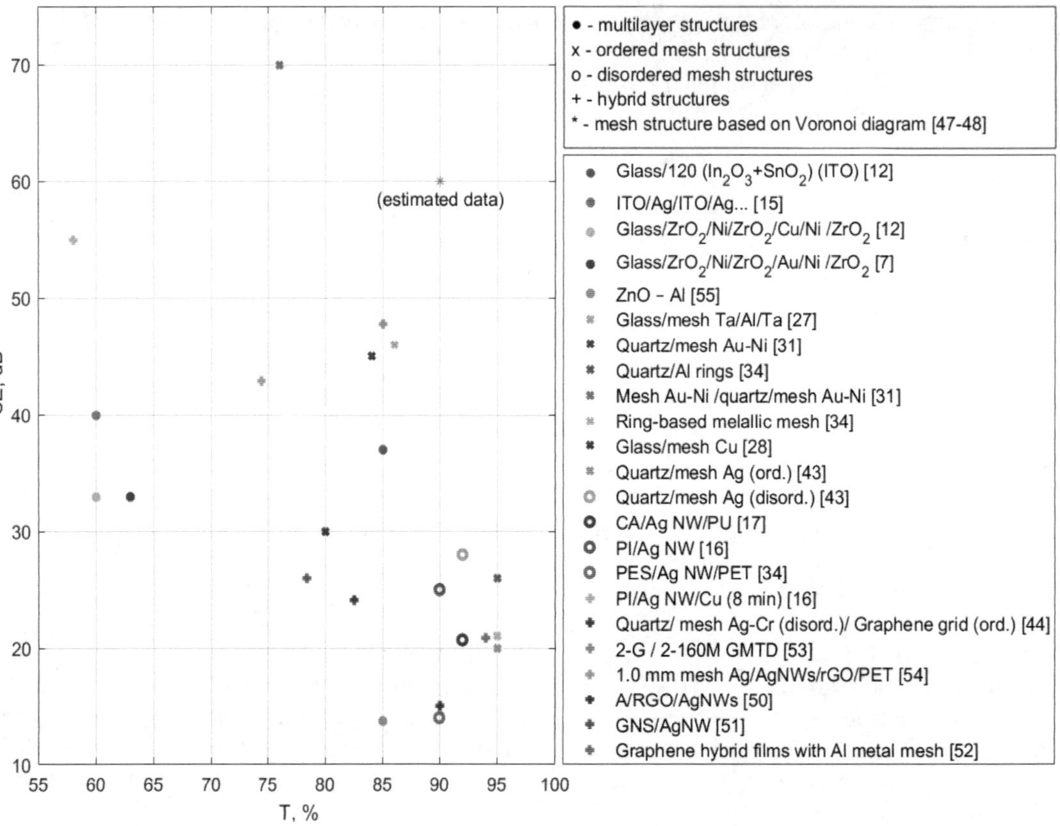

Figure 6. Comparison of the shielding efficiencies and light transmission coefficients of coatings with different structures

It is worth noting that the practical results obtained on multilayer continuous structures turned out to be generally lower compared to metal meshes. Such coatings, as a rule, lead to a decrease in transparency to 60 % or less when reaching SE up to 40 dB. In a number of studies, higher values SE were obtained, but in narrow frequency bands. At the same time, our theoretical calculations, which are beyond the scope of this article, show that the optimization of the thicknesses and chemical composition of heterostructure layers can significantly increase the functional characteristics of such coatings, bringing them closer to metal meshes. However, this issue requires further serious study, since in addition to structural parameters (layers thickness, chemical composition, electrical conductivity), technological aspects of manufacturing developed multilayer structures will affect the obtaining characteristics.

6. Conclusion

The results analysis allows to conclude that both approaches (fabrication of multilayer structures and mesh conductive structures) make it possible to obtain materials with high values of shielding efficiency in the radio range and transparency in the visible wavelength range. Currently, among practical results, the best characteristics have been achieved on ordered mesh structures obtained by lithographic methods. However, the need to use microelectronics technologies in their creation is a significant limiting factor of this approach. Moreover, it can be expected that the development of self-organization methods and additive technologies will lead to a significant simplification of the technological process, lower cost and the possibility of scaling it to large-area samples.

At the same time, the task of technology scaling in the case of multilayer structures can be solved much easier, since the technologies of applying thin-film coatings on roll materials are widely used in industry already now. There is a great potential for improving multilayer structures by optimizing their design and technological solutions, despite experimentally they have lower shielding efficiency and light transparency compared to metal grids. In this regard, the technological aspects can come to the fore, when an approach for the implementation of the considered materials with given characteristics are chosen, taking into account the possibility of subsequent scaling of the technology and economic indicators when choosing an approach for the implementation of the materials with the required characteristics.

Acknowledgments
This work was supported by the Ministry of Science and High Education of the Russian Federation (state task No. 0705-2020-0032).

References

[1] Wessapan T and Rattanadecho P 2018 *International Journal of Heat and Mass Transfer* **119** 65-76
[2] Pasquino N 2017 *Measurement* **109** 373-383
[3] Crupi I, Boscarino S, Strano V, Mirabella S, Simone F and Terrasi A 2012 *Thin Solid Films* **520**(13) 4432-4435
[4] Kim B R, Lee H K, Kim E and Lee S H 2010 *Synthetic Metals* **160**(17-18) 1838-1842
[5] Lee S H, Lee D H, Lee K and Lee C W 2005 *Advanced Functional Materials* **15**(9) 1495-1500
[6] Lee K, Cho S, Park S H, Heeger A J, Lee C W and Lee S H 2006 *Nature* **441**(7089) 65-68
[7] Zalesskiy V, Volochko A and Markov G 2017 *Nauka i innovatsiyi - Science and innovation,* **4**(170) (in Russian)
[8] Granqvist C G and Hultåker A 2002 *Thin solid films* **411**(1) 1-5
[9] Melnikov A A and Shur P A 2019 *Trudy VIAM - VIAM Proceedings,* **8** (80) (in Russian)
[10] Yun J, Park Y H, Bae T S, Lee S and Lee G H 2013 *ACS Applied Materials & Interfaces* **5**(1) 164-172
[11] Parshina L S, Novodvorsky O A, Khramova O D, Lotin A A, Khomenko M D and Shchur P A 2019 *Semiconductors* **53**(2) 160–164
[12] Volochko A T, Markov G V, Zelenin V A and Narushko E O 2017 *Izvestiya Nazional'noyi akademii nauk Belarusiyi. Seriya fiziko-technicheskih nauk - Bulletin of the National Academy of Sciences of Belarus. Physics and Technology Series,* **1** 25-30 (in Russian)
[13] Sarto M S, Sarto F, Larciprete M C, Scalora M, D'Amore M, Sibilia C and Bertolotti M 2003 *IEEE Transactions on Electromagnetic Compatibility* **45**(4) 586-594
[14] Wang H, Ji C, Zhang C, Zhang Y, Zhang Z, Lu Z, Tan J and Guo L J 2019 *ACS applied materials & interfaces* **11**(12) 11782-11791
[15] Zhao Y L, Ma F H, Li X F, Ma J J, Jia K and Wei X H 2018 *Chinese Physics* B **27**(2) 027302
[16] Kim D H, Kim Y and Kim J W 2016 *Materials & Design* **89** 703-707
[17] Jia L C, Yan D X, Liu X, Ma R, Wu H Y and Li Z M 2018 *ACS applied materials & interfaces* **10**(14) 11941-11949
[18] Jung J, Lee H, Ha I, Cho H, Kim K K, Kwon J, Won P, Hong S and Ko S H 2017 *ACS applied materials & interfaces* **9**(51) 44609-44616
[19] Hu M, Gao J, Dong Y, Li K, Shan G, Yang S and Li R K Y 2012 *Langmuir* **28**(18) 7101-7106
[20] Deng B et al. 2015 *Nano letters* **15**(6) 4206-4213
[21] Morag A, Philosof-Mazor L, Volinsky R, Mentovich E, Richter S and Jelinek R 2011 *Advanced Materials* **23**(37) 4327-4331
[22] Babych B, Borisova A, Machulyansky A, Machulyansky V, Rodionov M and Yakimenko Y 2017 *2017 IEEE 37th International Conference on Electronics and Nanotechnology (ELNANO)* 52-56
[23] Corredores Y, Castel X, Besnier P, Dupeyrat C and Foutrel P 2018 *The Journal of Engineering* **2018**(4) 239-241

[24] Lu Z, Wang H, Tan J and Lin S 2014 *Applied Physics Letters* **105**(24) 241904
[25] Van De Groep J, Spinelli P and Polman A 2012 *Nano letters* **12**(6) 3138-3144
[26] Khan A, Lee S, Jang T, Xiong Z, Zhang C, Tang J, Guo L G and Li W D 2016 *Small* **12**(22) 3021-3030
[27] Kumar P, Reddy P V, Choudhury B, Chowdhury P and Barshilia H C 2016 *Thin Solid Films* **612** 350-357
[28] Han Y, Zhong H, Liu N, Liu Y, Lin J and Jin P 2018 *Advanced Electronic Materials* **4**(11) 1800156
[29] Fujita Y, Okazaki K and Sasaki H 2007 *SID Symposium Digest of Technical Papers* **38**(1) 1213-1215
[30] Wang W, Bai B, Zhou Q, Ni K and Lin H 2018 *Optical Materials Express* **8**(11) 3485-3493
[31] Zhang Y, Dong H, Li Q, Mou N, Chen L and Zhang L 2019 *RSC advances* **9**(39) 22282-22287
[32] Zhu X et al. 2019 *Advanced Materials* **31**(32) 1902479
[33] Schneider J, Rohner P, Thureja D, Schmid M, Galliker P and Poulikakos D 2016 *Advanced Functional Materials* **26**(6) 833-840
[34] Wang H, Lu Z, Tan J, Zhang Y, Cao J, Lu X, Liu Y, Kong R and Lin S 2018 *Thin Solid Films* **662** 76-82
[35] Wang H, Lu Z and Tan J 2016 *Optics express* **24**(20) 22989-23000
[36] Wang H, Lu Z, Liu Y, Tan J, Ma L and Lin S 2017 *Optics letters* **42**(8) 1620-1623
[37] Spatz J and Lohmueller T 2009 *U.S. Patent Application No. 12/302,733*
[38] De La Vega F, Nir M M and Garbar A 2014 *U.S. Patent No. 8,633,474. Washington, DC: U.S. Patent and Trademark Office*
[39] Bormashenko E 2017 *Membranes* **7**(3) 45
[40] Ovcharov P V, Kryuchkov N P, Zaytsev K I and Yurchenko S O 2017 *The Journal of Physical Chemistry* **121**(48) 26860-26868
[41] Khrapak S A, Kryuchkov N P, Yurchenko S O and Thomas H M 2015 *The Journal of chemical physics* **142**(19) 194903
[42] Yakovlev E V et al. 2017 *Scientific reports* **7**(1) 1-10
[43] Han Y, Lin J, Liu Y, Fu H, Ma Y, Jin P and Tan J 2016 *Scientific reports* **6** 25601
[44] Han Y, Liu Y, Han L, Lin J and Jin P 2017 *Carbon* **115** 34-42
[45] Han B et al. 2014 *Advanced materials* **26**(6) 873-877
[46] S. Shen, Chen S Y, Zhang D Y and Liu Y H 2018 *Optics Express* **26**(21) 27545-27554
[47] Shiriaev P, Makeev M, Ryzhenko D and Popkov O 2019 *Materials Today: Proceedings* **19** 2179-2182
[48] Shiriaev P P, Ryzhenko D S, Makeev M O and Popkov O V 2019 *Journal of Physics: Conference Series* **1410** 012206
[49] Lloyd S 1982 *IEEE Transactions on Information Theory* **28**(2) 129-137
[50] Kim D G, Choi J H, Choi D K and Kim S W 2018 *ACS applied materials & interfaces* **10**(35) 29730-29740
[51] Zhang N, Wang Z, Song R, Wang Q, Chen H, Zhang B, Lv H, Wu Z and He D 2019 *Science Bulletin* **64**(8) 540-546
[52] Ma L, Lu Z, Tan J, Liu J, Ding X, Black N, Li T, Gallop G and Hao L 2017 *ACS applied materials & interface* **9**(39) 34221-34229
[53] Lu Z, Ma L, Tan J, Wang H and Ding X 2017 *2D Materials* **4**(2) 025021
[54] Huang C C, Gupta S, Lo C Y and Tai N H 2019 *Materials Letters* **253** 152-155
[55] Choi Y J, Kang K M, Lee H S and Park H H 2015 *Thin Solid Films* **583** 226-232

Study of the Destruction of Carbon Composite Panel using High-Speed Jet of Liquid

M I Abashin[1*], A L Galinovskiy[1], Kyaw Myo Htet[1], A V Denisova[2], S A Erokhin[2], TSzia Chzheniuan[1]

[1] Bauman Moscow State Technical University, Russia
[2] LLC Special Development and Technology Bureau "Plastic", Syzran, Russia

* abashin@bmstu.ru

Abstract. The article discusses the task of assessing the damage of carbon fibre plastics arising from the destructive effect of radiation using the diagnostic capabilities of ultra-jet technology. A technique is proposed for determining the thickness of a carbon fibre package that can protect an imaginary interior from the negative effects of radiation waves. As informative diagnostic parameters in the method, the geometrical dimensions of the cavern created on the surface of the sample following exposure to a high-speed jet of liquid (water) are used. Based on the results of the experiments, it was found that a decrease in the depth value indicates a decrease in the penetrating effect of radiation and the destruction of the binder in the structure of the composite material.

1. Introduction

In recent years, there has been a tendency to expand the use of Polymer Composite Materials (PCM) in the construction of spacecraft. Firstly, such interest in introducing composite materials in Russia and abroad is associated with their physical and mechanical properties, which can significantly (up to 40-50%) reduce the weight of the product [1-3]. It should be noted that some PCM have a lot of special properties. So, for instance, carbon plastics are characterized by such properties as: high strength and stiffness, wear resistance, resistance to the negative effects of aggressive environments, low thermal expansion coefficients, electrical conductivity, the ability to resist fatigue loads under static and dynamic loads, etc [2, 4-6].

Currently, carbon fibre reinforced plastics are widely used in the design of spacecraft and launch vehicles for the manufacture of fairings, antenna bodies, electrical wires and cables, *etc.* Nevertheless, the problems of the *behaviour* of carbon composite materials under the influence of adverse outer space conditions were not yet sufficiently studied [7]. The reasons for this are associated with the lack of well-developed methods for diagnosis and testing, and engineering methods for the existing control and diagnostic technologies [8-11]. Because of the need to ensure high levels of reliability of critical space products, extending capability for monitoring, diagnostics and testing is an urgent scientific and technical task. This prompts us to formulate the goal of the study, which in the present work is formulated as the *development of an engineering technique for ultra-jet PCM diagnostics for assessing carbon fibre damage* [12-15].

To assess the performance of a particular spacecraft design, it is necessary to evaluate and identify the totality of factors affecting the product life and its physical and mechanical properties and other characteristics. Among these factors, the following can be singled out: cosmic radiation fluxes with high-energy electrons and ions, cold and hot cosmic plasma, electromagnetic solar radiation, probable meteoroid impact [13, 15]. Each of these can not only disable working equipment and devices, but also reduce the strength characteristics of PCM structures as a result of various physical and chemical processes, leading, for example, to embrittlement of the binder and deterioration of adhesion in the binder-fibre structure [10, 16-17]. Moreover, some types of exposure, such as cosmic plasma and ultraviolet solar radiation, as a rule, affect the surface and subsurface layers of the material. This is particularly the case for the above mentioned high-energy charged particles [20-23].

These circumstances indicate that the existing variety of unfavourable factors of outer space, the variability and cyclic variations of the space environment, as well as the complex processes accompanying their influence and course, for example, wide spectra of corpuscular and electromagnetic radiation of various periodicities, significantly complicate the development of effective methods for predicting the resource and assessment of controlled quality parameters of PCM [16, 17]. In this regard, taking into account the experience in the development of technologies for Ultra-Jet Diagnostics (UJD) of materials, an attempt was made to adapt existing methods for diagnosing metallic materials and alloys with a high-speed fluid flow to the test sample [23-25].

2. The hypothesis of the study

The hypothesis of the study was the phenomenological position on the change in the parameters of hydro-erosive destruction of the surface layer of the sample exposed to radiation. As a result of the very likely embrittlement of the binder material, the propagation of cracks in it, and the decrease in adhesion with carbon fibre, a high-speed fluid flow will wash it out of the material volume, providing a deepening of the jet into the material thickness under conditions of low resistance to its dynamic high-energy impact. In this case, the parameters of surface hydro-erosion are the main informative data of ultra-jet diagnostics by which it is possible to judge the current state of the material, its degree of damage, internal defects [20]. Moreover, the defectiveness of the material can be either: present initially, associated with the imperfection of the manufacturing process; or acquired as a result of the negative influence of radiation.

3. Experimental technique

Figure 1 shows a schematic diagram of the UJD process, the implementation of which in this work was carried out on a machine for waterjet cutting Flow Mach 3 3131b (USA) at the Centre for Hydro-physical Research at Lomonosov Moscow State University.

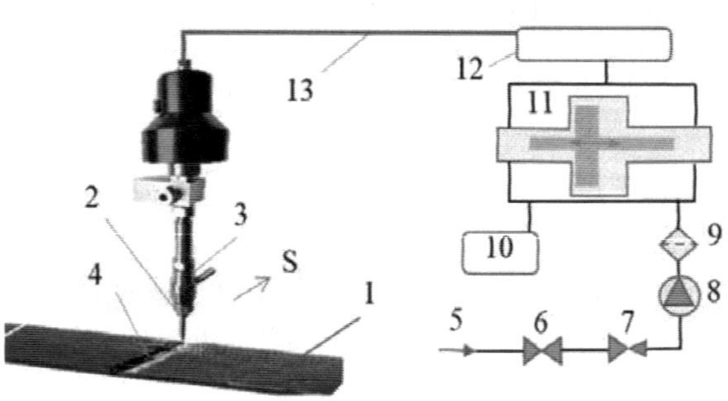

1 – Sample; 2 – Focusing tube;
3 – Head nozzle block; 4 – Cavern;
5 – Water supply; 6 – Way valve; 7 – Gear;
8 – Booster pump; 9 – Block filters;
10 – Control system; 11 – Multiplier;
12 – Vibration damper;
13 – High pressure pipeline

Figure 1. Schematic diagram of ultra-jet diagnostics with the composition of the main elements of hydraulic equipment.

An engineering technique was developed to implement the idea of PCM UJD for assessing the level of damage/defectiveness. The first stage of this was the production of a batch of carbon-plastic panel samples in accordance with the standard technological process specified by the LLC Special Development and Technology Bureau "Plastic". The samples were manufactured using carbon tape LU-P/0.1, impregnated with a binder ENFB, and layup (0°/45°/135°/90°/90°/135°/45°/0°). The resulting panel had the following properties: modulus of elasticity 59.166 MPa; tensile strength 263 MPa; and thickness 3.6 mm.

At the second stage, samples of 150 x 20 mm in size cut from the panel were collected in a bag and placed in the chamber of the industrial XT H 450 fluoroscopy and computed tomography system, see figure 2(b), to exert an exposure to radiation in accordance with the parameters presented in table 1. Preliminarily, the samples were collected in a bag, as shown in figure 2(a).

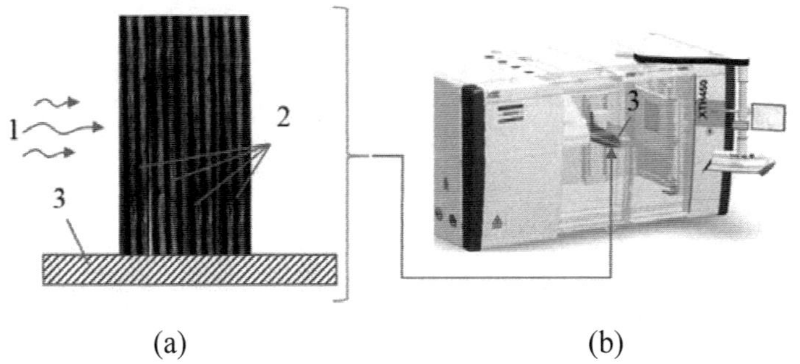

1 – Radiation
2 – Parallel stacked carbon fibre samples
3 – Desktop monitoring system

Figure 2. UJD implementation scheme:
(a) – 1st stage
(b) – 2nd stage

Table 1. Parameters set in the system of fluoroscopy and computed tomography for the destruction of carbon fibre samples

№	Voltage, kV	Current, μA	Experiment time (working time), min
1	376	185	37
2	360	185	37
3	360	185	35
4	300	100	5
5	310	105	37
6	350	145	105
7	375	185	74
8	360	100	37
9	360	100	37
10	325	120	57
11	300	105	14
12	300	105	14
13	300	105	14
14	300	105	14
15	300	105	14
16	265	100	14
17	380	200	26

Following a series of test trials, the following diagnostic mode was selected by the experimental method: feed of the cutting head 7500 mm/min; water pressure 130 MPa; distance from the focusing tube cut to the sample surface 160 mm (figure 1). With the choice of these parameters, it was possible to ensure surface destruction of the binder without noticeable destruction of the fibre. An increase in pressure, a decrease in distance and a feed rate were observed penetrating destruction of the material with rupture and damage to the fibres (figure 3). When conducting experiments on the UJD, the authors assumed that, when moving away from the radiation source, the damage to the samples in the packet (see figure 2a), especially the binder, decreases.

The UJD parameters were selected as follows: 160 mm from the cut of the focusing tube to the surface of the sample (figure 1); the feed rate of the cutting head is 7500 mm/min; the working fluid pressure in the hydraulic system is 130 MPa. The selection of these parameters was carried out empirically according to the criterion of not breaking through the sample and the formation of a local notch – a cavern.

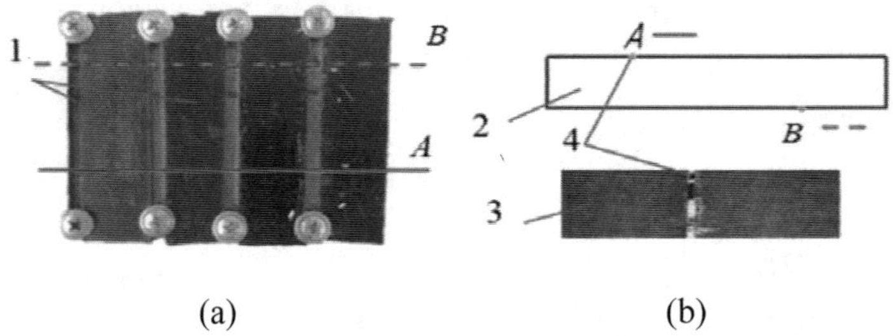

1 – Samples of carbon fibre 1-4, where A is the scanning path of the ultrasound from the front and B-back side
2 – View of the end surface of the sample
3 – Top view of the sample
4 – Caverns

Figure 3. (a) Carbon fibre samples on the substrate, and (b) areas for exposure to ultra-jet on the surface of the samples.

4. Experiment results

Figure 4 shows 3D images of hydro-caverns and their profilograms, measured after the experiment was carried out using a Keyence VHX-6000 digital microscope (Japan). Statistical processing of the results of measurements of depths of hydro-caverns made it possible to construct a dependence of their depth on the distance to the radiation source. Figure 5 shows this dependence. In the docking areas of the samples, the average values of the depth of the cavern were calculated.

5. Discussion

In total three samples of each type of material were examined. Table 1 shows the average values. The spread of values was less than 5%. The research demonstrates that the UJD method proposed has potential for implementation in production for the assessment of PCM damage in carbon plastics arising from the destructive effect of radiation using the diagnostic capabilities of ultra-jet technology. According to the results of simulation and accelerated tests, the control valve can be used to assess the influence of certain adverse environmental factors, such as climatic, various types of radiation, temperature, thermal cycling, alternating loads, and others, on material damage. Damage resulting from the action of a single or combination of factors can manifest itself in the appearance of various defects, such as delamination, pore formation or poor adhesion.

6. Conclusion

The UJD method has shown its versatility and flexibility in the example of solving various diagnostic problems of mechanical engineering, which made it possible to adapt the previously obtained results to assess the damage to carbon fibre panels and determine the depth of its penetration into the thickness of the package.

The proposed engineering method will allow to determine the rational thickness of the composite material, for example, according to the criterion of protecting the internal space from the negative effects of radiation, for example, protecting personnel working with radiation equipment, NPP rooms and other critical human life objects.

The method is distinguished by the breadth of possibilities for controlling the technological parameters of the UJD, which allows it to be used both for surface diagnostics and for studying subsurface layers of a composite material, the presence of defects, damage, and violation of adhesive bonds in them.

Because the method has a small but non-negligible effect on the surface layer of the material, this limits the site of application, which should therefore be focused on selective inspection of products, diagnostics of test specimens, and large structures for which this damage will not affect performance, reliability and resource.

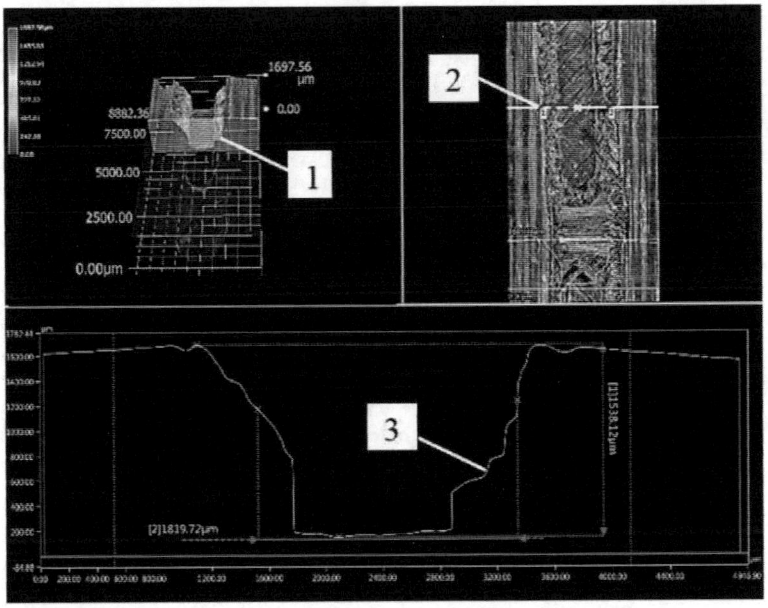

1 – Profile of the cavern
2 – Section with tracing
3 – Profile.

Figure 4. Image of a cavern as a result of a study using a digital microscope.

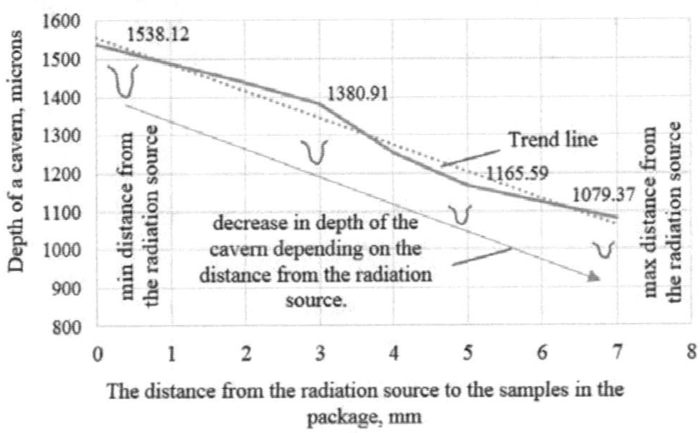

Figure 5. Dependence of the depth of the cavern on the distance to the radiation source.

Acknowledgments

The studies were carried out under Russian Foundation for Basic Research grants 18-29-18081 and 19-38-90228\19, a grant of the President of the Russian Federation for state support of leading scientific schools of the Russian Federation NSh-3778.2018.8, and grants from the Fund for the Promotion of Innovations under the UMNIK-18 program (in) following the contract No. 14727GU/2019 and No. 14549GU/2019.

References

[1] Bhagwan D A, Lawrence J B and Chandrashekhara K 2017 *Analysis and Performance of Fiber Composites* 4th ed: John Wiley & Sons

[2] Peebles L H 2018 *Carbon Fibers: Formation, Structure, and Properties* Boka-Raton: *CRC Press*

[3] Rahul V, Alokita S, Jayakrishna K, Kar V R, Rajesh M, Thirumalini S, Manikandan M, 2019, Structural health monitoring of aerospace composites, 33-52, in *Structural Health Monitoring of Biocomposites, Fibre-Reinforced Composites and Hybrid Composites.* Ed. Jawadi M, Thariq M, Saba N, (Amsterdam: Elsevier)

[4] Baurova N I, Zorin V A, 2016, *The use of polymer composite materials in the manufacture and repair of machines* (Moscow: MADI [Moscow Automobile and Road Institute])

[5] Ahmad F, Choi H S and Park M K 2014 A Review: Natural Fiber Composites Selection in View

of Mechanical, Light Weight, and Economic Properties *Macromol. Mater. Eng.* **300**(1) 10-24

[6] Fleischer J, Teti R, Lanza G, Mativenga P, Möhring H-C and Caggiano A 2018 Composite materials parts manufacturing *CIRP Annals* **67**(2) 603-26

[7] Gholizadeh S 2016 A review of non-destructive testing methods of composite materials," *Procedia Structural Integrity* **1** 50-7

[8] Gorman M R 2011 Modal AE analysis of fracture and failure in composite materials, and the quality and life of high pressure composite pressure vessels *J. Acoust. Emiss.* **29** 1-28

[9] Rana S and Fangueiro R 2016 *Advanced Composite Materials for Aerospace Engineering: Processing, Properties and Applications*, (Cambridge: Woodhead Publishing)

[10] Sause M G R 2016 *In Situ Monitoring of Fiber-Reinforced Composites: Theory, Basic Concepts, Methods, and Applications* (Berlin: Springer)

[11] Troitsky V A, Karmanov M N and Troitskaya N V 2014 Non-destructive quality control of composite materials *Technical diagnostics and non-destructive testing* **3** 29-33.

[12] Larin A A 2013 *Methods for assessing the health of products from composite materials by computed tomography*: Diss. Cand. Tech. Sci., Research Institute of Introscopy MNPO "Spectrum", Moscow

[13] Murashov V V and Slyusarev M V 2016 Revealing Cracks in Polymer Composite Parts and in Multilayered Glued Constructions by a Low-Frequency Acoustic Method *Russian Journal of Nondestructive Testing* **52**(6) 324-3.

[14] Lammering R, Gabbert U, Sinapius M, Schuster T and Wierach P 2017 *Lamb-Wave Based Structural Health Monitoring in Polymer Composites* (Berlin: Springer)

[15] Figlus T and Koziol M 2016 Diagnosis of early-stage damage to polymer - glass fibre composites using non-contact measurement of vibration signals *J. Mech. Sci. Technol.* **8** 3567-76

[16] Murashov V V 2014 Identification of Areas of Absence of Adhesive Bonding between Layers in Multilayer Structures *Polymer Science. Series D. Glues and Sealing Materials* **7**(1) 46-8

[17] Liu G, Cheng L and Luan X 2019 Radiation damage behavior of carbon/carbon composite in Low Earth Orbit environment *Ceramics International* **45**(13) 16088-96

[18] Herath H. *et al.* 2019 Carbon fibre reinforced shape memory polymer composites for deployable space habitats *Engineer* **52**(1) 1-9

[19] Karger-Kocsis J, Mahmood H, and Pegoretti A 2015 Recent advances in fiber, matrix interphase engineering for polymer composites *Prog. Mater. Sci.* **73** 1-43

[20] Niu Y, Yang Y and Wang X 2018 Investigation of the interphase structures and properties of carbon fiber reinforced polymer composites exposed to hydrothermal treatments using peak force quantitative nanomechanics technique *Polym. Compos.* **39**(S2) 791-6

[21] Adamov A A, Adamov A A, Laptev M Yu and Pot-va E G 2012 Analysis of domestic and foreign regulatory framework for mechanical testing of polymer composite materials *Konstuktsii iz kompozitsionnykh materialov* [Composite materials Constructions] **3** 72-7

[22] Xueyan Li, Semashko V V and Abashin M I 2018 Development of a method of ultra-jet diagnostics of cutting tools for the disposal of polymer composite materials in the construction of rocket and space technology *Politekhnicheskiy Molodezhnyy Zhurnal* [Polytechnic Youth Journal] **8** 1

[23] Sleight D W 2019 Development of Composite Sandwich Bonded Longitudinal Joints for Space Launch Vehicle Structures *AIAA Scitech 2019 Forum*. 0236

[24] Htet K M, Galinovsky A L and Provatorov A S 2020 Prospects for the development of ultra-jet dispersion technology for nanocontaining suspensions *IOP Conf. Ser.: Mater. Sci.* Eng. **709** 044092.

[25] Galinovskiy A L, Htet K , and Provatorov A S 2020 Ultra-Jet as a Tool for Dispersing Nanosuspensions *Polym. Sci. Ser. D* **13** 209-13

Development of a new diagnostic method for materials abrasion resistance

A A Barzov[1], A L Galinovskiy[2] and A S Vyshegorodtseva[2*]

[1] Lomonosov Moscow State University, Moscow, 119991, Russia
2 Bauman Moscow State Technical University, 105005, Russia

E-mail: * anastaciamarchenko@mail.ru

Abstract. The article deals with the development of a new diagnostic method for materials abrasion resistance. It lists the advantages of the proposed ultra water-jet abrasive diagnostics method and the disadvantages of existing ones. The implementation diagram of the method, the results of the experiments and the samples images with traces of abrasive jet interaction are presented. The phenomenological ideas of the nature of erosion of interacting materials are highlighted.

1. Introduction

One of the main requirements for some materials and products is their abrasion resistance, *e.g.* working surfaces of nodes and mechanisms. The surface damages like scratches, chips, sore from rubbing, loss of color (or dullness) may prevent product acceptance or require its removal from service. To ensure testing and control of abrasive effect prone surfaces during process design, the special devices called abrasers are used. The application area of such devices is extremely wide. We can use them to control painted surfaces, anodized and powder coatings, laminated material, fabrics, glass and rubber, metals, lacquered surface, leather and leather-like material, plastics, paper and cardboard, linoleum, natural and faux stone, ceramics.

Traditionally, abrasers are used to test the wet or dry abrasion resistance of various films, coatings and paint materials. Testing is carried out by simulating daily wear resulting from general use or cleaning processes. It is also possible to determine the minimum number of cycles for which the coating is destroyed (or loses its appearance), to conduct various tests for scratching and abrasion.

Some designs of abrasers, *e.g.* TQC Taber 5135 (rotational), are designed to test the wear resistance and abrasion of various materials, such as plastics, ceramics, metal, leather, rubber. Testing methods may vary based on the purpose of the product, for example, abrasion using abrasive particles.

The analysis showed that the existing diagnostic methods and the implementing equipment have mainly an abrasive effect on the material, and as a result, we can talk about the mechanical form of abrasive wear [1]. However, we do not always face abrasive wear as a result of frictional contact of two or more bodies in practice [2]. Elements of hydraulic equipment for pumping oil products, coal, and ore experience not only frictional contact with the working medium, but dynamic loads [3]. An example of such equipment is centrifugal slurry pumps for pumping abrasive fluids and cement mortars, abrasive fluids in mine drainage systems [4]. Separate abrasive particles are able to accelerate in the fluid flow and have a quick dynamic effect on the surface of the nodes and assemblies of pumping systems [5]. Another problem could be the assessment of the abrasion resistance of coatings and materials subjected to abrasion. The issues of assessing the abrasive wear impact on equipment performance are also very important, especially in airspace engineering. Here, a special place is taken by the abrasive wear of the outer contours of aircraft units, which are hit by solid sand particles carried away by the air stream [6]. As a result, erosive wear of the working surfaces occurs, including the formation of a surface with densely located "holes" or cavities, and also the numerous scratches and chips of the paint coatings [7]. And we are talking about important units of airplanes and helicopters, such as nozzles, gas turbine blades, walls of combustion chambers, *etc.* This requires the formation of new approaches and the development of simulation diagnostic methods to assess the quality of the surface layer of units under the dynamic flow of abrasive particles [8].

2. The development of a new simulation diagnostic method

The goal to assess the adaptability of the method of ultrasonic diagnostics for the quality assessment of materials and coatings with an abrasive-liquid ultra-jet was set to the SM-12 department (Special Mechanic Engineering) "The Technology of space rocket engineering" of Bauman Moscow State Technical University [9].

The difference of the new method is in the usage of suspension containing abrasive powder for diagnostics instead of previously used water. The method was developed for metallic materials and coatings. It can be assumed that it will also be effective for composite polymer materials. There can be some problems with the implementation of the method if the applied coating has very weak adhesion with the base and the coating can fall off from the surface only because of the centrifugal forces. In this case, it is necessary to implement other kinematic schemes. As a result of preliminary experiments on the effect of abrasive-liquid flow on the surface layer of materials, this effect was found not to provide an opportunity to analyze erosion parameters. Moreover, a negative result was obtained even when choosing processing methods with maximum technically feasible axis velocity, minimum pressure in the hydraulic system and the maximum permissible distance of the focusing tube to the sample surface [10]. As a result of this experiment, assumptions about the need to impose an additional velocity component were made. Such velocity can be created due to the rapid movement of the sample. Since linear drive units will obviously not give a noticeable result, an option with the implementation of the rotation of the sample has been proposed. With the need to vary the velocity to select the optimal velocity indicators for the experiments, we used Sturm AG9012TE angle grinder (manufactured in China) with the following technical characteristics – power is 1000 W, voltage is 220 V. A distinctive feature of this angle grinder model is the ability to change the speed of rotation. This technical solution opens up additional possibilities for choosing technological parameters that can be adjusted, depending on the material being studied [11].

Figure 1 shows a principle diagram of ultra water-jet abrasive diagnostic process. Experiments of the implementation of this process were carried out at Moscow State University, Physics Department, Hydrophysical Research Center. The formation of abrasive-liquid suspension was carried out using water-jet cutting machine Flow Mach 3 3131b (manufactured in the USA).

Figure 2 shows the machine-tool attachment of angle grinder to the water-jet machine. The attachment with the angle grinder fixed in it was fastened to the worktable of the water-jet cutting machine.

3. Experimental testing

The experiment was carried out with the following technological parameters: 1) pressure in the hydraulic system 100 MPa, axis velocity of the jet forming head S = 212 mm / s (the maximum possible for this water-jet machine); 2) the distance from the cut of the focusing tube to the surface is 150 mm (the maximum possible, taking into account the height of the tool with the angle grinder fixed in it); 3) the abrasive concentration is 100 g / min; 4) the spindle speed is 10000 rpm. After all the preparatory operations for placing and securing the machine-tool attachment with the angle grinder to the worktable and after selecting the indicated technological parameters, the abrasive-liquid flow directly affected the sample. After turning on the angle grinder and reaching its stable operating mode, the hydraulic head section began to move to the center of the disk in a horizontal plane. High pressure water supply stopped as soon as the hydraulic head section was in the closest possible position to the center of the sample. The concentration of particle impacts increases as it approaches the center of the disk. This fact has a positive feature, since it seems possible to evaluate how the material or coating will behave at different particle impact densities. This data can be obtained on one single sample and there is no need to make additional test samples. As a result of this experiment, a surface with the presence of multiple impacts of abrasive particles was formed (figure 3).

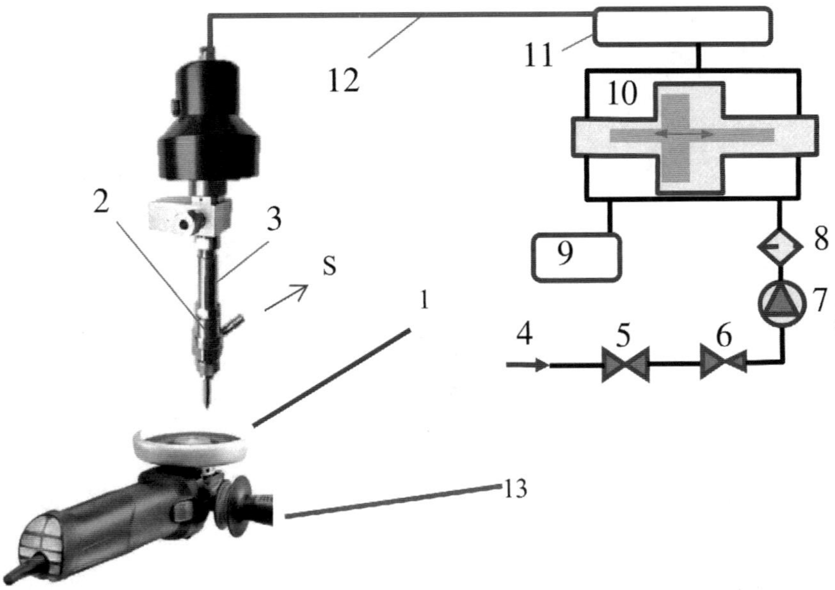

Figure 1. The Principle Diagram of ultra water-jet abrasive diagnostics.

1 – the sample; 2 – focusing tube; 3 – head section of nozzle unit; 4 – water supply; 5 – lock valve; 6 – reduction gearbox; 7 – priming pump; 8 – filter unit; 9 – control system; 10 – intensifier; 11 – vibration damper; 12 – pipeline of high pressure; 13 – angle grinder.

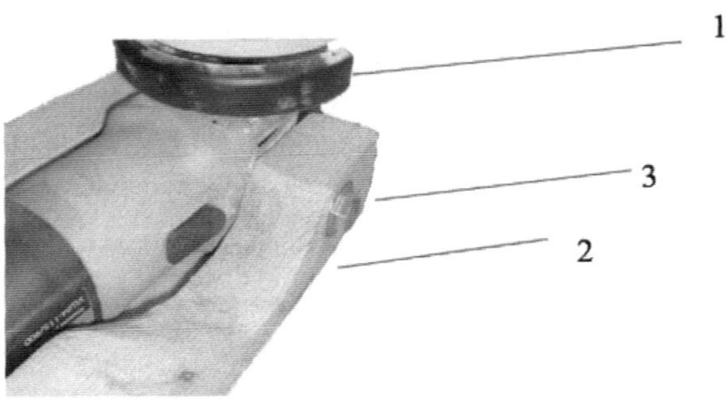

1 – angle grinder
2 – support assembly
3 – fixing screw

Figure 2. The machine-tool attachment of angle grinder

Aluminum alloy was chosen as the diagnosed material [12]. A round sample was cut from a sheet, corresponding in size to standard cutting discs for angle grinders with external diameter of 125 mm, internal bore diameter of 22 mm, and thickness of 2 mm.

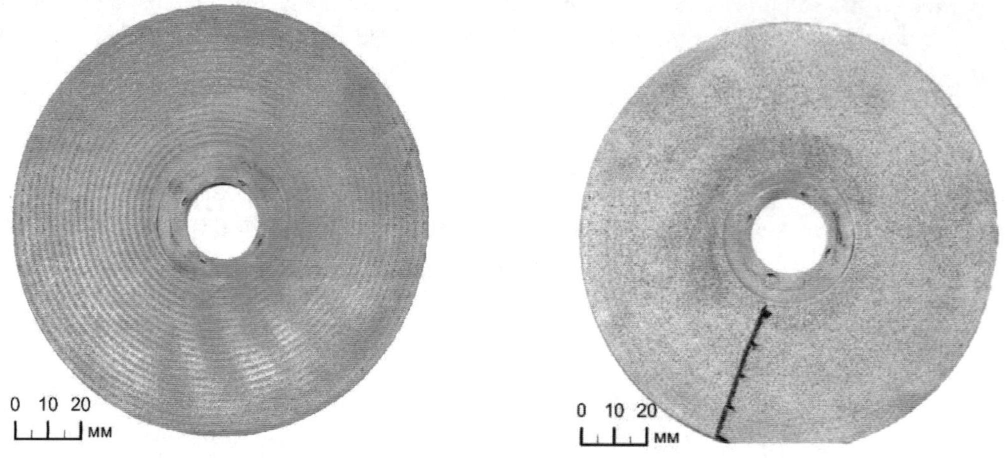

(a) (b)

Figure 3. Samples after water-jet machining: (a) at 4000 rpm, (b) at 10000 rpm.

The results presented in Figure 3(a) do not allow assessing the degree and nature of the penetration of single abrasive particles into the material, but they can be used to assess the overall level of total erosion resistance of a particular material [13].

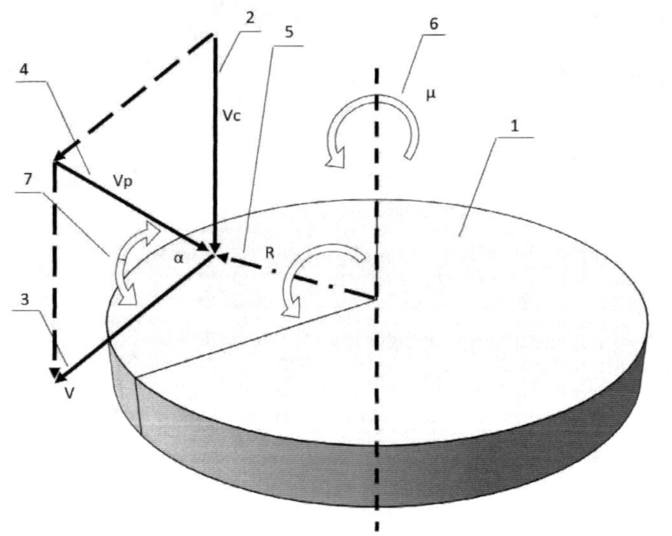

1 – Rotating sample
2 – Velocity vector of water-jet suspension (V_c)
3 – Velocity vector of sample rotation at the point of water-jet impact (V)
4 – Vector of resulting velocity (V_p) of water-jet and sample interaction
5 – Radius (R) of water-jet and sample interaction point
6 – Angular rotational velocity of sample
7 – Angle between velocity vector of water-jet suspension (V_c) and velocity vector of sample rotation (V)

Figure 4. The scheme of the water-jet effect of abrasive suspension on a rotating sample

The sample is of great interest since it can be used to evaluate the penetration depth of the abrasive into the material, the geometric parameters of the prints and their shape [14]. The images of the surface of the sample have been taken by using the MBS-10 Microscope (manufactured in Russian Federation), and 100 times the power of zoom (Figure 3(b)). Figure 5 shows a surface area with holes formed as a result of impacts of abrasive particles.

The measurement of the depths of cavities can be carried out using optical microscope, optical profilometer or profilograph-profilometer. The best results were obtained using optical profilometers of the SURFIEW GLtech series (South Korea).

1, 2 – Cavities on the sample surface.

Figure 5. Photo of the sample surface with traces of interaction with the abrasive-liquid flow.

Obviously, disks made from a wide variety of materials, including composite materials, samples with various coatings, ceramic disks, etc. can be used as a sample [15]. There is no doubt that the nature of the erosion of their surface formed as a result of the application of the described diagnostic method can show a number of operational indicators of materials and products made from these materials [16]. The analysis and map of particle impacts on the surface is of great interest since the parametric characteristics of the water jet and the specifics of the movement of the abrasive powder in it are revealed. Figure 6 presents the expected surface profiles of samples from various materials.

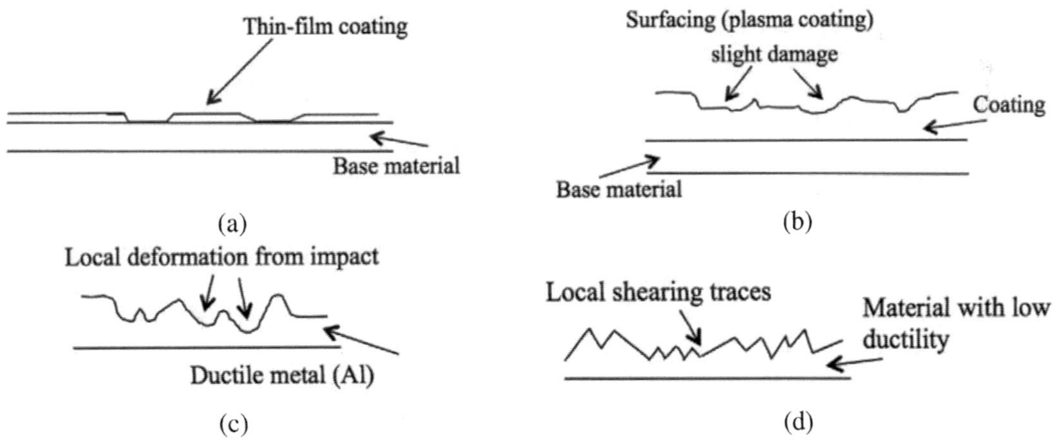

Figure 6. The assumed damage on the surface of various materials as a result of the abrasive-liquid ultra water-jet machining:
(a) – Model of a thin-film coating, *e.g.* TiN, TiCN
(b) – Minor damage during surfacing (plasma coating)
(c) – Assessment of the dynamic ductility of metal and alloys
(d) – Local impact cracking determination of ceramics and composites).

Such parameters as the depth, width, length of the cavities, as well as the height of the plastic displacement rollers can give important diagnostic information useful for choosing a particular material, hardening technology, or surface modification according to the criterion of abrasion resistance [17]. The use of this method in the future can take its rightful place in the development of aerospace engineering products and manufacturing technology development.

The rational implementation of the functional and physical capabilities of discrete ultra water-jet diagnostics in the life cycle of rocket and spaceship equipment items is given below [18].

- Controlled simulation of local impact action on facial layer of details
- Erosive wear of blades and propellers of spaceship power plants
- Rapid recognition of erosive resistance of functional coatings
- Analysis of erosive change in solid-propellant rocket engine elements profile
- Rapid estimation of functional quality parameters of surface by means of locally dynamic indentation
- Quality control test (QCT) of hardening techniques of various materials and parts made from them
- Recognition of adhesive interaction of coatings and protected surfaces
- Rapid estimation of physical and mechanical properties of various powdery materials
- Experimental verification of finite-element models of impact hydrodynamic interaction of dispersed particles with solid target
- Functional properties of composite and ceramic materials
- Reliability support of getting rocket and space equipment parts in additive manufacturing

4. Conclusion

To sum up, there are brief general conclusions:

- Preliminary processing methods are established. They allow to obtain both quantitative and qualitative characteristics of surface erosion as a result of a single high-speed interaction of particles and investigated materials surface.
- According to a set of preliminary experiments, the method of ultra water-jet abrasive diagnostics is found to be the development of the already known technology of ultra water-jet diagnostics and replenishes abrasive materials control means.
- The phenomenological ideas of various materials erosion nature during the interaction between the materials and the abrasive flow of fluid are defined. The verification of the ideas requires a set of additional experimental studies.
- Areas of potential use of the ultra water-jet diagnostics method for solving various technological problems (primarily in airspace engineering) are defined.

Acknowledgments

The studies were carried out within the framework of grants of the Russian Foundation for Basic Research 18-29-18081 and 19-38-90228\19, a grant of the President of the Russian Federation for state support of leading scientific schools of the Russian Federation NSh-3778.2018.8 and grants from the Fund for the Promotion of Innovations under the UMNIK-18 program in accordance with the contracts No. 14727GU/2019 and No. 14549GU/2019.

References

[1] Barzov A A, Galinovsky A L, Provatorov A S 2019 Information physical mechanism of ultra water-jet diagnostics of the functional coatingsquality *Innovative mechanical engineering* pp. 51-57.
[2] Tischenko L A 2014 Development of a methodology for assessing the durability period of a nozzle exploring tube in order to ensure a given performance of water-jet cutting (Moscow: Bauman Moscow State Technical University Press) p169
[3] Korolkov U V 2011 The reliability improvement of the centrifugal-rotational machining process in an abrasive medium (Rostov-on-Don: Don State Technical University Press) p 186
[4] Rozhnenko O A The efficiency improvement of shaped parts surfacesprocessing with a free

abrasive (Rostov-on-Don: Donetsk State Technical University Press) p 21
[5] Krukov S A, Baydakova N V, Mirosedi A I, Revzina E M 2018 Grinding tools with fillers and mixtures of abrasives of different shapes and grit *High Technology(Saint Petersburg)* p 154
[6] Khmelevskaya V B, Myakonkov M B, Moseyko E S 2014 Friction and wear in mechanical engineering mechanisms (Saint Petersburg: Peter the Great St. Petersburg Polytechnic University Press) p 190
[7] Barabanschikov U G, Chulkin S G 2011 Friction and wear during the formation of aqueous dispersed mixtures (Saint Petersburg: Peter the Great St. Petersburg Polytechnic University Press) p 217
[8] Galinovsky A L, Abashin M I, Khafizov M V 2013 Analysis of hydrophysical processes and their capabilities in ultra water-jet processing of materials of rocket and space technology (Moscow: Bauman Moscow State Technical University Press)
[9] Abashin M I 2013 Accelerated determination of the quality parameters of the surface layer of the products' material based on the results of exposure to a supersonic jet of liquid (Moscow: Bauman Moscow State Technical University Press) p 16
[10] Khazifov M V 2015 Technological support for ultra water-jet machining of mechanical engineering units by using acoustic emission method (Moscow: Bauman Moscow State Technical University Press) p 130
[11] Gui-Lin Yang 2013 Forecast surface quality of abrasive water jet cutting based on neural network and verified by experiments *J. Sensors & Transducers* **156**(9):379-383
[12] Moravčíková J, Delgado Sobrino Daynier R, Košťál P 2018 *Analisis of the surface morphology of the S235JRG1 steel after an abrasive water jet cutting process* (Trnava: Faculty of Materials Science and Technology. Slovak University of Technology)
[13] Duspara M, Samardžić I, Starčević V, Horvat M 2017 Optimization of abrasive waterjet machining process parameters *J. Tehnički Glasnik* **11**(4):143-149
[14] Anasiewicz K, Kuczmaszewski J 2017 Effect of abrasive waterjet peening surface treatment of steel plates on the strength of single-lap adhesive joints *J. Adv in Sci and Tech* **11**(3):270-276
[15] Ishfaq K, Ahmad N, Mufti N A, Saleem M Q, Al-Ahmari A M 2019 Abrasive Waterjet Cutting of Clad Composite for Achieving Minimal Cut Quality Difference Between Constituent Layers *J Metals* **9**(7):754 DOI: 10.3390/met9070754
[16] Mital G, Dobransky J, Ruzbarsky J, Olejarova S 2019 Application of Laser Profilometry to Evaluation of the Surface of the Workpiece Machined by Abrasive Waterjet Technology *J Applied Sciences* **9**(10):2134 DOI: 10.3390/app9102134
[17] Liu H T 2019 Advanced waterjet technology for machining curved and layered structures *J Curved and Layered Structures* **6**(1):41-56 DOI: 10.1515/cls-2019-0004
[18] Kmec J, Harnicarova M, Borzan C, Borzan M, Valicek J, Kriz J, Kusnerova M 2019 Proposal for a method of measurement and control of surface quality in the course of abrasive waterjet cutting of material *J MATEC Web of Conf* 299:02003 DOI: 10.1051/matecconf/201929902003

Lightweight nodal parts for large space truss structures enabled by SLM: modelling, manufacturing, and testing

E Golubev[1,2], M Arkhipov[1], A Galinovsky[2] and A Klem[1,2]

[1] Astro Space Center of Lebedev Physical Institute of Russian Academy of Sciences, 53 Leninskiy prospect, Moscow, 119991, Russia
[2] Bauman Moscow State Technical University, 5/1 Baumanskaya 2-ya str., Moscow, 105005, Russia

E-mail: golubev.ev@asc.rssi.ru

Abstract. Selective laser melting (SLM), as a method of metal additive manufacturing, has developed rapidly over the last years to become a breakthrough technology capable of upending the way that products from aerospace and other various industrial sectors are not only fabricated but also designed. The purpose of this paper is to investigate the potential benefits and drawbacks of SLM implementation in the frame of design and development of components for large space lightweight structures. The three types of the most common structural engineering solutions are compared, based on the different manufacturing techniques and materials. The paper describes the results of the measurement of the shape of fittings after fabrication and cryogenic temperature-cycle tests. An extra option for the weight reduction of structures is given via the implementation of the topology optimization method. Three challenges have been identified, including quality of surfaces, deviation of geometry, residual internal stresses. This work shows the possibility to solve this topical issue by using a waterjet inspection technique and represents one of the first attempts for its implementation to perform a fast evaluation of mechanical properties of structural materials manufactured using SLM.

1. Introduction

Lattice or truss frameworks are widely utilized in the design of contemporary large space lightweight structures thanks to their excellent structural strength to weight ratio. Fittings that connect truss members in nodes have rather complex cross-sections and hence are difficult to manufacture. Connecting fittings for such structures made of metal alloys can be developed and manufactured using conventional machining methods [1]. However, fittings can represent up to 60% of the total truss weight. Therefore, the application of additive technologies allows to significantly expand development options of creating more efficient fitting structures [2].

The use of additive technologies for manufacturing fittings was studied based on the frame structure of the primary mirror of the space telescope "Millimetron" [3]. Its load-bearing frame is a complex 3D truss consisting of carbon fibre tubes and connecting fittings. A general view of the primary mirror's structural engineering model is shown in figure 1. This structure is defined by a series of requirements, such as maximum rigidity and shape stability, high requirements for manufacturing and assembly accuracy, extremely low operating temperatures (up to 20K), a large number of part form factors, and their total quantity. All these searches for new engineering and design solutions a very topical issue.

This paper presents a study of the possibilities of using SLM for manufacturing truss fittings, a comparison with other competing technologies, and the results of experiments to verify key quality parameters.

Figure 1. Primary Mirror Frame Structure of Millimetron Space Observatory (engineering mock-up).

2. Literature Review

2.1. Benefits and drawbacks of selective laser melting

Selective laser melting (SLM) is one of additive technologies, which has been actively developed during the last two decades. However, while other additive technologies used for prototyping, cast pattern and mould manufacturing have been firmly established in the industry, the list of the potential applications of SLM technologies as the suitable manufacturing method is still only being studied [2, 4]. This is caused both by the benefits and drawbacks of this approach.

Firstly, there is a prohibitively high level of geometrical complexity of parts, including thin-walled structures with interior cavities. Secondly, there is a wide range of metal powders: heat resistant and corrosion-resistant steels, aluminium, and titanium alloys. Thirdly, the selective laser melting is characterized by zero waste manufacturing (excluding a certain number of removable manufacturing supports), because unused powder can be reused. Furthermore, the physical and mechanical properties of manufactured material match those of conventional semi-finished products (rolled and forged products) [5, 6]. In addition, SLM tools with the capacity for building larger products (up to 500x500x500 mm) are now available, also there are many ongoing research projects aimed at increasing dimensions of manufactured parts. What is more, several parts can be built simultaneously, so long as they fit in the build chamber. Finally, many authors see this approach as a way to reduce the cost of parts and assemblies [7, 8].

However, nowadays there are quite a few drawbacks. Firstly, that is the high probability of shape warping as-fabricated SLM-built parts. This issue is caused by high thermal gradients during SLM process that leads to thermal deformations and residual stresses [9, 10]. Then, the modern manufacturing method as SLM requires the rapid and reliable quality control methods of produced parts. Further, we address both of these issues in this paper.

2.2. Waterjet inspection technique as the quality control method of SLM produced parts

Papers [11-16] substantiate the need for adequate information and diagnostics support using non-destructive inspection methods to control the quality of parts manufactured using additive technologies, in particular, selective laser melting (SLM). However, the issue of rapid determination of heterogeneity of mechanical properties of such materials remains unsolved. Therefore, the problem of fast evaluation of the heterogeneity of mechanical properties of structural materials manufactured using SLM is very topical, since the quality of work surface of products made of such materials usually defines operating parameters.

In this study, waterjet inspection technique (WIT) was chosen as a non-destructive testing method. It is based on the scouring along some path impact of a high-speed water jet on the surface of the controlled object [17]. Upon that, the geometrical parameters of the formed erosive cavity serve as the informative attributes of diagnostics. The studies [18-19] present some of the results of experiments that demonstrate the capabilities of assessing homogeneity of the material. It shows the dependence between the depth of the cavity formed after waterjet impact and the hardness, impact viscosity of a material. This opens up the possibility of performing a detailed analysis of mechanical properties of SLM manufactured products, where the heterogeneity of mechanical properties often limits operating and structural reliability [20, 21]. Thus, we experimentally sampled this technique based on manufactured prototypes.

3. Experimental design

3.1. Search for structural engineering solutions

This paper analyses potential structural engineering solutions (figure 2). The carbon fibre fitting is characterized by low weight and is manufactured from a material with qualities close to those of the tube material (because another reinforcement method is used), which is especially important for coefficients of thermal expansion of various materials due to the deep cooling of the structure and the need to minimize thermal deformations. However, a significant drawback of this option is an extremely high price because of, primarily, the price of tooling required for a composite structure with complex geometry.

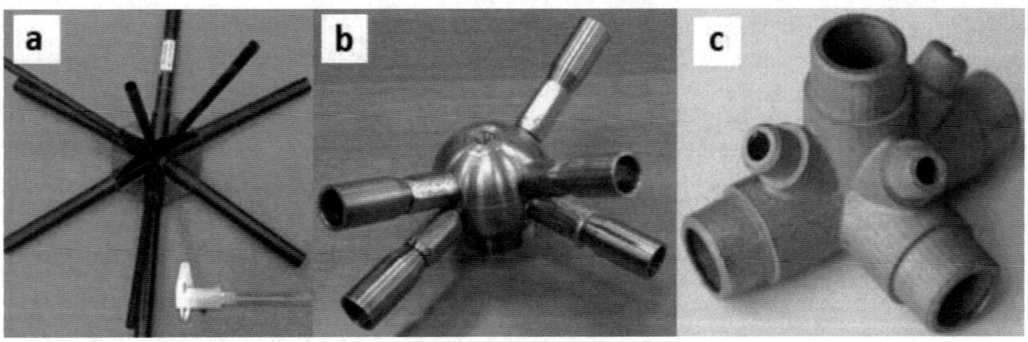

Figure 2. Examined Structural Engineering Solutions for Fittings:
(a) carbon fibre fitting; (b) machined fitting; and (c) fitting manufactured using SLM.

The machined fitting (figure 2, b) was used to create a structural engineering mock-up (figure 1). The fitting has a modular design. Its threaded elements allow for a certain adjustment of the length of truss members during assembly to improve accuracy. The key drawback of this structure is large weight of the fitting, which leads to the combined weight of fittings in the frame structure reaching 60% of the total weight. It also negatively influences the level of thermal deformations of the structure during deep cooling.

Fitting No. 3 (figure 2, c) is manufactured using the SLM method with the Concept Laser M2 Cusing printer out of corrosion-resistant steel dust powder (the domestic analogue is 12X18H10T).

3.2. Manufacturing prototypes

The SLM-built fitting prototype proved acceptable manufacturability and cost of production. Thus, five fitting prototypes were made from Ti-6A-l4V titanium alloy using a Concept Laser M2 SLM machine (figure 3). Figure 4 shows four fittings connected in a closed member of the frame. The fifth prototype was designed for cryogenic testing. Due to manufacturing restrictions, the thickness of fitting walls was increased to 1 mm, while the thickness limited by strength requirements is 0.8 mm. The weight of finished parts was 14-22 g, while the weight of the machined fitting (figure 2, b) was 79 g.

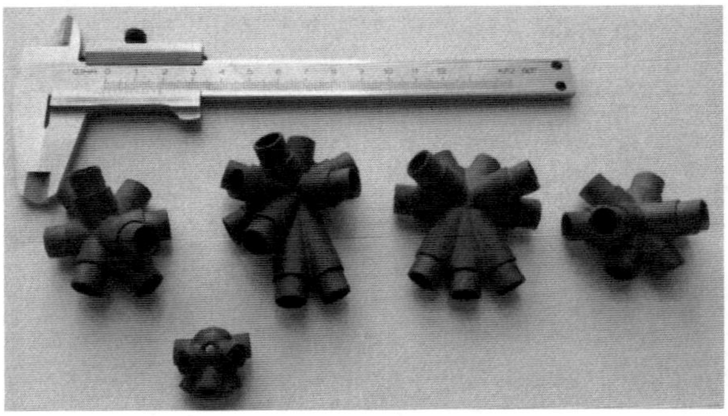

Figure 3. SLM-built fittings from Ti-6Al-4V alloy.

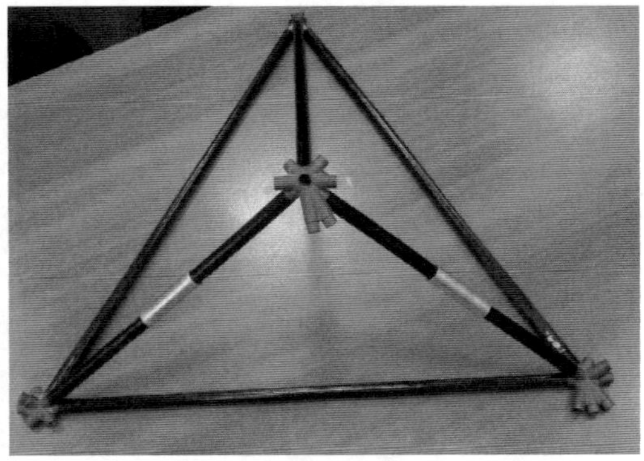

Figure 4. The prototype of the assembled petal frame segment.

3.3. Measurement of manufacturing errors

In order to measure the manufacturing errors, high precision geometry scanning was used [22]. The geometry of the produced fittings was measured at room temperature using the optical scanner ATOS III, with an accuracy of ±6 μm. The scan data was overlaid on the reference CAD model using a Gaussian best fit approach realized in the commercial software Geomagic Control®. The deviation analysis provided clarity that the accuracy of the SLM-built fittings was within ±0.1 mm of the reference geometry. Figure 5 shows the deviation colour map and table 1 gives calculated root-mean-square deviations (RMS) for the surface and the deviations of the axes of cylindrical surfaces (interfaces for truss members from CFRP tubes).

Table 1. Fitting Geometry Measurements after Manufacturing.

	RMS (mm)	Axes deviation (degrees)
Fitting No. 1	0.074	≤1.2°
Fitting No. 2	0.082	≤1.5°
Fitting No. 3	0.084	≤2.4°

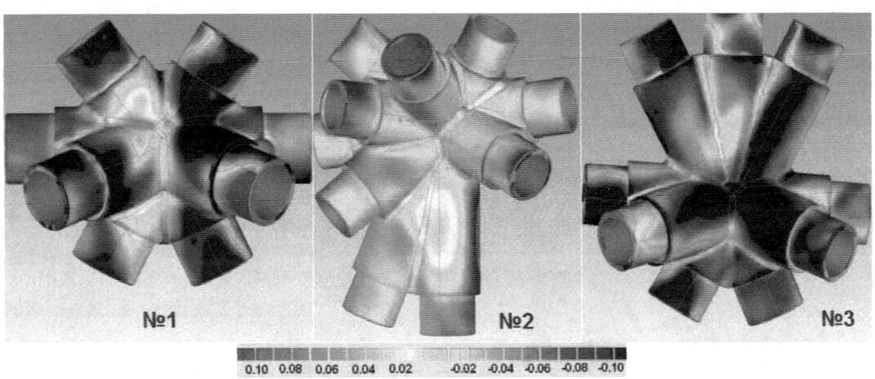

Figure 5. Deviation Map (mm, relative to the reference CAD-geometry).

Given the fact that the considered frame structure operates at the temperature of 20K and exposed to almost 300 K temperature range, the issue of thermal dimensional stability is essential. According to the requirements, the thermal distortions of the primary mirror frame structure should not exceed 2 mm. In order to measure the residual thermal distortions of the frame structure, a cryogenic test was performed. For this test the specimen, which represents a section of the frame structure, was assembled from fitting No. 5 and CFRP tubes by an adhesive of cryogenic operation. During the first cryogenic thermal cycling, the specimen was cooled down to -195°C. Geometry was measured before and after cooling. Figure 7 shows measurement results – comparison of tube axes position before and after the cryogenic test. The maximum deviation of the tube axis was 0.12 degrees that fully meets the requirements for the dimensional stability of the truss.

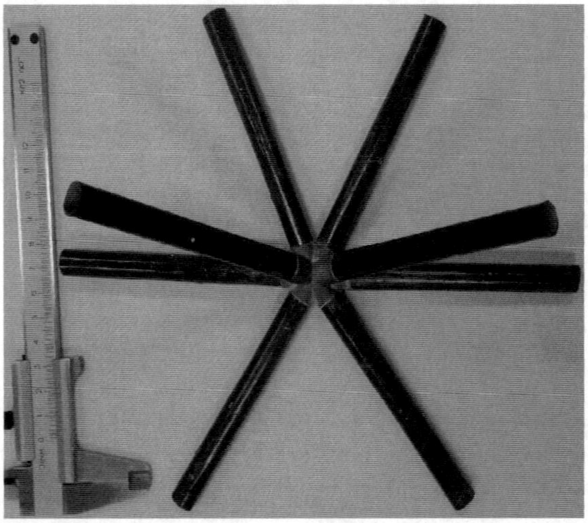

Figure 6. Specimen for the Cryogenic Test.

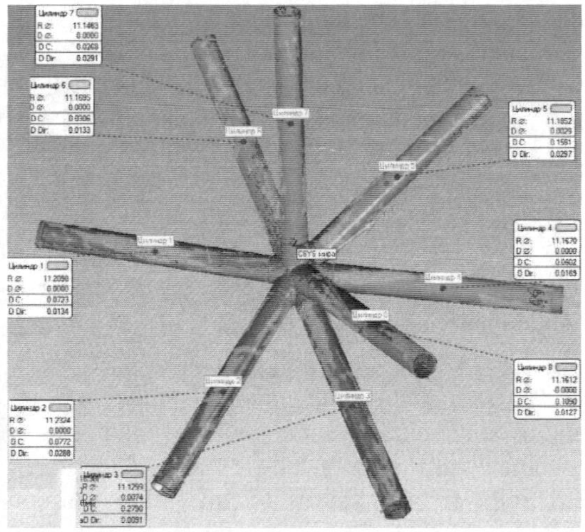

Figure 7. Comparison of Tube Alignment Before and After Cooling.

3.4. Using topological optimization methods to minimize fitting structural weight

Using SLM to manufacture the fittings helped reduce the structural weight to less than 3 times the weight of the machined fitting and save 57% of the overall weight for the complete assembly. Additionally, the possibility of structural weight reduction using topological optimization based on computational modelling was investigated [23]. The optimization objective function was the minimum weight of the fitting with the constraints of required mechanical strength and rigidity. The boundary conditions for the static analysis were the loads emerging in the frame when it is delivered to outer space on a rocket. Figure 8 compares the initial geometry of fitting No. 1 and the so-called "biomorphic" optimized structure. It appeared that topological optimization enabled a weight reduction of the optimized area by an extra 34%, while the total weight of the fitting was reduced by 9%. As a result, the level of maximum stress in the structure fell by 21%.

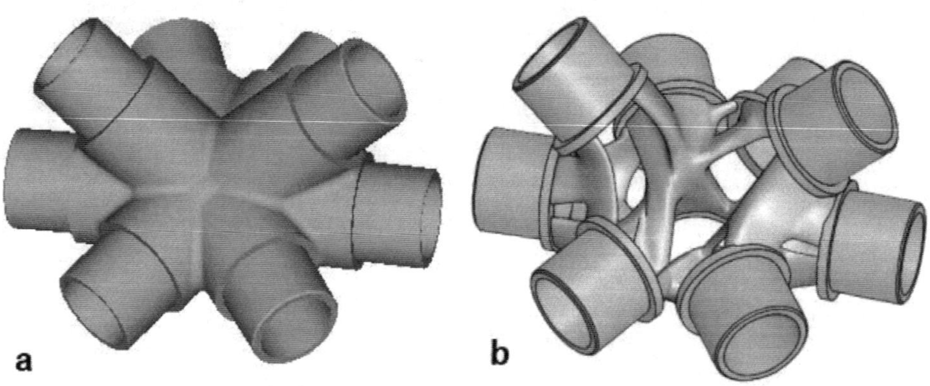

Figure 8. Comparison of two fitting variants:
(a) reference variant; (b) optimized variant.

3.5. Waterjet inspection of fitting

The WIT method for identifying the degree of the heterogeneity of an SLM-built material is based on the assumption that the mechanical properties change with a period equal to the step of the laser beam or the thickness of the growth. It follows that the regions of a material having lower mechanical properties as hardness or strength are subjected to greater erosion. A diagram of the formation of a microrelief of the cavity on the studied surface under the action of the waterjet is shown in figure 9.

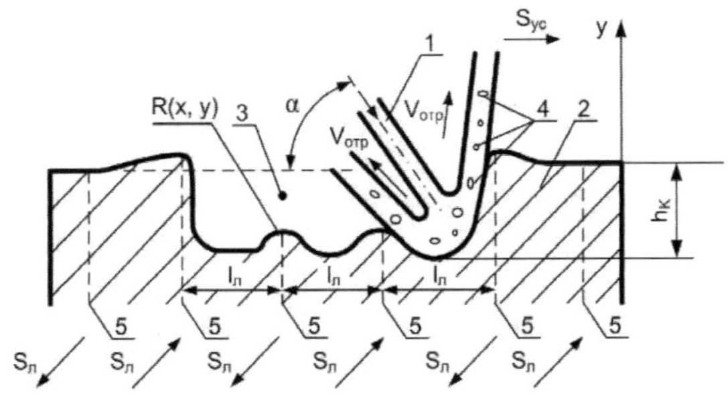

1 – Waterjet
2 – Surface layer of SLM-built materials
3 – Cavity
R(x,y) – Microrelief of cavity
hk – Current depth of the cavity
Syc – Direction of waterjet movement
SL – The step between the laser beam passes

Figure 9. Scheme of WIT of the surface of SLM- built material.

4. Results and discussion

The WIT procedure was carried out using the abrasive waterjet cutting equipment Multicam WaterJet (US). One of the fittings has been exposed to the waterjet along its growth direction (figure 10). The following operating parameters were used: pump pressure – 350 MPa, waterjet traverse speed – 2 mm/s, stand-of distance – 5 mm. Pure water-jet was applied at the right angle to the surface. The image of the waterjet footprint was obtained using the laser scanning confocal microscope Carl Zeiss LSM 700.

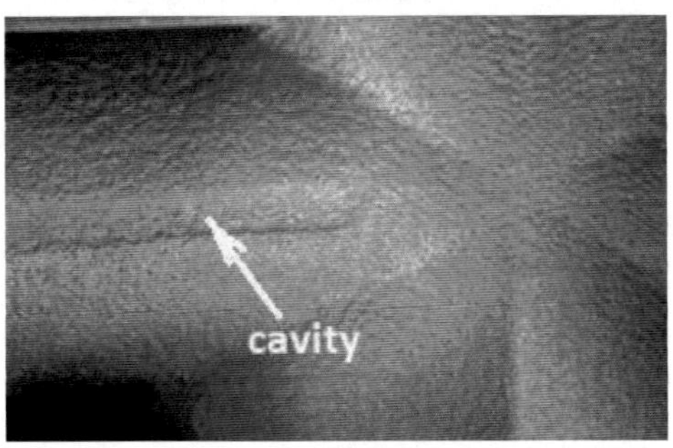

Figure 10. Fitting after WIT test.

Figure 11 shows the results of the laser scanning of microrelief of waterjet footprint (or cavity) and its post-processing analysis. The standard Fourier analysis method was used via MATLAB for a detailed consideration of the informative parameters of the cavity. Fourier analysis spectrogram identifies dominant harmonic, its amplitude, and period. These are the characteristic of geometric deviations of waterjet footprint, which are due to periodic changes in mechanical properties along the cavity. It follows from the data in figure 11 that the dominant harmonic of the Fourier analysis of the cavity microrelief has a similar character to the geometry of laser melting of the material (the step between melt pools).

At the next step of experimental study, in order to verify the results of WIT, measurements of the surface microhardness were carried out parallel to the formed cavities. Based on the data obtained on the microrelief of the waterjet footprint and on the approximation of the microhardness measurement in the selected areas (600 μm in length), their correspondence and satisfactory correlation were established.

Thereby, the waterjet inspection technique shows sensitivity to variations in the mechanical properties (i.e. microhardness) of the SLM-built material with periodic dependence. That makes this method is perspective to develop for rapid determination of the degree of heterogeneity of materials and parts produced by SLM. Undoubtedly, it requires further study and comparison with other non-destructive inspection methods.

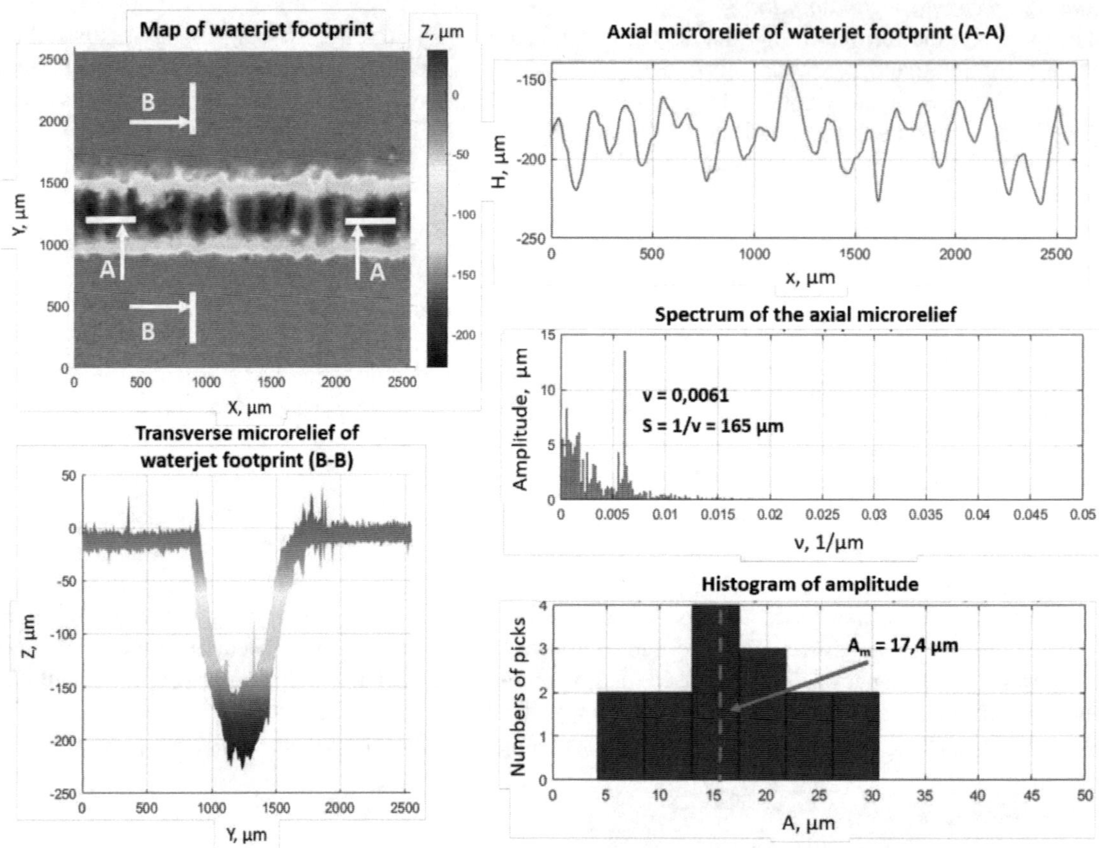

Figure 11. Waterjet footprint data for the specimen.

5. Conclusions

Based on the study results outlined above, we can draw the following conclusions:

1. The layered selective laser melting technology is efficient for manufacturing parts with complex profiles, such as truss fittings. Conventional technologies (casting, machining) cannot provide the same level of structural strength and low weight.
2. Satisfactory manufacturing accuracy was achieved. Accuracy can be improved by different building orientation and using manufacturing supports.
3. The possibility to use SLM manufactured products in cryogenic equipment together with carbon fibre parts was confirmed.
4. Topological optimization methods efficiently complement the potential of additive technologies in reducing the structural weight of load-bearing elements. However, they require significant computational costs and time to post-process obtained "biomorphic" geometry.
5. Waterjet inspection technique has shown itself to be sensitive to variations in the mechanical properties of the SLM-built material with periodic dependence. It requires further study and the comparison with other non-destructive inspection methods.

References

[1] Atkinson Ch, Texter S, Keski-Kuha R and Feinberg L, 2014, "Status of the JWST optical telescope element," *Proc. of SPIE 9143 Space Telescopes and Instrumentation.*

[2] Kumar L J and Krishnadas Nair C G, 2017, "Current Trends of Additive Manufacturing in the Aerospace Industry," in *Advances in 3D Printing & Additive Manufacturing Technologies* ed. Wimpenny D *et al.*, Springer, Singapore, 39-54.

[3] Smirnov A V, Baryshev A M, Pilipenko S V, Myshonkova N V, Bulanov V B, Arkhipov M Yu, Vinogradov I S, Likhachev S F and Kardashev N S, 2012, "Space Mission Millimetron for Terahertz Astronomy," *Proc. of SPIE 8442 Space Telescopes and Instrumentation.*

[4] Gibson I, Rosen D and Stucker B, 2015, *Additive Manufacturing Technologies: 3D Printing, Rapid Prototyping, and Direct Digital Manufacturing,* 2nd ed., Springer-Verlag, New York.

[5] Alcisto J, Enriquez A and Garcia H, 2011, "Tensile Properties and Microstructures of Laser-Formed Ti-6Al-4V," *J. Mater. Eng. Perform.,* **20**, 203-212.

[6] Vilaro T, Colin C and Bartout J D, 2011, "As-Fabricated and Heat-Treated Microstructures of the Ti-6Al-4V Alloy Processed by Selective Laser Melting," *Metall. Mater. Trans. A Phys. Metall Mater Sci.,* **42**, 3190-3199.

[7] Quarshie R, MacLachlan S, Reeves P, Whittaker D and Blake R, 2012, *Shaping our national competency in additive manufacturing,* Additive Manufacturing Special Interest Group, Technology Strategy Board, UK.

[8] Durakovic B, 2018, "Design for Additive Manufacturing: Benefits, Trends and Challenges," *Period. Eng. Nat. Sci.,* **6**, 179-191.

[9] Blackford B, Gene Z and Kim I, 2020, "The effect of scan path on thermal gradient during selective laser melting," *Int. J. Adv. Manuf. Technol.,* **110**, 1-14.

[10] Yingli L, Zhou K, Tan P, Tor S B and Leong K F, 2017, "Modeling temperature and residual stress fields in selective laser melting," *Int. J. Mech. Sci.,* **136**, 24-35.

[11] Aleshin N P, Murashov V V, Evgenov A G, Grigoriev M V, Simakov N A, Vasilenko S A and Krasnov I S, 2016, "The Classification of Flaws of Metal Materials Synthesized by Selective Laser Melting, and the Capabilities of Non-Destructive Testing Methods for their Detection," *Defektoskopia,* **52**, 48-55.

[12] Babentsova L P and Antsiferova I V, 2017, "Quality and Ecology of the Selective Laser Technology," *Master's Journal,* **1**, 87-92.

[13] Kosushkin P A, 2017, "Quality Control in Additive Manufacturing." *Vector of High Technology,* **2**(31)**,** 28-31.

[14] Koester L, Taheri H, Bigelow T, Collins P and Bond L, 2018, "Nondestructive Testing for Metal Parts Fabricated Using Powder-Based Additive Manufacturing," *Mater. Eval.,* **76**, 514–524.

[15] Senck S, Happl M, Reiter M, Scheerer M, Kendel M, Glinz J and Kastner J, 2020, "Additive manufacturing and non-destructive testing of topology-optimised aluminium components," *Nondestruct. Test. Evaluation,* **35**(3), 315–327.

[16] Charalampous P, Kostavelis I and Tzovaras D, 2020, "Non-destructive quality control methods in additive manufacturing: a survey," *Rapid Prototyp. J.,* **26**(4), 777-790.

[17] Abashin M I, 2013, "Accelerated Determination of the Quality Parameters of the Surface Layer of the Material of Products by the Results of the Influence of Supersonic Liquid Jet on It," *Abstract diss. Cand. Tech. Sci.. Moscow: Bauman Moscow State Technical University.*

[18] Abashin M I, Bochkarev S V, Tsaplin A I and Kobernik N V, 2015, "Ultra-Jet Diagnostics of the Quality of Weld," *Izvestiya Vysshih Uchebnyh Zavedenij. Mashinostroenie,* **12**(669), 52-61.

[19] Bochkarev S V, Tsaplin A I, Galinovskii A L, Abashin M I and Barzov A A, 2017, "Ultrajet Diagnostics of the Microstructure of a Material under Heat Treatment," *Met. Sci. Heat Treat.,* **6**(744), 58-63.

[20] Kok Y, Tan X, Wang P, Nai M, Loh N H, Liu E, and Tor S B, 2018, "Anisotropy and Heterogeneity of Microstructure and Mechanical Properties in Metal Additive Manufacturing: A Critical Review," *Mater. Des.,* **139**, 565–586.

[21] Popovich A A, Sufiiarov V S, Borisov E V, Polozov I A, Masaylo D V and Grigoriev A V 2017 Anisotropy of Mechanical Properties of Products Manufactured Using Selective Laser Melting of Powdered Materials *Russian Journal of Non-Ferrous Metals,* **58**(4), 389-395.

[22] Chua C K, Wong C H and Yeong W Y, 2014, *Standards, Quality Control, and Measurement Sciences in 3D printing and additive manufacturing,* Elsevier Ltd.

[23] Eschenauer H A and Olhoff N, 2001, "Topology optimization of continuum structures: a review," *Appl. Mech. Rev.,* **54**, 331-390.

IOP Conf. Series: Materials Science and Engineering 1060 (2021) 012011 doi:10.1088/1757-899X/1060/1/012011

Influence of the profile of longitudinal grooves of various depths on increasing static characteristics of radial gas bearings

I V Vishtak[1*], O V Petrov[2], V I Savulyak[3], S I Sukhorukov[2]

[1] Department of Life Safety and Safety Pedagogy, Vinnytsia National Technical University, Khmelnytske shoes 95, 21021, Vinnytsia, Ukraine
2 Department of Technology and Automation of Machine Building, Vinnytsia National Technical University, Khmelnytske shoes 95, 21021, Vinnytsia, Ukraine
3 Department of of Industrial Engineering, Vinnytsia National Technical University, Khmelnytske shoes 95, 21021, Vinnytsia, Ukraine

E-mail: innavish322@gmail.com

Abstract. Radial gas bearings with one line for supplying compressed gas to the working gap with blind microgrooves of constant width and variable depth of the different profiles are considered. Such as with constant depth microgrooves; triangular or stepwise with decreasing depth of grooves along the course of compressed gas. For them, the radial lifting force, the recovery moment for angular displacements of shaft in the range of constancy of radial and angular stiffness and consumption of compressed gas for the bearing's work are found. Calculations and studies have shown that a radial bearing with longitudinal microgrooves whose depth decreases linearly in the direction of gas flow has a large radial lifting force and a recovery moment for angular displacements of shaft compared to a bearing with microgrooves of constant depth. Bearings with stepped longitudinal microgrooves with greater depth at the entrance to working gap are operable in the range of practical use of bearings and have greater dimensionless radial and angular stiffness and are more economical compared to bearings with microgrooves of constant depth. In practical activities (when designing spindle units), it is not recommended to use gas bearings with a variable depth of microgrooves having a minimum depth at the entrance to working gap. With optimal design parameters, the flow rate of compressed gas for bearings with longitudinal microgrooves of different transverse profiles with alignment of shaft and sleeve is almost the same (the difference is about 7, 8%).

1. Introduction

One of important problems of modern engineering is to achieve high accuracy, rigidity and vibration resistance of equipment. One of the most critical assembly units of machine is its spindle, which is always involved in shaping movements and is subjected to operational loads [1-3]. Already at the design stage, the problem arises of creating such assembly units and elements of metal-cutting machines, which during the entire operational period would provide the specified processing accuracy. Analysis of the design of high-speed spindles on gas-lubricated bearings shows that they include radial and thrust gas-static bearings. The most important operational characteristics of such bearings are the stiffness of the lubricant layer, the recovery moment from the skew axis of the spindle and the load capacity. Therefore, the problem of creating gas bearings to ensure high performance spindles is of great importance in industry. Gas bearings can operate reliably at low and high temperatures and humidity, their use eliminates environmental pollution, reduces the level of vibration and noise. Such bearings are practically free of wear; therefore, high characteristics of spindle rotation accuracy are maintained for almost the entire life of the machines and equipment.

Spindle units with gas bearings are widely used in modern domestic and foreign production in machines and equipment for various purposes [4, 5]. They the ability to work reliably and remain durable at the required high shaft rotation speed (150 thousand min-1 and more). they have a number of advantages among other types of supports (bearings). Firstly, they can operate in a wide range of temperatures and pressures (the viscosity of gases is practically independent of temperature and pressure), as well as in the zone with increased radiation (gases are not subject to phase changes). Secondly, virtually no wear their work surfaces. Third, they have minimal friction losses, and therefore, negligible heat dissipation and can work with minimum sliding speed. Among the various types of radial gas bearings, the most technologically advanced and simplest in design are bearings with longitudinal microgrooves.

2. Literature Review

Studies have shown [1, 6] that for symmetrical gas bearings in the central position, the linear lifting force depends on the relative eccentricity in the range $-0.5 \leq \varepsilon \leq 0.5$. Therefore, the lifting force of bearing will also linearly depend on the relative radial movement of the movable sleeve. The problem of optimizing the structural parameters of the gas bearing [6] was the search for a criterion in which dimensionless rigidity itself was maximum at the minimum flow of gas. Calculations showed that for a fixed value of relative length of microgroove there always exists a value of the microgrooves parameter at which dimensionless stiffness reaches a maximum, and the closer α1 ($0 \leq \alpha_1 \leq 1$) to a maximum, the greater bearing stiffness. But as the relative length approaches unity, the gas consumption rises even faster. Therefore, bearings with a large distance between the gas supply lines have a low ratio of stiffness to gas flow.

Researches and calculations [1] showed that the dimensionless lifting force F^* and the dimensionless recovery moment M^* linearly depend on the relative radial eccentricity, $\varepsilon = \frac{e}{c}$, and angular $\theta = \frac{e_\theta}{c}$ ($e_\theta = l_0 \cdot \upsilon$ – eccentricity on the shaft end face at $e = 0$, $\upsilon \neq 0$) displacements of shaft within: $-0.5 \leq \varepsilon \leq 0.5$; $-0.5 \leq \theta \leq 0.5$.

For gas bearings, the effect of radial displacements on the recovery moment M and angular displacements on the lifting force F is practically zero [1].

Calculations and studies have shown that a bearing with longitudinal microgrooves whose depth decreases in the direction of gas flow (figure 1 b, c) [7, 8] has a significantly larger radial lifting force and recovery moment compared bearings with microgrooves constant depth. Since the gas pressure in bearings with microgrooves with a variable depth of microgrooves are functions of the longitudinal coordinate, that for calculation of characteristics of bearings was used spline methods (approximating the gas pressure in the grooves by cubic polynomials) [7,8] and cyclic sweep [9-11].

Radial bearings with stepped longitudinal microgrooves with greater depth at the entrance to the working clearance (figure 2, d) [9] are operable in the range of practical use of bearings and have greater dimensionless radial stiffness and are more economical compared to bearings with microgrooves constant depth. In perspective of further research, it is planned to optimize the design parameters of the microgrooves whose depth decreases in the direction of gas flow and stepped microgrooves with a greater depth in the zone of compressed gas supply to the working clearance of radial gas bearings according to the criteria [1].

An important characteristic of the operation of gas bearings with stepwise longitudinal microgrooves is their response to the angular displacements of the shaft (figure 1). But in scientific journals there are no calculations and studies of such bearings.

The shaft of a radial bearing with stepped longitudinal microgrooves is rotated through an angle $\upsilon = e_0/l_0$ under the action of an external load (figure 2). Working clearances in the radial bearing at $\upsilon \neq 0$:

$$h_1 = (c+\sigma)(1 - \nu \cdot \vartheta \cdot \xi \cdot \cos\varphi); \quad h_2 = (c+\sigma_1)(1 - \nu_1 \cdot \vartheta \cdot \xi \cdot \cos\varphi);$$
$$h_3 = c \cdot (1 - \nu_1 \cdot \xi \cdot \cos\varphi),$$

where $\nu = \frac{c}{c+\sigma}$, $\nu_1 = \frac{c}{c+\sigma_1}$ are parameters for changing gap as a result of applying microgrooves and $\xi = \frac{z}{l_0}$, is the dimensionless axial coordinate.

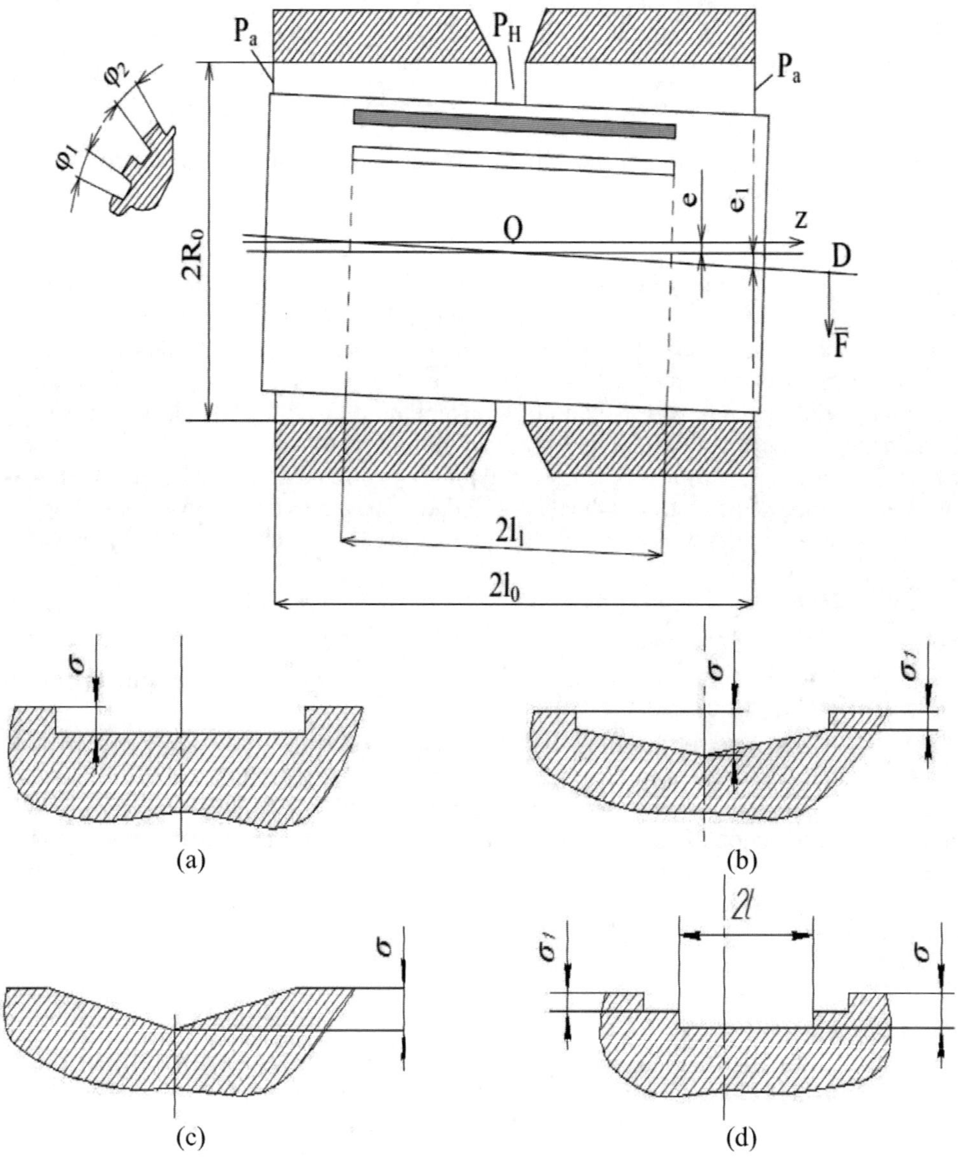

Figure 1. Radial bearing with longitudinal microgrooves for radial and angular movements of shaft under the action of force *F*:
(a) – Microgrooves of constant depth
(b), (c) – Microgrooves depth is decreases according to a linear law along the gas flow
(d) – Stepped microgrooves with a maximum depth at the inlet from an external source of compressed gas.

According to the methodology [1], the dimensionless pressure squares U_1, U_2, U_3 – the square of dimensionless gas pressure in the profiled (U_1, U_2) and smooth zones (U_3) of support with longitudinal stepwise grooves (figure 2, b) of the gas bearing are found from the expressions:

$$\left.\begin{aligned} U_1 &= P_H^2 + \chi_7 b_{02}\xi + \vartheta((\lambda\chi_7 - c_2)e^{\lambda\theta\xi} + c_2 e^{-\lambda\theta\xi} - \frac{3\beta}{\lambda^2\theta^2}\cdot\chi_7))\cdot b_{02}\cdot\cos\varphi\,; \\ U_2 &= 1 + b_{02}(c_1+\xi) + \vartheta\left(ce^{-\lambda\theta_1\xi} - (B_4 + cB_3)e^{\lambda\theta_1\xi} - \frac{3\beta_1}{\lambda^2\cdot\theta_1^2}\right)\cdot b_{02}\cdot\cos\varphi\,; \\ U_3 &= 1 - \chi_5\cdot b_{02}(1-\xi) + \vartheta\left(c_4 e^{\lambda\xi} + c_3 e^{-\lambda\xi} - \frac{3}{\lambda^2}\cdot\chi_5\right) b_{02}\cdot\cos\varphi\,. \end{aligned}\right| \qquad (1)$$

The restoring moment of a radial gas bearing with stepped microgrooves with angular displacements of the shaft (figure 2):

$$M = 4p_a R_0^3\cdot M^*.$$

where M^* – the dimensionless recovery moment.

$$M^* = \lambda^2(\int_0^{\alpha}\xi d\xi\int_0^{\pi}\sqrt{u_1}\cdot\cos\varphi\cdot d\varphi + \int_{\alpha}^{\alpha_1}\xi d\xi\int_0^{\pi}\sqrt{u_2}\cos\varphi\, d\varphi + + \int_{\alpha_1}^{1}\xi d\xi\sqrt{u_3}\cos\varphi\, d\varphi.$$

Given the squares of dimensionless pressure U_1, U_2, U_3 (1) in the working sections of bearing, we obtain:

$$M^* =$$

$$= \lambda^2\int_0^{\alpha}\xi d\xi\int_0^{\pi}\left(\sqrt{\begin{gathered}P_H^2 + \chi_7 b_{02}\xi + \vartheta((\lambda\chi_7 - c_2)e^{\lambda\theta\xi} + c_2 e^{-\lambda\theta\xi} - \frac{3\beta}{\lambda^2\theta^2}\cdot\chi_7)) \\ \cdot b_{02}\cdot\cos\varphi\times\cos\varphi\cdot d\varphi\end{gathered}}\right)$$

$$+\int_{\alpha}^{\alpha_1}\xi d\xi\int_0^{\pi}(\sqrt{\begin{gathered}1 + b_{02}(c_1+\xi) + \vartheta\left(ce^{-\lambda\theta_1\xi} - (B_4 + cB_3)e^{\lambda\theta_1\xi} - \frac{3\beta_1}{\lambda^2\cdot\theta_1^2}\right)) \\ \cdot b_{02}\cdot\cos\varphi\times\cos\varphi\cdot d\varphi\end{gathered}}$$

$$+\int_{\alpha_1}^{1}\xi d\xi\int_0^{\pi}\sqrt{1 - \chi_5\cdot b_{02}(1-\xi) + \vartheta\left(c_4 e^{\lambda\xi} + c_3 e^{-\lambda\xi} - \frac{3}{\lambda^2}\cdot\chi_5\right) b_{02}\cdot\cos\varphi\cdot\cos\varphi\cdot d\varphi}\,.$$

In the working range θ, the actual expression $M^* = \vartheta k_{\upsilon}^*$ [1, 10], where k_ϑ^* is the derivative of M^* with respect to θ at $\theta = 0$, that is, the angular stiffness in the linearity range M^*.

$$\left(\frac{dM^*}{d\vartheta}\right)_{\vartheta=0} = k_\vartheta^* =$$

$$= \lambda^2\int_0^{\alpha}\xi d\xi\int_0^{\pi}\frac{(\left(\lambda\chi_7 - c_2 e^{\lambda\theta\xi} - \frac{3\beta}{\lambda^2\theta^2}\cdot\chi_7\right))\cdot b_{02}\cdot\cos^2\varphi\cdot d\varphi}{2\sqrt{P_H^2 + \chi_7 b_{02}\xi}} +$$

$$+\int_{\alpha}^{\alpha_1}\xi d\xi\int_0^{\pi}\frac{(\left(-(cB_3 - B_4)e^{\lambda\theta_1\xi} + ce^{-\lambda\theta_1\xi} - \frac{3\beta_1}{\lambda^2\theta_1^2}\right))b_{02}\cdot\cos^2\varphi\cdot d\varphi}{2\sqrt{1 + (c_1+\xi)\cdot b_{02}}} + \qquad (2)$$

$$+\int_{\alpha}^{1}\xi d\xi\int_0^{\pi}\frac{(c_4 e^{\lambda\xi} + c_3 e^{-\lambda\xi} - \frac{3}{\lambda^2}\cdot\chi_5)\cdot b_{02}\cdot\cos^2\varphi\cdot d\varphi}{2\sqrt{1 - \chi_5(b_{02}(1-\xi))}}.$$

IOP Conf. Series: Materials Science and Engineering 1060 (2021) 012011 doi:10.1088/1757-899X/1060/1/012011

The dimensionless flow rate of compressed gas Q^* for operation of the radial gas bearing are found at $\theta = 0$ [1]:

$$Q^* = \frac{2(P_H^2 - 1)}{\lambda(1 - \alpha + \frac{\alpha v^3}{a_e + v^3(1 - a_e)})}.$$

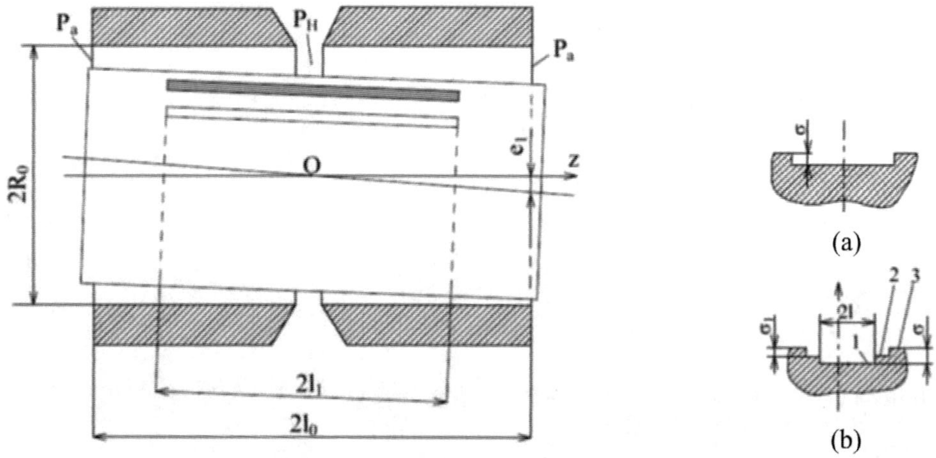

Figure 2. Radial bearing with longitudinal microgrooves at angular shaft movements

3. Results

In radial bearings with longitudinal microgrooves of constant depth (figure 2, a), for a fixed value of the width of microgrooves a_e, there are optimal values of the depth and length of microgrooves (figure 3, 4) at which function K_ε^*/Q^* reaches its maximum. But the design parameter of microgrooves a_e is not included in the values by which the optimization is performed. Originality of choice a_e is complicated by fact that as the microgrooves width decreases their optimum depth increases, and the dimensionless radial rigidity of the bearing increases closer to maximum value (figure 3).

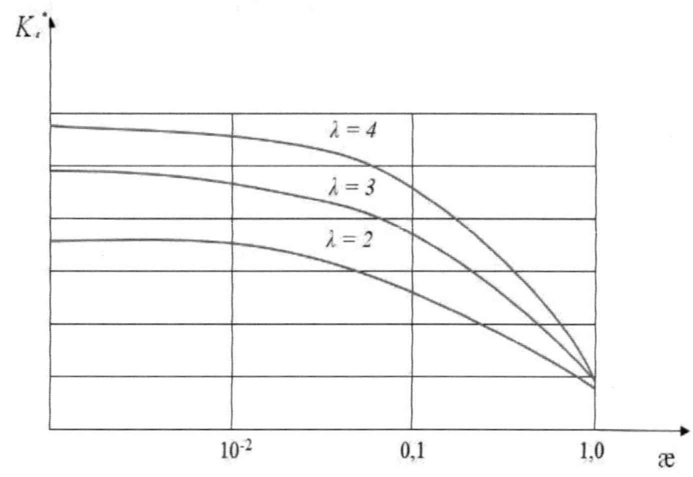

Figure 3. Dependence of dimensionless radial rigidity, K_ε^*, of a radial bearing with longitudinal microgrooves at $P_H = 3$ on the relative width of microgrooves a_e.

Depending on the optimum values of parameters ν and α on a_e (fig. 4) at the dimensionless $P_H = 3$ gas pressure supplied to the input of working clearance of the radial bearing, it is possible to draw an incorrect conclusion that longitudinal microgrooves should be tried to be made narrow and deep. But the Reynolds equations [1] are valid only if the gas flow is laminar in the working clearance of the bearings and depth of microgrooves must at least two orders of magnitude (Critical R_e = 500 to 1000) should be less than their width. Therefore, the actual width of microgrooves during calculations [1, 10] and use in spindle units and in production [1, 12] is always limited by reasonable limits $a_e = 0.25\ to\ 0.5$.

For a fixed value of the relative microgroove length α for a radial bearing with longitudinal microgrooves, the depth of which decreases in the direction of gas flow (figure 1, b), there are values ν and β (microgroove depth parameter) for which the dimensionless stiffness K_ε^* of the radial bearing reaches a maximum ($K_\varepsilon^* = 1.884$), which is significantly larger (by 54%) than at bearing with microgrooves of constant depth.

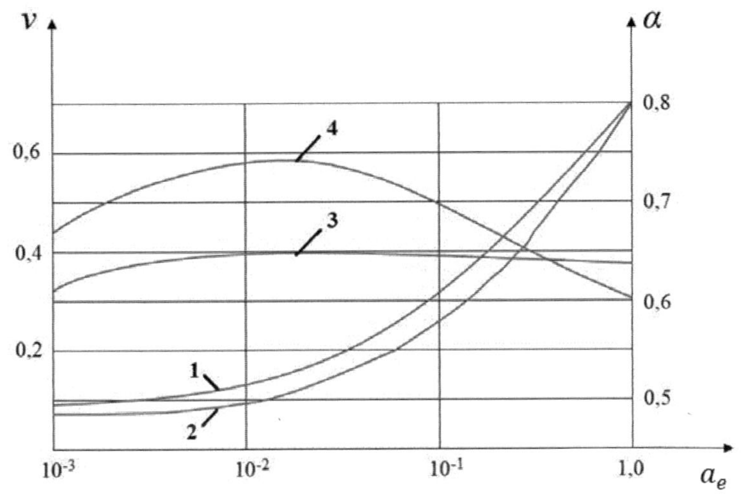

Figure 4. Dependence of P_H = 3 on the parameters ν (curves 1, 2) and the relative length α (curves 3, 4), on the relative width of microgrooves a_e.

A radial bearing with a maximum depth of microgrooves on the gas boost line (figure 1 b, c) has extremes of dimensionless radial stiffness at the coefficient of depth of microgrooves β and the clearance smoothness parameter ν and according to the criteria [1], no static instability zones were found for such bearings at $0 \leq \beta \leq 1$ and $0.2 \leq \nu \leq 1$.

4. Discussion

Calculations showed that with increasing only the relative step length α_1, without changing the depth of longitudinal microgrooves, the dimensionless radial and angular stiffness of the bearing increases for the profile of microgrooves, in which the step depth decreases in the direction of gas flow (figure 1, d, figure 2, b).

In radial gas bearings with stepped microgrooves with a maximum depth on the compressed gas supply line to the working clearances (figure 2, b), dimensionless radial stiffness are larger (about 12%), angular stiffness and the ratio K_ε^*/Q^* are large (25%) in comparison with a bearing in which microgrooves has constant depth and design parameters are optimal (maximum K_ε^*/Q^*).

5. Conclusions

A radial bearing with longitudinal microgrooves whose depth decreases linearly (figure 1 b, c) in the direction of gas flow has a large radial lifting force and a recovery moment for angular displacements of the shaft compared to a bearing with microgrooves of constant depth (figure 1, a).

Bearings with stepped longitudinal microgrooves with greater depth at the entrance to the working clearance (figure 2, b) are operable in the range of practical use of bearings and have large dimensionless radial and angular stiffness and more economical ($\Phi = K_\varepsilon^*/Q^*$) in comparison with bearings in which microgrooves of constant depth.

The decrease in radial lifting force (about 4,67 %) of bearing with stepped microgrooves (figure 1, d) in comparison with bearing with longitudinal microgrooves whose depth decreases linearly (figure 2, b, c) is explained by an increase in the accuracy of approximate calculations, since the length of section with microgrooves are $0 \leq \alpha_1 \leq 0{,}669$ (figure 2, b, c), and in the stepped are approximated by cubic polynomials at $0{,}4 \leq \alpha_1 \leq 0{,}669$ and $0 \leq \alpha_1 \leq 0{,}4$.

In practical activities (when designing spindle units), it is not recommended to use gas bearings with microgrooves of variable depth, which have a minimum depth at the entrance to the working gap.

With optimal design parameters, the flow rate of compressed gas for bearings with longitudinal microgrooves of different transverse profiles at $\varepsilon = 0$ (figure 2) is almost the same (the difference is about 7.8%).

References

[1] Fedotov V O and Fedotova I V 2010 *Gas bearings of spindle unit* (Vinnytsia: Vinnytsia Nationat Technical University) p 244 http://ir.lib.vntu.edu.ua/handle/123456789/8315.

[2] Loginov V N, Kosmynin A V and Shirokova Z V 2012 *Analytical solution of the problem of determining the characteristics of a cylindrical gas bearing. Modern problems of science and education 5*. http://www.science-education.ru/ru/article/view?id=7188.

[3] Fedotov V O and Vishtak I V 2016 *Calculation of the characteristics of the gas suspension with variable external inductor in a complex misalignment (Bulletin of National Technical University "KhPI")* vol 5 (Technologies in mechanical engineering: Kharkiv, Ukraine) pp. 21 - 24 http://repository.kpi.kharkov.ua/handle/KhPI-Press/23662.

[4] Fedorynenko D, Sapon S and Boyko S 2018 *Accuracy of spindle units with hydrostatic bearings* SCImago Journal Rank (SJR) 2018: 0.243. Source Normalized Impact per Paper (SNIP) 2018: 0.615. 10 (2) pp 117-124 DOI 10.1515/ama-2016-0019.

[5] Guihua Han, Jianying Li, Yuhong Dong and Junpeng Shao 2009 *Control Method of Heavy Hydrostatic Thrust Bearing (Intelligent Human-Machine Systems and Cybernetics 2009)* vol 2 (IHMSC '09. International Conference) pp. 62-65 DOI: 10.1109/ICAL.2008.4636314

[6] Xiaodong Yu, Yanqin Zhang, Junpeng Shao, Chao Yin, Bo Wu, Zhimin Shi, Yan Ni, Shuyan Zhao, Hui Jiang, Xuemei Chang and Changqing Yang 2009 *Simulation Research on Gap Flow of Circular Cavity Multi-pad Hydrostatic Thrust Bearing (Intelligent Human-Machine Systems and Cybernetics)* vol 2 (IHMSC '09. International Conference) pp 41-44 DOI: 10.1109/IHMSC.2009.136

[7] Yoshimoto S, Kume T and Shitara T 1998 *Axial load capacity of water-lubricated hydrostatic conical bearings with spiral grooves for high speed spindles: Comparison between rigid and complaint surface bearings* vol 31 (Tribology international: Elsevier) pp 331-338 DOI.org/10.1016/S0301-679X(98)00043-7

[8] Stefano Morosi Ilmar F Santos 2011 *Active lubrication applied to radial gas journal bearings (Modeling. Tribology International)* Part 1 (Tribology International) pp 1949-1958 DOI: 10.1016/j.triboint.2011.08.007

[9] Fabian Pierart Ilmar F Santos 2015 *Active lubrication applied to radial gas journal bearings. (Modelling improvement and experimental validation)* Part 2 (Tribology International) pp 237-246 DOI: 10.1016/j.triboint.2015.12.004

[10] Belforte G, Colombo F, Raparelli T, Trivella A and Viktorov V 2013 *High Speed Rotors on Gas Bearings: Design and Experimental Characterization)* chapter 6 (Tribology in Engineering Haşim Pihtili, Intech Open) pp 81-108 DOI: 10.5772/50795

[11] Petrov O, Kozlov L, Lozinskiy D and Piontkevych O2019 *Improvement of the Hydraulic Units Design Based on CFD Modeling (Lecture Notes in Mechanical Engineering XXII)* pp 653-660 DOI: org/10.1007/978-3-030-22365-6_65

IOP Conf. Series: Materials Science and Engineering 1060 (2021) 012012 doi:10.1088/1757-899X/1060/1/012012

Polymer composite materials based on ultra-high molecular weight polyethylene

A A Okhlopkova[1], S N Danilova[1], A A Dyakonov[1,2], A P Vasilev[1], A N Ivanov[1] and S A Sleptsova[1]

[1]North-Eastern Federal University, 46 Kulakovskogo St., Yakutsk, 677013, Russia
[2]Institute of the Physical-Technical Problems of the North, Siberian Branch of the RAS, 1 Oktyabrskaya St., Yakutsk, 677980, Russia

E-mail: okhlopkova@yandex.ru, dsn.sakhayana@mail.ru

Abstract. Ultra-high molecular weight polyethylene (UHMWPE) has a number of unique properties: low friction coefficient, high strength and toughness, frost resistance, and resistance to aggressive media. This determines the use of UHMWPE as a polymer matrix for the manufacture of high-strength technical products. The use of nanoparticles as polymer fillers, which has emerged over the last two decades, has made it possible to significantly improve the characteristics of nanocomposites while reducing their degree of filling, achieving an improvement in properties that could not be achieved using traditional fillers and modifiers. Due to the high specific surface area, nanofillers have high activity for structuring the polymer matrix. However, this becomes a problem in the development of polymeric materials with nanofillers, due to the processes of agglomeration, as well as the difficulty to distribute them uniformly in the volume of the polymer matrix. We have demonstrated an efficient mixing process for UHMWPE and nanofillers, consisting of mechanical activation in a planetary mill and the technology of liquid-phase compounding under the continuous action of ultrasonic vibrations. Despite the large number of studies aimed at enhancing the interfacial interaction between UHMWPE and inorganic fillers, the problem of the weak interaction between the polymer matrix and the inorganic filler remains unsolved. Thus, the research work is aimed at obtaining new composite materials based on UHMWPE strengthened with organically modified montmorillonite and compounds containing 2-mercaptobenzothiazole, ZnO and sulfur.

1. Introduction

Issues of developing durable and wear-resistant polymer composite materials (PCM) pertain to highly prospective and intensively developing areas of modern material engineering, which are determined by the necessity of replacing traditional and costly materials [1]. One of the prospective polymer matrixes for creating high-strength technical items is an ultra-high molecular weight polyethylene (UHMWPE). UHMWPE possesses a whole range of advantages such as a low friction ratio, increased mechanical strength, self-lubrication properties, chemical resistance to acids and bases, and durability in terms of cracking and wear [2]. In spite of its unique properties UHMWPE is characterized by low values of hardness, Young's modulus, and susceptibility to constant load tensile creep. In order to improve the complete set of performance properties, various fillers of micro- and nanoscale sizes, such as fibres (glass, basalt, carbonic, aramid, etc.), as well as dispersed matters (silicon dioxide, quartz, wood powder, metal oxides, metal powders, etc.), are introduced into UHMWPE [3, 4]. Using various matters as fillers we can intentionally alter the performance characteristics of materials.

Interfacial interaction between the filler and the polymer matrix plays and important role in determining qualities and properties of created composite materials. Weak adhesion between the PCM components brings about defects, which reduces the physical and mechanical, as well as tribological properties of materials. The process of combining PCM components is still a complex technological issue today [5]. At present, various approaches and modification methods are used in order to increase the interfacial interaction and facilitate an even dispersion of filler in a polymer. Methods of ultrasonic processing, mechanic activation, high-speed mixing of components of a composite material and other methods are used to upgrade the properties of PCMs [6-8]. Besides, various specific methods of preparing and modifying fillers, mixtures of fillers and polymers in different thinners, as well as adding surface active agents [9-11]. And CVD method modification for fillers – that is a chemical vapor deposition [12]. Also, utilization of functionalized fillers and other methods are used, too [13, 14].

Therefore, the given work contains a comparison of montmorillonite (MMT) without modifications to one with organo-modification. Such organo-modified clay (organoclay) shows good dispersion in the polymeric matrix and enhances interaction with polymer macromolecules compared to plain clay [15]. Besides, the work suggests an original technology of modifying UHMWPE in order to adjust the interfacial interaction between the components of PCM. A distinctive peculiarity of this technology lies in utilizing reactive compounds (2-mercaptobenzothiazole), which forms free radicals during processing the PCM, thus influencing the interaction of polymers with solid fillers.

2. Materials and Methods

For the research we used UHMWPE GUR 4022 (obtained from Celanese Corporation, Germany). The salient characteristics of UHMWPE are presented in table 1.

Table 1. Characteristic of the UHMWPE GUR-4022.

Properties	Data
Molecular weight, g/mol	5.3×10^6
Average particle size, μm	145
Density, g/cm^3	0.93

And as a filler we used MMT of MONAMET 1H1 brand without modifications and MONAMET 1P3 brand ("Metaclay", Russia) the surface of particles in which was modified with resorcinol resin with presence of a surfactant containing aliphatic radicals ranging from 14 to 18 methylene groups. Combinations of fillers such as zinc oxide, sulphur powder and 2-mercaptobenzothiazole (MBT).

Samples for research were obtained by hot pressing at a temperature of 175 °C and a pressure of 10 MPa holding for 20 minutes, followed by cooling to room temperature. The hardness (H) of the samples was measured by pressing the ball Russian standard ISO 2039/1-87 on the UMT-3 universal testing tribometer. Testing parameters of the hardness PCM shown in table 2.

The mechanical properties of UHMWPE and PCM were studied using the Autograph AGS-J tensile testing machine (Shimadzu, Japan). The tensile strength, elongation at break and Young's modulus were tested according to ASTM D3039/D3039M-14 at the moving gripper speed of 50 mm/min, the number of samples was 6. The tribological characteristics of UHMWPE and PCM were determined according to ASTM G99–17 using a universal testing machine UMT-3 (CETR, USA). The pin-on-disc testing method was used for tribological characterization.

Table 3 illustrates the testing parameters of the tribological properties of the PCM. The counter-body was made of tempered unalloyed carbon steel AISI 1045.

Supramolecular structure of PCM was characterized by scanning electron microscopy (SEM) JSM-7800F (JEOL, Japan) with the X-MAX-20 attachment (Oxford Instruments plc, UK) in the secondary electrons mode at an accelerating voltage of 1–1.5kV.

Table 2. Data of the testing hardness.

Characteristics	Data
Diameter of the steel ball indenter, mm	5.0±0.05
Accuracy of indentation depth measurement, mm	±0.005
Indentation time, s	0.1
Sample size, mm	30×30
Thick of the sample, mm	4.0±0.1

Table 3. Data of the tribological testing.

Parameters	Data
Diameter of the sample, mm	10.00±0.02
Height of the sample, mm	20.0±0.2
Hardness of the counter-body, HRC	56–58
Roughness of the counter-body, μm	0.06–0.08
Load, MPa	1.9
Linear slipping velocity, m/s	0.5
Test time of each sample, h	3

3. Results

3.1. Comparison influence with and without organically modified MMT on the mechanical and tribological properties of the PCM

Using of organo-modified filler greatly simplifies the technological process of creating PCMs and solves the problem of combining the hydrophilic surface of the filler with non-polar or low-polarity polymer matrix [16]. The mechanical and tribological properties of a PCM filled with MMT with and without the organo-modification are presented at figure 1. A polymer filled with MMT without the modification leads the increased by 20 % tensile strength compared to those of initial UHMWPE. In the cases of polymer containing a modified MMT, there was registered an increase of the yield strength by 26 %. Nonetheless, the coefficient of friction of the PCM with low amount of MMT remains unchanged from that of the original UHMWPE. The further increase of filler content up to 5 wt.% is followed by some decrease of strength PCM, which is related to the formation of defect areas initiating material failure from external stresses [17].

Tribological investigation of PCM revealed that the addition of 5 wt.% of organo-modified MMT decreases the wear rate by 17 times and reduced the friction coefficient by 18 % compared with the initial polymer. These are results of the increase in wear resistance of a PCM due to the participation of the organo-modifier in the intensification of tribo-chemical reactions. The organic component takes part in forming labile structures on the friction surfaces of the composite material which bring about the localization of shear deformations and facilitate the surface sliding processes of the steel counter-body. The use of an MMT without a modifier results in some decrease of weight wear rate by 16 %, for filler quantities of 1 wt.% and 2 wt.%. The coefficient of friction is reduced by 10 % for a an MMT filling quantity of 0.5 wt.% and it is unchanged by the increase of a filler content to the level of the original UHMWPE. Hence, MMT without modifications fails to bring about a significant change in the friction coefficient.

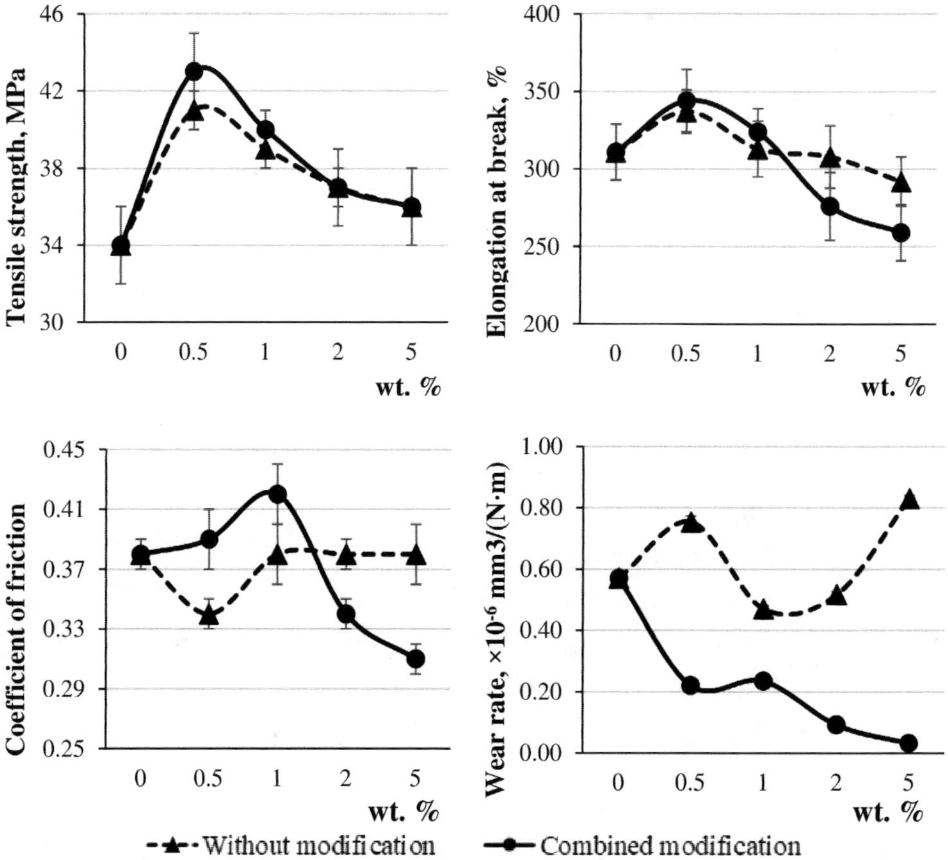

Figure 1. Dependence of mechanical and tribological properties of PCM on the content of fillers

The composite with organo-modified MMT shows better mechanical and tribological properties compared with those of the composites/MMT without modifier. The observed results can be explained by the strengthening effect of the polymer matrix due to its interaction with the modified surface of the organoclay.

In order to explain the acquired mechanical results microstructures of composites were studied. Besides, to compare the interfacial interaction between the polymer matrix and MMT samples filled with non-modified and organo-modified clay were drawn for comparison (figure 2). It is obvious that the supramolecular structure of UHMWPE transforms lamellar to spherulitic at the introduction of the fillers. Along with this, the composite with an organo-modified MMT is distinctive for its formation of a dense microcrystalline packing with a uniform supramolecular structure which confirms the enhancement of strength properties of PCM. Such a phenomenon can be explained by structural lability of small spherulites having enough time to get reorganized at the outside stretching action while the large-sized spherulitic structure starts to disintegrate [18]. As observed in figure 2, PCM containing non-modified clay is notable for containing large agglomerates and defects as well as for the uneven mixing with the clay ("a structural fragmentation" of the system in whole). Besides, when filling unmodified clay into the polymer, the resulting composition contains interlayer nano-spaces. For the case that the PCM is filled with organoclay agglomerates of filler this is not observed.

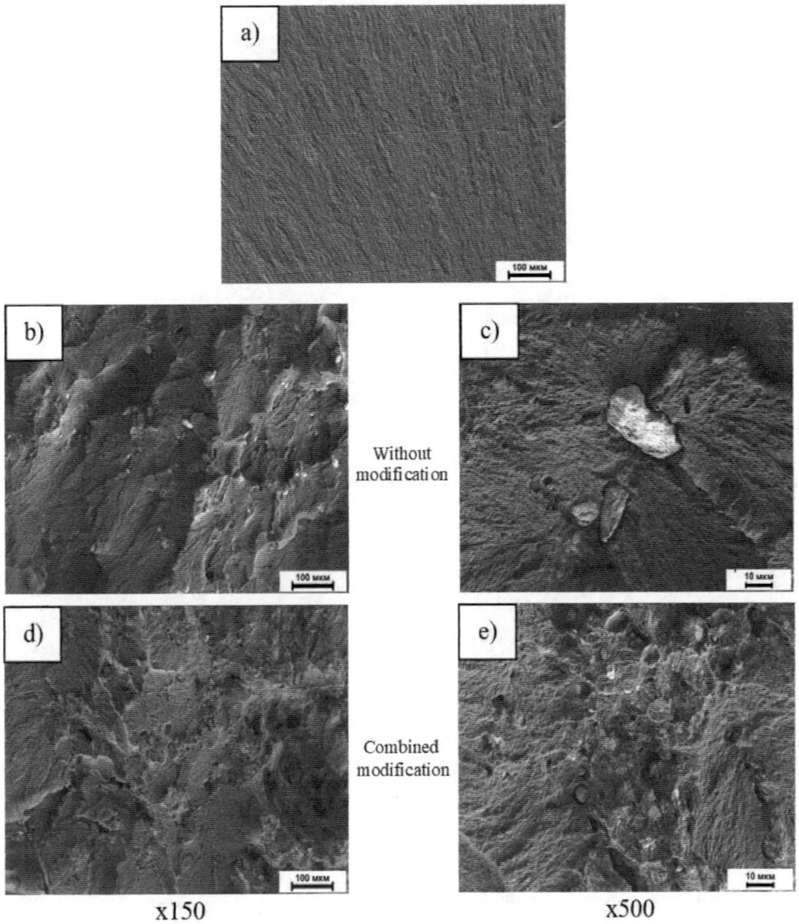

Figure 2. Micrographs of the supramolecular structure of: (a) UHMWPE, (b-c) PCM filled with unmodified MMT, and (d-e) organo-modified MMT

3.2. Investigation of the effect sulphur, zinc oxide and 2-mercaptobenzothiazole on the characteristic of PCM

Further studies concerned the use of combined filler in the system of ZnO/S/MBT. The advantage of using combined fillers for enhancing the properties of PCM consists in the fact that each of the fillers introduced there had different influences on the polymer matrix. For example, it was shown in work [19] that a combined filler represented by carbon fiber and vermiculite greatly enhanced the mechanical properties and wear resistance the PCM based on polytetrafluorethylene. Thus, combined use of fillers having different nature allows for achieving the necessary performance properties of PCM.

Figure 3 shows the results of studies of mechanical properties of PCM filled with sulphur, 2-mercaptobenzothiazole and zinc oxide depending on their content. The analysis of the acquired data in the studies of the physical and mechanical properties of PCM showed that the fillers added to it in the proportion of 1 wt.% ZnO / 1 wt.% S / 1 wt.% MBT to the polymer brings about a substantial increase in the tensile strength by 44 % and Young's modulus by 70 % as compared to the original UHMWPE.

Similar results were obtained in work [20], in which the authors greatly enhanced the polymer properties using sulphur as a filler. It was shown that the mechanical characteristics of the developed materials exceed those of characteristics of UHMWPE filled with nanosized particles of CuO, AlN, SiO_2, as well as those of basalt and carbon fibers [21-23]. Figure 3 shows that the hardness of the composite containing 5 wt.% ZnO / 5 wt.% S / 5 wt.% MBT increased by 13 % compared to UHMWPE. The registered enhancement of mechanical properties of PCM can be explained by changes in the supramolecular structure of UHMWPE at the addition of complex fillers together with the formation of clearly expressed spherulitic structures without defect areas (figure 5).

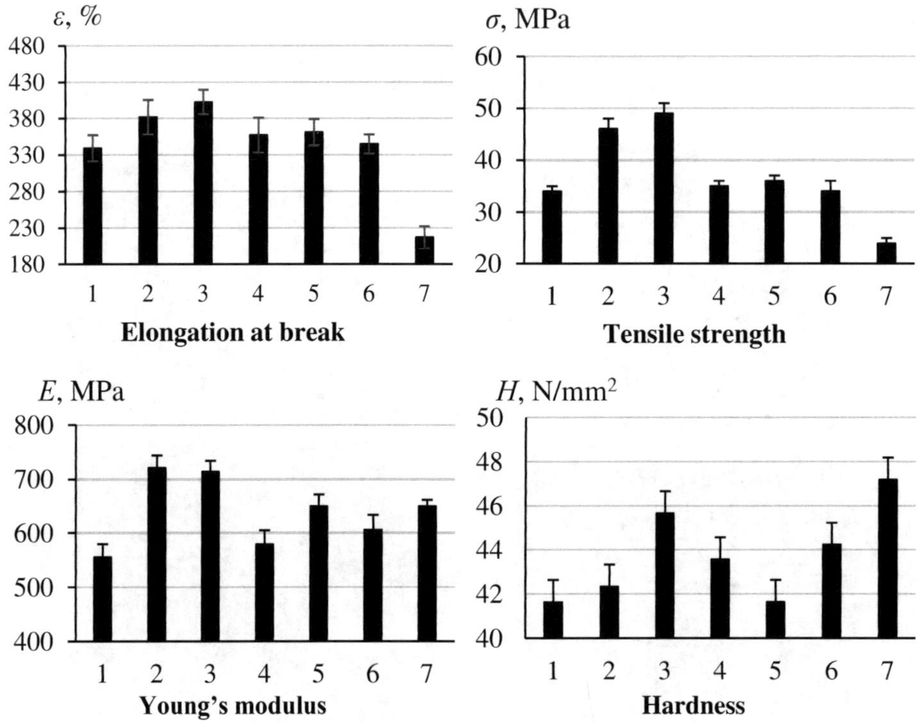

Figure 3. The dependence of the mechanical characteristics of PCM on the ratio of fillers:
1 – initial UHMWPE;
2 – UHMWPE/ 0.5 wt.% ZnO/ 0.5 wt.% S/ 0.5 wt.% MBT;
3 – UHMWPE/ 1 wt.% ZnO/ 1 wt.% S/ 1wt.% MBT;
4 – UHMWPE/ 2 wt.% ZnO/ 2 wt.% S/ 2 wt.% MBT;
5 – UHMWPE/ 2 wt.% ZnO/ 5 wt.% S/ 2 wt.% MBT;
6 – UHMWPE/ 2 wt.% ZnO/ 5 wt.% S/ 5 wt.% MBT;
7 – UHMWPE/ 5 wt.% ZnO/ 5 wt.% S/ 5 wt.% MBT

It was observed in the tensile tests for PCM that the stress characteristics of the modified (1 wt.% ZnO / 1 wt.%. S / 1 wt.% MBT) UHMWPE are 1.75 higher compared to those of the original polymer (figure 4). Having analyzed the curves a) and b) it was found out that the stress-strain curve possesses two values of Yield stress point, and it is probably connected with the rearrangement of macromolecules UHMWPE during the deformation. Such character of the curve demonstrates resistance to tensile strength with a structure that is different from the original.

In order to define the mechanism of the interaction between the UHMWPE and the used additions structural research for PCM was conducted using the SEM method. Figure 5 shows microphotographs of supramolecular structure of UHMWPE and PCM. As is shown in figure 5, the supramolecular structure of the composite with combined fillers is characterized by the formation of spherulites. It was also registered that particles of zinc oxide initiate the crystallization growth. The further increase of the filler content in UHMWPE brings about fragmentation of the supramolecular structure which explains the reduction of the mechanical characteristics of PCM. According to works [24-27], the

IOP Conf. Series: Materials Science and Engineering 1060 (2021) 012012 doi:10.1088/1757-899X/1060/1/012012

spherulite structure of PCM increases the rigidity and bearing capacity of the material. A decrease in the size of the structural components of the composite helps to reduce the proportion of defective areas and pores, as well as stress concentrations.

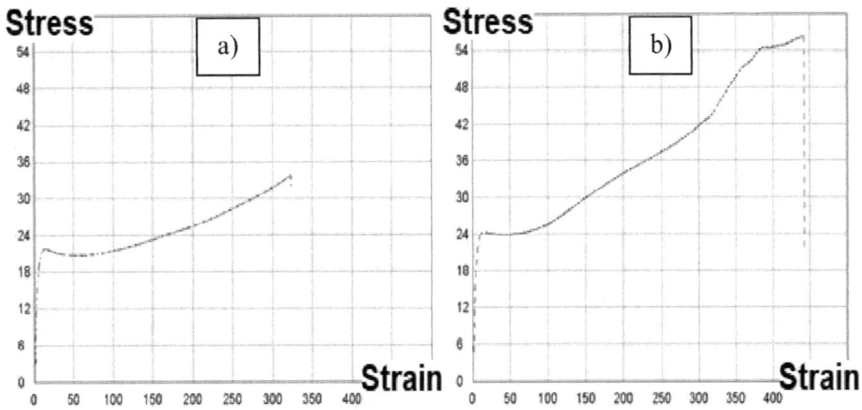

Figure 4. Stress-strain curve a) of the initial UHMWPE and b) PCM filled with S, ZnO and MBT

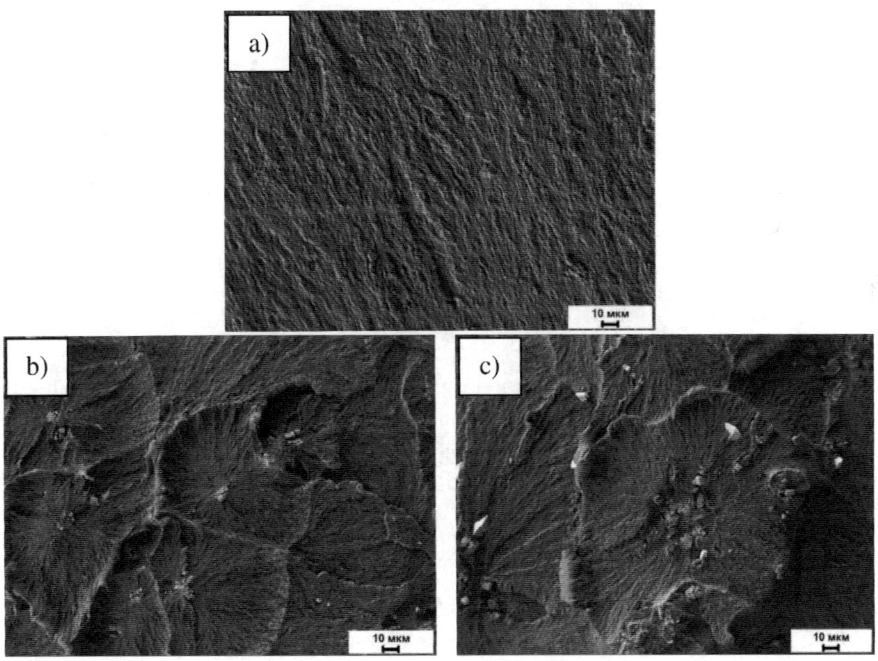

Figure 5. Micrographs of the supramolecular structure of the PCM at x500 magnification: (a) - initial UHMWPE; (b) - UHMWPE + 0.5% ZnO + 0.5% S + 0.5% MBT; (c) - UHMWPE + 1% ZnO + 1% S + 1% MBT

4. Discussion

4.1. Mechanical and tribological properties of PCM filled by MMT

As you know, powdered MMT is usually represented by agglomerates, each of which contains up to several tens of silicate plates of the clay mineral. To ensure the dispersion of the clay, MMT powders are treated with a surfactant. Quaternary ammonium salts (QAS) are usually used as surfactants for the treatment of clays. Positively charged NH_4 interact with the negatively charged surface of the aluminosilicate, displacing the interlayer cations of alkali and alkaline earth metals, and thereby increasing the distance between the silicate plates, facilitating the intercalation of polymer macromolecules or their fragments into the interlayer space (figure 6) [28]. The main disadvantage of using QAC is their low thermal stability - when processing UHMWPE (about 200 °C), the destruction of surfactants begins. Therefore, along with the modification of the QAC, the clays are additionally treated with heat stabilizers. For this, a resorcinol resin is used, which is an aromatic ether that contains polar bridging ether groups (figure 7) [29].

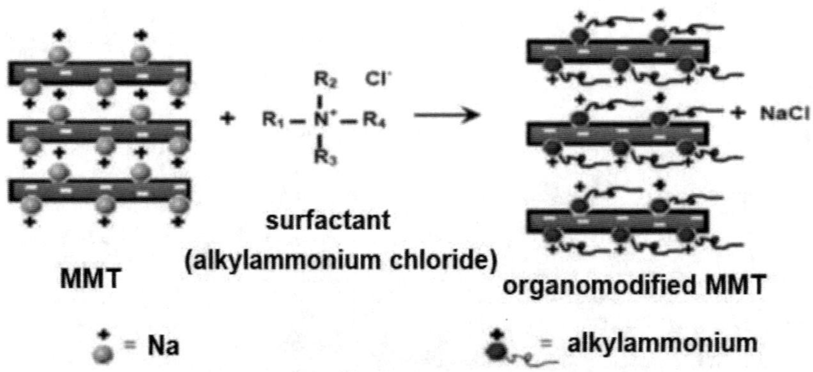

Figure 6. The reaction of ion exchange between Na+ MMT and a cationic surfactant to obtain "organoclay" [28]

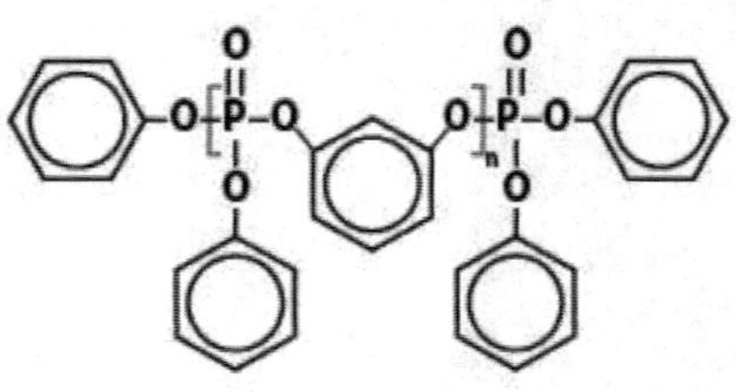

Figure 7. Resorcinol bis(diphenyl phosphate) (rdp) [29]

The increase in the mechanical properties of PCM may be explained in the wettability of the MMT surface by non-polar UHMWPE. Also, due to the intercalation of surfactant particles into the interlayer space of the MMT plates, the dispersion of agglomerates is ensured. Structural studies in figure 2 shows that the presence of agglomerated filler particles was not recorded in the supramolecular PCM containing organo-modified MMT. It is shown in [30] that organo-modified MMT “Cloisite C15A” promotes better delamination of layered clay components, which leads to a decrease in the wear rate by 41 % and the friction coefficient by 38 % compared to unfilled UHMWPE. The research results are confirmed by structural studies of the SEM friction surface. It is shown that the formation of a smooth transfer film on the conjugated counter-body is observed on the friction surface of such a composite. Based on these studies, it can be assumed that a secondary structure is formed on the friction surface of the composites from the tribo-degradation products of UHMWPE and organo-modified MMT. Such a layer during frictional contact protects the material from wear.

4.2. Enhance mechanical properties of UHMWPE with combined fillers

Improvement of the deformation and strength properties of PCM with combined fillers is associated with the plasticizing effect of 2-MBT and sulfur on UHMWPE. This facilitates the processes of a reorganization of UHMWPE macromolecules during stretching. At the same time, zinc oxide particles act as reinforcing elements of the composite material. It was revealed that 2-MBT interacts with UHMWPE macromolecules within the amorphous phase. Based on the results of investigations of the

IR spectra of PCM (figure 8), the appearance of new peaks related to oxygen-containing functional groups appeared. For example, in the 1595 cm^{-1} region, which corresponds to the stretching vibrations of –C=O of the carbonyl group. And in the area of 1180-1400 cm^{-1}, corresponding to the vibrations of –C–O–C– groups [31]. The emergence of new active oxygen-containing functional groups indicates the occurrence of oxidative processes during the processing of UHMWPE composites, which suggests their participation in the formation of the polymer structure. The reactive formed groups from the polymer chain can interact with filler particles and/or decomposition products, thereby enhancing interfacial interaction [32-33]. As a result of these factors, there is an increase in the tensile strength of PCM by 44%, elongation at the break by 19 %, and elastic modulus by 70 %.

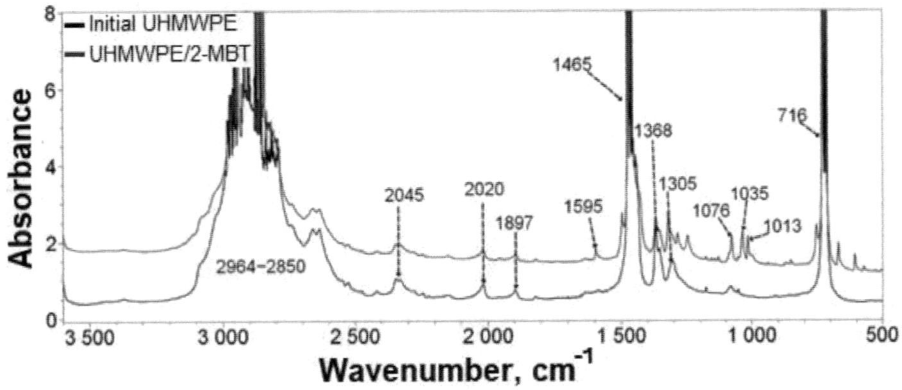

Figure 8. IR spectra of films of the initial UHMWPE and UHMWPE/2-MBT composite

5. Conclusion

As a result of this work, an enhancement of interfacial interaction between UHMWPE fillers added to it has been found, which has been proven by the increase of the whole set of mechanical characteristics of composites. The results of studies of the PCM microstructure have shown that the supramolecular structure of composites with organo-modified MMT has no defects, such as hollows and pores at the phase boundary. Besides, a uniform distribution of organoclay is also observed in the polymer matrix without the presence of agglomerated particles.

The prospective use of combined fillers in the future as functional additives into the UHMWPE has been proven. A substantial enhancement of the tensile strength characteristics of PCM by 44 % of the relatively unfilled UHMWPE has been observed.

The studies of the structure revealed that adding fillers into UHMWPE is capable of bringing about the formation of a spherulitic structure of composites where particles of non-organic fillers serve as centers of crystallization. The energy of the destruction of such a spherulitic structure exceeds that of lamellar structure, which explains the increase in the strength and wear resistance of PCM compared to the original UHMWPE.

As a result, the optimal formula of PCM with a complex enhancement of performance properties has been worked out. The developed materials are intended for the manufacture of parts for friction units of machinery and equipment operated in extreme climatic conditions in regions with a cold climate. For structural materials where higher strength and hardness values are required, a combination of zinc oxide, sulfur and 2-mercaptobenzthiazole fillers is preferred. For example, as a lining material used for facing mining and processing and mining equipment.

Acknowledgments

The work was carried out with support of the Ministry of Science and Higher Education of the Russian Federation Scientific Research Work № FSRG-2020-0017

References

[1] Kurdi A and Chang L 2018 *Lubricants* **7**(1) 2

[2] Kurtz S 2009 *UHMWPE biomaterials handbook: ultra-high molecular weight polyethylene in total joint replacement and medical devices* (Burlington: Academic Press)

[3] Šupová M, Martynková G S, Barabaszová K 2011 *Science of advanced materials* **3**(1) 1-25

[4] Rajak D K, Pagar D D, Menezes P L, Linul E 2019 *Polymers* **11**(10) 1667

[5] Bryk M T 1989 *Destruction of filled polymers* (Moscow: Chemistry) 192

[6] Okhlopkova A A, Borisova R V, Nikiforov L A, Spiridonov A M, Okhlopkova A A, Jeong D Y, Cho J H 2016 *Bulletin of the Korean Chemical Society* **37**(4) 439-44

[7] Yin X, Li S, He G, Feng Y, Wen J 2018 *Ultrasonics sonochemistry*.**43** 15-22

[8] Zhang H, Wang L, Chen Q, Li P. Zhou A, Cao X, Hu Q 2016 *Materials & Design* **92** 682-9

[9] Zec J, Tomić N Z, Zrilić M, Lević S, Marinković A, Heinemann R J 2018 *Composites Part B: Engineering* **153** 36-48

[10] Wang Y, Qiao X, Wan J, Xiao Y, Fan X 2016 *RSC Advances* **6**(83) 80262-7

[11] Feng C P, Chen L, Wei F, Ni H Y, Chen J, Yang W 2016 *RSC Advances* **6**(70) pp 65709-13

[12] Silva C R, Lago R M, Veloso H S, Patricio P S 2018 *Journal of the Brazilian Chemical Society* **29**(2) 278-84.

[13] Chen R, Ye C, Xin Z, Zhao S, Xia J, Meng X 2018 *Journal of Polymer Research* **25**(6) 135

[14] Yeh J, Wang C K, Tsai C C, Ren D H, Chiu S H 2016 *Textile Research Journal* **86**(16) 1768-87.

[15] Mikitaev A K, Kaladzhyan A A, Lednev O B, Mikitaev M A 2004 *Investigated in Russia* **7** 1365-90

[16] Demidyonok K V, Bahov F N, Cherkina U Yu, Solomkin I A, Novikov A I, Zlobin V B, Aleshinskaya S V, Hilazhdinov V V 2013 *Vestnik evrazijskoj nauki* **5**(18)

[17] Danilova S N, Okhlopkova A A, Sleptsova S A, Ivanov A N, Grigoreva L A, Spiridonov A M 2019 *IOP Conf. Series: Earth and Environmental Science* **320** 012059

[18] Sutyagin V M, Bondaletova L I 2003 *Chemistry and physics of polymers* (Tomsk: TPU Publishing) 208

[19] Vasilev A P, Struchkova T S, Nikiforov L A, Okhlopkova A A, Grakovich P N, Shim E L, Cho J H 2019 *Molecules* **24**(2) 224

[20] Jena K K and Alhassan S M 2016 *Journal of Applied Polymer Science* **133**(9) 43060

[21] Okhlopkova T A, Borisova R V, Nikiforov L A, Spiridonov A M, Sharin P P, Okhlopkova A A 2016 *Russian Journal of Applied Chemistry* **89**(9) 1469-76

[22] Panin S V, Alexenko V O, Kornienko L A, Buslovich D G 2018 *AIP Conference Proceedings: AIP Publishing* **2007**(1) 040014

[23] Panin S V, Kornienko L A, Alexenko V O, Qitao H, Ivanova L R 2016 *IOP Conference Series: Materials Science and Engineering: IOP Publishing* **156**(1) 012026

[24] Panin S V, Shilko S V, Kornienko L A, Chernous D A, Aleksenko V O, Buslovich D G, Korchagin M A, Chaikina M V 2017 *MOJ Applied Bionics and Biomechanic* **1**(5) 192-9

[25] Khalil Y, Hopkinson N, Kowalski A, Fairclough J P A 2019 *Materials* **12**(21) 3496

[26] Tanniru M, Yuan Q, Misra R D K 2006 *Polymer* **47**(6) 2133-46

[27] Way J L, Atkinson J R, Nutting J 1974 *Journal of materials science* **9**(2) 293-9

[28] Tretyakova V D, Bahov F N and Demidyonok K V 2011 *Vestnik evrazijskoj nauki* **4**(9) 8

[29] Bahov F N 2013 *Vestnik evrazijskoj nauki* **3**(16) 8

[30] Ali bin A, Mohammed A S, Merah N 2018 *Polymer composites* **39**(7) 2224-31

[31] Tarasevich B N 2012 *The IR spectra of the major classes of organic compounds* (Moscow: MGU (Moscow State University) 54

[32] Xanthos M 2010 *Functional fillers for plastics* (John Wiley & Sons) 531

[33] Karaseva Yu S Bashkatova T V Cherezova Ye N Khusainov A D 2006 Bulletin of the Technological University **5** 57-62

Ultralight structurally optimized carbon fibre reinforced polymer composite wing designing based on parametric modelling and topology optimization

S V Baranovski and K V Mikhailovskiy

Bauman Moscow State Technical University, Moscow, Russian Federation

serg1750@mail.ru

Abstract. The passenger aircraft wing is an extremely complex object, especially in design, in which it is necessary to take into account many different factors, physical processes and phenomena both in individual areas of science and engineering, and at their junction. The main enlarged areas and issues that are solved during the development of wings include aerodynamics, aeroelasticity, strength, stability, and manufacturability. All this causes significant time and labour costs, especially at the stage of design calculations. In this connection, the urgent task of compiling a universal integrated methodology for designing a wing of polymer composite materials, which allows to accelerate and facilitate the selection of design parameters at the stage of outline design. The methodology being developed is tested on the passenger aircraft wing and takes into account the choice of the geometric shape of the wing, taking into account optimal aerodynamics and minimizing the operating loads, determining the position of the main power elements providing the necessary structural strength, as well as calculating the main structural parameters of individual polymer composite elements.

1. Introduction

Currently, in connection with the design and manufacture of a new generation of passenger aircraft, the proportion of the use of polymer composite materials (PCM) with carbon fibre in the structure is increasing (up to 50%). In this connection, the design and justification of PCM structures and parts, with the required reinforcement scheme, plays a key role in ensuring the operability of the critical load-bearing power elements of the aircraft.

One of the most important elements of the aircraft body is a wing designed to create aerodynamic lift and ensure lateral stability. Currently, carbon fibre reinforced polymer composite wings, which allow for mass efficiency without compromising performance.

Work in the field of wing design techniques is actively carried out by both international and domestic experts, however, they are disparate in nature and are aimed at certain tasks, in particular, optimization of existing structures using various methods, studying aerodynamics and wing loads, and obtaining design structural and structural arrangement (SA).

Some studies are focused on the study of aerodynamics and wing loads [1], which are not limited only to experiments and process modelling using computer-aided design (CAD) systems, but also combine these results to obtain the most accurate data [2]. Some works [3] also include strength calculation, however, the results are limited to obtaining the design SA of the product. Much attention is paid to methods for optimizing an existing design using various methods. In particular, the use of genetic algorithms [4, 5], Kriging models [6]. Particular attention is paid to the calculation and optimization of the weight of the wing structure [7], including the use of curved force elements [8, 9]. In addition, PCM panels [10] with integrated stringers [11, 12] are considered as an object of optimization. Work is underway to study the properties of PCM stacks other than quasi-isotropic [13, 14]. For a rational layout that takes into account multivariate analysis, methods of multidisciplinary optimization are used [15, 16], optimization of non-traditional aerodynamic layouts of an aircraft is considered [17, 18]. Separate studies are carried out not only in the field of rational arrangement of the engine to minimize power consumption and create efficient traction, reduce fuel consumption required for the flight of a subsonic aircraft, but also the effect on the aeroelasticity of the aircraft [19, 20]. Much attention is paid to various physical processes, their description during the formation of ice on the surface of the apparatus and related phenomena [21], for example, the effect of ice roughness [22], modelling of airborne droplets and the processes of moisture loss on a streamlined surface [23]. Also, when designing aircraft, topological optimization is applied using volumetric modelling methods [24] using finite elements in conjunction with the basic algorithms for structural design and engineering analysis [25] to determine the layout scheme and dimensions of the future product [26].

In general, it is logical to state that the work associated with the development of a comprehensive methodology that takes into account all the stages of designing a wing and its elements from PCM, starting with primary geometry and ending with the parameters of parts, is fragmented. Design stages carried out at aviation enterprises are a trade secret. Based on the foregoing, the work is devoted to the urgent task of designing a wing made of carbon fibre reinforced polymer composite.

2. Goals and General Structure of the Methodology

The purpose of the work is to develop a design technique for a carbon fibre reinforced polymer composite wing, combining both designing of structural elements and individual power elements, using topological optimization methods, taking into account pronounced anisotropy of material properties, taking into account operational loads [27-30].

In the framework of the goal, a number of problems are solved in connection with the design of an ultralight PCM wing:

- Development of geometric models of the wing with an internal power set
- Determination of operational loads at various flight modes
- Determining the degree of influence of the calculation characteristics and the model on the results
- Optimization of geometric parameters of wing PCM elements
- Topological optimization of wing elements
- The use of structurally adapted curved layouts of the material in the wing structure

These tasks are solved in accordance with the proposed methodology, which includes a number of successive stages, and allow you to develop the primary geometry of the ultralight PCM wing:

- The choice of primary geometric shape options, i.e. aerodynamic profile, scope, area
- Carrying out calculations to obtain loads for a given flight mode (speed, angle of attack, altitude)
- Design of the SA (selection of the number and location of power elements: spars, ribs, stringers)
- Designing of individual power elements of the PCM structure (selection of material, reinforcement scheme, wall and belt thicknesses, geometric profile, *etc.*)

The methodology was tested on the carbon wing of a medium–range passenger aircraft.

3. Rationale for Geometric Dimensions and Calculation of Wing Aerodynamic Load

3.1. Source data

Based on the analysis of the fleet of existing medium-range aircraft, a number of wing variants with a wingspan of 35 m were simulated. For the calculation, the absolute flight altitude of 11500 m was adopted, the atmospheric parameters were selected in accordance with standard values. We considered the cruising flight mode at a speed of 870 km / h, and the angles of attack, respectively: 0°, +11°, –7°.

The parametric calculation for determining the aerodynamic load on the wing was carried out in the ANSYS software package using the CFX module. Based on existing aircraft, 12 variants of wing geometric models were built (Figure 1). When choosing the geometric shape of the wing, the following varied: sweep angle along the leading edge (χ); the angle of the transverse V wing (ψ); plan view of the wing (with a straightened section and without a straightened section); aerodynamic profile.

3.2. *Results*

As a result of the calculation, the aerodynamic loads on the wing were determined: wing pressure, temperature as a result of aerodynamic heating, flow directions near the wing surface. In Figure 2, as an example, individual calculation results are presented, data on the swept wing, with a symmetrical aerodynamic profile, with a sweep angle along the leading edge $\chi = 29°$, a transverse angle V, $\psi = +6°$, and an angle of attack $\alpha = 0°$ are presented.

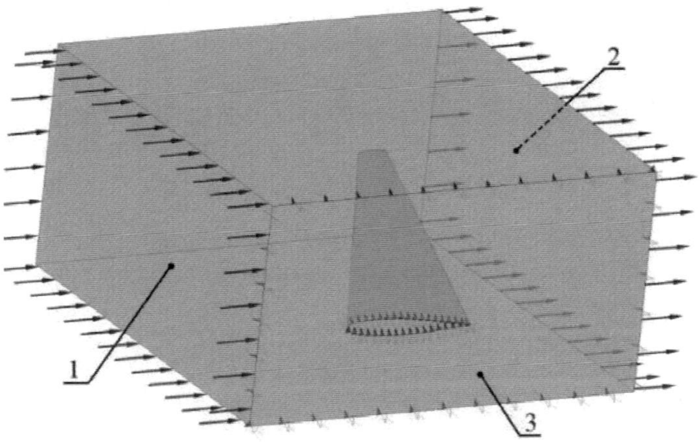

Figure 1. Type of computational model of the medium flowing around the wing with boundary conditions: 1 – inlet, 2 – outlet, 3 – symmetry.

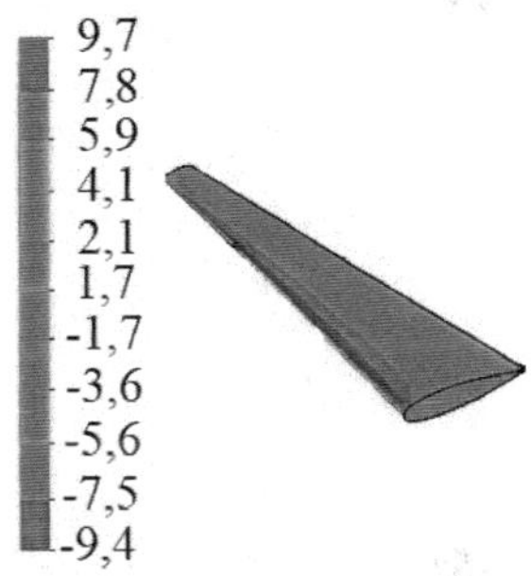

Figure 2. Pressure distribution along the wing profile [Pa 10³].

Figure 3 shows the individual results of calculating the pressure distribution over the wing profile from the relative coordinate of the point ($\overline{x} = x/c$), where x is the coordinate of the point in the profile section and c is the profile chord length in this section.

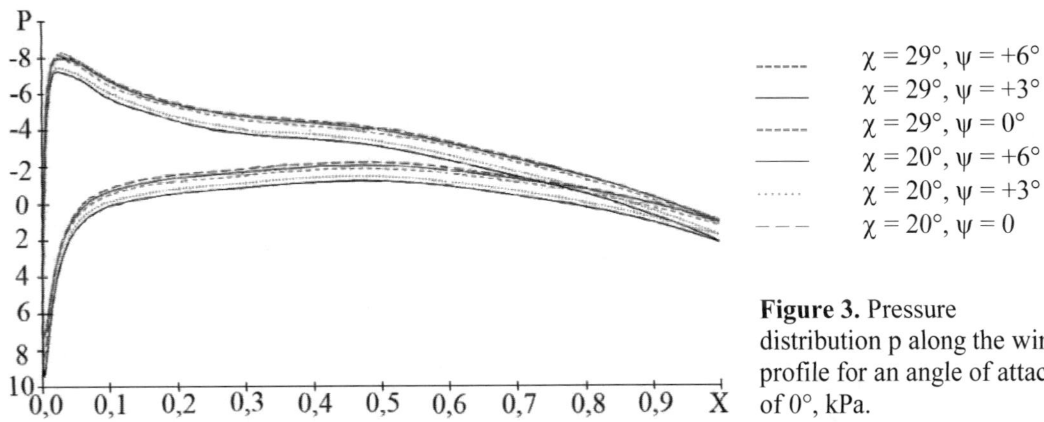

Figure 3. Pressure distribution p along the wing profile for an angle of attack of 0°, kPa.

After analysing the calculation results, the option of the swept wing with an asymmetric aerodynamic profile, with a sweep angle along the leading edge χ = 29°, a transverse angle V, ψ = +6°, with a straightened section was selected.

4. Selection of Wing Structural–Arrangement

4.1. Source data

The calculation was carried out for three calculated cases: angles of attack 0°, +11°, –7°. The load values were obtained when simulating the problem of external aerodynamics and were imported into the calculation. In the side section, the wing is limited in movements and turns about the axes. The elements of the SA were modelled flat, with the assignment of the corresponding thickness in groups. 90 variants of the SA were modelled. When choosing the SA, the pitch of the spars (% of the chord), stringers, ribs (% of the average aerodynamic chord), the direction of installation of the ribs (in flight, perpendicular to the spar) were varied.

Taking into account the varied parameters, 210 calculated cases of SA are considered in the work.

4.2. Results

As a result of mathematical modelling, the stress-strain states are determined and the values of the wing displacement under the action of aerodynamic loading are obtained. The Pareto set is determined by two criteria – the minimum mass and the minimum deflection corresponding to the maximum load-bearing capacity of the structure (Figure 4). The choice of the best option was made by determining the shortest distance to the ideal centre.

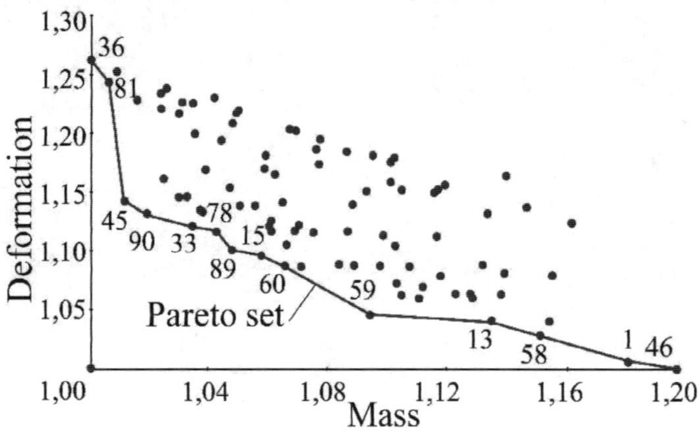

Figure 4. The distribution of values in relative units.

After analysing the calculation results, the SA option was selected with the parameters presented in Table 1, where MAC is the Mean Aerodynamic Chord.

Table 1. The values of the selected wing SA geometric parameters.

Front spar pitch (% of chord)	Rear spar pitch (% of chord)	Ribs pitch (% MAC)	Ribs mounting orientation	Stringer pitch (m)
20	70	40	Perpendicular	0,2

5. Influence of Airliner Core Elements and Icing on Aerodynamic Load Definitions

5.1. Source data

The influence of the main elements of the aircraft on the aerodynamic loads from high-speed pressure was studied by examining a separate aerodynamic surface of the wing as part of the aircraft, a wing with an engine nacelle mounted on a pylon below it, and separately the aerodynamic surface of the wing

(Figure 5). The influence of the size of the computational domain and the grid of finite volumes on the surface of the objects of study was considered on several options with different dimensions. When considering the dimensions of the computational domain, the variable distance from the surface of the object under study to the boundary varied from 3 to 30 chord lengths. Several flight modes were considered in water and mixed clouds at different altitudes.

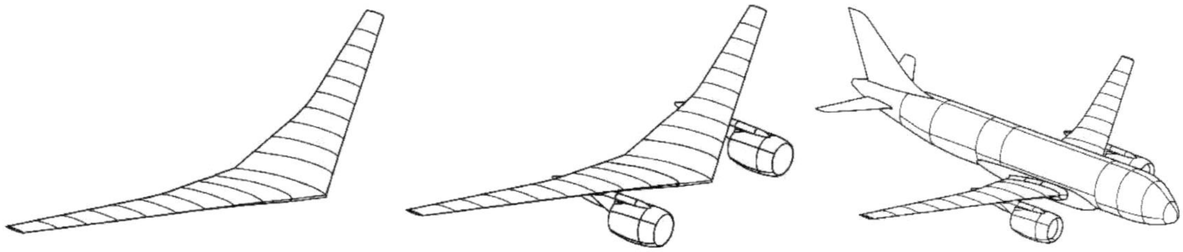

(a) (b) (c)

Figure 5. Geometric model variants: (a) – wing; (b) – wing with engine nacelle; (c) – aircraft.

5.2. Results

As a result of parametric modelling, the values of pressure along the wing, temperature as a result of aerodynamic heating, the nature of the flow around the wing, and the time dependences of the duration of preparation and calculation for the various options considered are determined and analysed. It was established that at the initial stages of design it is rational to use only the aerodynamic surface of the wing. In this case, the difference will be no more than 7–10%, and time costs will decrease by 15–20%. To obtain the result in the shortest possible time, roughening the grid of finite volumes will reduce the calculation time by 10 times, and the loss in accuracy will be no more than 10%. Taking into account the diffuser and bypass of the engine at this stage of design affects the design characteristics insignificantly, both when considering separately a wing (2%) and the entire aircraft (1.5%). The mass, thickness, geometric shape of ice on the wing surface, and the aircraft were determined depending on the phase composition of the air, flight mode, angle of attack and altitude. It is shown that at the initial stages of the design of a wing made of carbon fibre reinforced polymer composite, the consideration of icing from the standpoint of strength has an indirect effect, a detailed consideration of this effect in the framework of the task gives an insignificant result.

6. Determination of Carbon Fibre Wing Elements Parameters

6.1. Source data

For the selected SA, the basic parameters of the carbon elements were optimized: the choice of thicknesses and the direction of layers. The SA consists of finned panels (skins reinforced by eleven stringers), two spars and fourteen ribs (Figure 6). The angles of attack varied: 0°, + 11°, –7°. The simultaneous action of the following concentrated and distributed loads (the weight of structural elements, the weight of the engine, fuel, aerodynamic load) was considered. The total number of estimated cases exceeded 11,000.

6.2. Results

As a result of mathematical modelling for the choice of the SA element thicknesses, the deflection and mass of the structure and the optimal values of laying are determined.

7. Topological Optimization

7.1. Source data

The geometric model was a skin forming the theoretical surface of the wing and SA of the wing box: two spars, fifteen ribs, eleven stringers (for each skin). The SA of the leading-edge assembly and tail parts of the wing were not taken into account. The root and end chords are 6 and 1.3 m, respectively. The effect of the loads was considered as when determining the parameters of the wing elements. Several design models are considered, both from an isotropic material (aluminium alloy) and from anisotropic (carbon fibre). Structures made of isotropic material were considered for: analysis of the selected SA, comparison with a wing made of anisotropic material, determination of critical zones, as well as zones of structural reinforcement during further wing design.

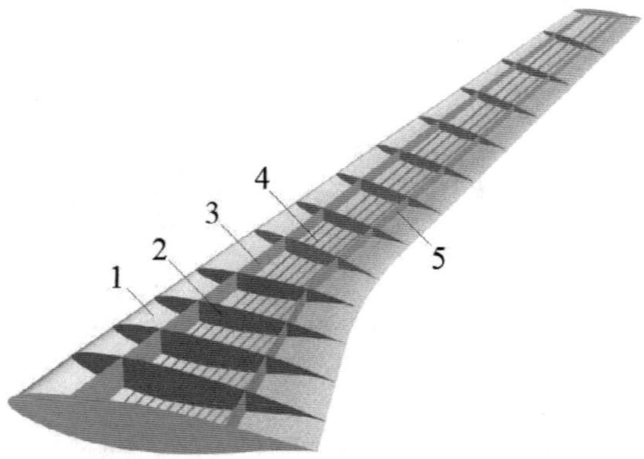

Figure 6. The main elements of the SA wing for selecting reinforcement schemes (upper skin and stringers not shown): 1 – lower skin; 2 – rib; 3 – front spar; 4 – stringer; 5 – rear spar.

7.2. Results
As a result of optimizing the design of the wing and the power caisson under the action of operational loads at several angles of attack, the following were obtained and analysed:

- Material distribution over the volume of an isotropic material wing and individual power elements (Figure 7)
- Geometric characteristics of PCM wing elements with stackings different from quasi-isotropic with different pitch orientations of the layers
- The most loaded areas of the structure
- The geometric shape of the zones of reinforcement of structural elements

The use of piling with a large number of layers of different orientations allows reducing the weight of the finished product by adapting the reinforcement circuit to the existing loads while maintaining the structural displacement but can complicate the manufacturing technology.

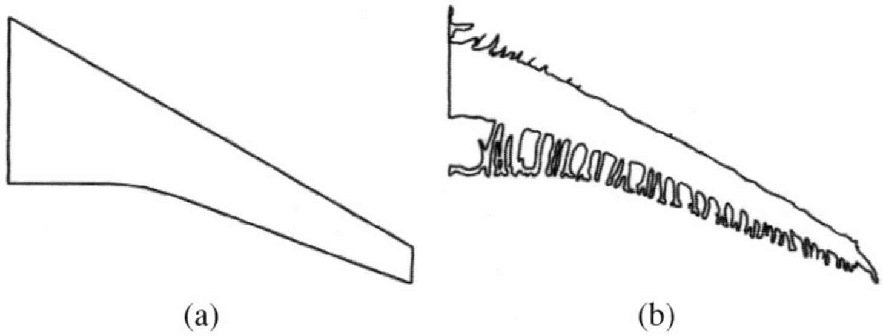

Figure 7. The results of an ultralight carbon fibre wing topological optimization to determine rational reinforcement zones: (a) – original wing geometry; (b) – material distribution after optimization.

8. Tow-Steered Composites

8.1. Source data
For the previously obtained wing with the SA, the possibility of using directional laying of fibres in the layers was considered.

The choice of reinforcement directions was carried out in two ways: zone and direction methods. For the zone method, the casing is divided into zones, the laying is determined for each individually.

For the direction method, the direction of action of normal stresses is determined, based on which calculation paths are modelled.

8.2. Results

As a result of parametric modelling, the following are defined:

- Angles for laying a prepreg based on a carbon tape in a layer in two different ways
- Stress-strain states under the action of operational loads for each option
- Optimal options for laying directions (Figure 8)

The section-by-section modelling allows obtaining layering, which gives a gain both in deflection (up to 8%) and in the damage ratio (on average, 1.5 times). Layering modelling based on the direction of action of normal stresses for finite elements can reduce the deflection of the structure by 12%, and the damage ratio by 1.3 times

This part of the study can be considered the initial stage of research in the field of structurally optimized polymer composite wing structure, aimed at creating a universal design methodology. The results obtained will be used in further studies to determine the rational directions of the power structural layout elements, to create a method for choosing rational ways to lay out the PCM layers, aimed at improving the strength characteristics and aeroelastic properties of the wing.

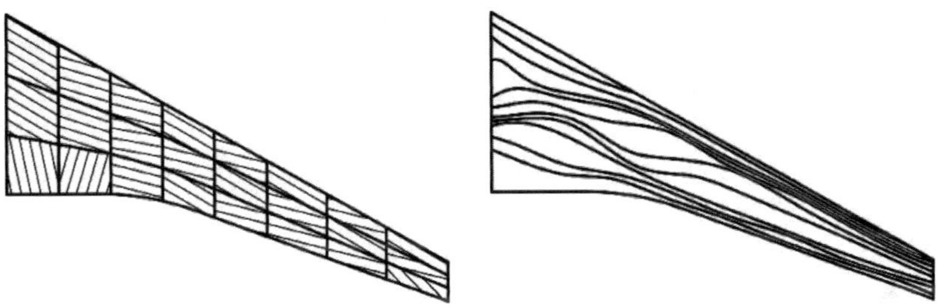

(a) (b)

Figure 8. Laying in a layer:
(a) – zone method;
(b) – referral method.

9. Conclusion

In accordance with the proposed universal methodology, the following individual tasks were completed:

- Aerodynamic loads were determined, based on parametric modelling of the external aerodynamics problem when flying at cruise, for 12 geometric wing models at different angles of attack. The following were established: rational geometric parameters for the wing taking into account sweep angles, transverse V and aerodynamic profile, and aerodynamic loads taking into account the possible formation of turbulent flows.
- The influence of the main elements of the aircraft in determining aerodynamic loads on the wing was analysed.
- The primary geometric model of the ultralight wing of a medium-range carbon fibre aircraft which meets the level of the outline design was determined.
- On the basis of the parametric modelling of 90 geometric models of SA wing, made of carbon fibre and at different angles of attack, the following were established: rational geometric parameters of SA wing - installation step and the number of ribs, stringers, spars, the shape and material of the power element of the panel – stringer.
- Using parametric optimization for elements of SA the following were defined: the rational minimum possible number of layers, optimal layer orientation angles, design deflection values for the selected optimal styling, and safety factors of structural elements.
- A topological optimization of the wing structure and individual elements was carried out.
- A variant of directional laying of fibres in layers was proposed.

IOP Conf. Series: Materials Science and Engineering 1060 (2021) 012013 doi:10.1088/1757-899X/1060/1/012013

The work will be expanded and supplemented by expanding the computational studies, experimentation and studies of additional factors affecting the wing.

References

[1] Cummings R M, Schütte A 2008 *46th AIAA Aerospace Sci. Meet. and Exh. (Reno)* **2008–396**, 1–24.

[2] Barber T J, Doig G, Beves C, Watson I, Diasinos S 2012 *Proc. Inst. of Mech. Eng., Part G: J. of Aerospace Eng.* **226**(6), 602–19.

[3] Caixeta P R, Marque F D 2010 *51st AIAA/ASME/ASCE/AHS/ASC Struct., Structural Dynamics, and Mat. Conf. (Orlando)* **2010–2762**, 1–10.

[4] Reznik S V, Prosuntsov P V, Mikhaylovskiy K V 2018 *MATEC Web of Conf.* **224 03019**, 5.

[5] Kumano T, Jeong S, Obayashi S, Ito Y, Hatanaka K, Morino H 2006 *44th AIAA Aerospace Sci. Meet. and Exh. (Reno)* **2006–932**, 1–13.

[6] Chen Bai, Luo Mingqiang, Shen Zhong, Wu Zhe, Man Yiming and Fang Lei 2014 *IJASS* **15**, 383–95.

[7] Locatelli D, Mulani S and Kapania R 2011 *J. of Aircraft* **48**, 1671–84.

[8] Gorodetsky M A, Mikhaylovsky K V, Reznik S V 2019 *IOP Conf. Series: Mat.Sci. and Eng.* **683 012075**, 8.

[9] Khani A, IJsselmuiden S, Abdalla M and Gürdal Z 2011 *Comp. Part B: Eng.* **42**, 546–52.

[10] Stanford B, Beran P and Bhatia M 2014 *J. of Aircraft* **51**, 938–44.

[11] Reznik S V, Prosuntsov P V, Sapronov D V 2015 *EPJ Web of Conf.* **82**(01034), 6.

[12] Mitrofanov O V and Ognyanova T S 2013 *Nat. and tech. sci.* **6**, 261–65.

[13] Peeters D, Abdallay M 2017 *J. of Aircraft* **54**(4), 1454–64.

[14] Petersson O, Daoud F 2012 *Deutscher Luft- und Raumfahrtkongress* [German Aerospace Conference, Berlin], 1–7.

[15] Reznik S V, Prosuntsov P V, Mikhailovsky K V, Shafikova I R 2016 *IOP Conf. Ser.: Mat. Sci. and Eng.* **153**(1), 10.

[16] Gagnon H, Zingg D W 2014 52nd Aerospace Sci. Meet., *AIAA SciTech Forum (National Harbor)* **2014–0908**, 1–18.

[17] Reznik S V, Prosuntsov P V, Mikhailovskii K V 2015 *J. Eng. Phys. Thermophy* **88**(3), 594-601.

[18] Pezhman M 2013 *PhD thesis,* Georgia Institute of Technology, Atlanta, p 133.

[19] Vasil'ev V I, Lavruhin G N, Lazarev V V, Noskov G P, Talyzin V A 2014 *TsAGI Sci. J.* **45**(3), 45–52.

[20] Deiler C 2018 *J. of Aircraft* **55**(1), 145–61.

[21] McClain S T, Vargas M, Tsao J 2016 *8th AIAA Atm. and Space Env. Conf. (Washington)* **2016–3592**, 1–14.

[22] Prihodko A A, Alekseenko S V 2013 *Aerospace techniq. and technol.* **4**(101), 59 – 67.

[23] Lukomskij D K, Ripeckij A V, Frejlehman S A, Petrakova E V 2018 *Modern sci.: Act. Prob. Of theory and pract. Ser: Nat. and tech. sci.* **11**, 79–83.

[24] Aleksandrin Ju S, Tuktarov S A, Chedrik V V 2017 *TsAGI Sci. J.* **48**(1), 72–85.

[25] Yang W, Yue Z, Li L, Wang P 2016 *Eng. Optimiz.* **48**(1), 94–114.

[26] Baranovski S V, Mikhailovskiy K V 2019 *TsAGI Sci. J.* **50**(3), 325-39.

[27] Mikhailovskiy K V, Baranovski S V 2016 *Proc. Higher Educational Inst. Mach. Build.* **11**(680), 86-98.

[28] Mikhaylovskiy K V, Baranovski SV 2018 *Vestn. Mosk. Gos. Tekh. Univ. im. N.E. Baumana, Mashinostr* [Herald of the Bauman Moscow State Technical University Series Mechanical Engineering] **5**, 15–28.

[29] Mikhailovskiy K V, Baranovski S V 2019 *Proc. Higher Educational Inst. Mach. Build.* **3**, 61–70.

IOP Conf. Series: Materials Science and Engineering 1060 (2021) 012014 doi:10.1088/1757-899X/1060/1/012014

Prediction and detection of primary microcracks in carbon fiber reinforced polymer under load

K V Mikhaylovskiy and S V Reznik

Bauman Moscow State Technical University, Moscow, 105005, Russian Federation

E-mail: konst_mi@mail.ru

Abstract. Carbon fiber reinforced polymers (CFRP) are widely used in the manufacturing of critical parts and structures in the aerospace industry due to the combination of low density, high strength, and stiffness. Production of parts and structures with pre-predicted properties is a difficult task provided significant anisotropy of physical properties and complex microstructure. The solution to this problem has to be based on the correct mathematical model describing the behavior of parts and structures made of CFRP under operational loads. Moreover, experimental data on physical, mechanical, and thermophysical properties and their change depending on the number of loading cycles has to be implemented. In view of the above, prediction of the occurrence and development of microcracks in the material becomes significant. The aim of this article is to develop algorithms for prediction and detection of primary microcracks in CFRP under loading using fiber Bragg grating (FBG) sensors and multiscale mathematical modeling. The results of measuring the primary residual deformations in the plates made of CFRP at the manufacturing stage by embedded FBG sensors and testing its performance under loading was presented. Multiscale mathematical modeling of a numerical experiment performed to evaluate the occurrence of areas of primary microcracks in CFRP under loads. It is demonstrated that splitting of peaks of resonant wavelengths of embedded FBG sensors indicates the occurrence of primary damage and microcracks.

1. Introduction

High-strength and high-modulus carbon fiber reinforced polymers (CFRP) are widely used to create light, strong structures of aviation and rocket space vehicles [1 – 5]. CFRP exposed to significant thermal and type of other types of loading during operation. CFRP have high stiffness, strength properties, low coefficients of temperature expansion, and low density. Due to microstructural features and different fiber laminates, CFRP are anisotropic materials. In addition, CFRP characteristics depend on the adhesive strength at the fiber – matrix interface.

Due to these features, it is a very important part of scientific work, to study the behavior of CFRP and identify the conditions which lead to the occurrence and development of microcracks. Traditionally, various strength criteria such as Mises, Tsai-Wu, Goldenblat – Kopnov, Tsai-Hill, Hashin, Hoffman, maximum deformations, maximum stresses are used to evaluate the parts and different structures made of CFRP. However, the main forms of the damage mechanism of CFRP laminates are fiber-matrix debonding, intralaminar and interlaminar matrix cracks, and fiber breakage. In view of the above, prediction of the occurrence and development of microcracks in the material becomes significant. The research in this area involved scientists in the USA, Japan, Germany, Great Britain, France, China, the Russian Federation, and other countries [6 – 9]. However, those studies are devoted to mathematical modeling of delamination in polymer composites or modelling of representative volume elements at the level of «fiber – matrix», where the process of occurrence of microcracks is caused by the development of critical stresses and deformations in the components of the material.

That is why many fatigue life prediction models, which are based on experimental data, statistical approach, damage mechanics, or numerical modeling, have been proposed so far. However, there have been few studies on the prediction of crack initiation in CFRP laminates under fatigue loading because it is difficult to evaluate experimentally or analytically the initiation behavior of a crack under fatigue loading. A different model was proposed to predict transverse crack initiation in cross-ply and quasi-isotropic CFRP laminates under fatigue loading based on the Walker model [10, 11]. The paper [11] proposed a probabilistic slow crack growth model to predict the initiation and multiplication behavior of transverse cracks in CFRP laminates under fatigue loading. The influence of fiber bridging on delamination failure in multidirectional composite laminates with different thickness scales is characterized, and the dependence of fiber bridging significance on laminate thickness as well as loading regime is investigated in paper [12]. Furthermore, different numerical models have been developed to predict fiber-bridged quasi-static delamination growth in composite laminates [13–15].

At the paper [16] a multi-scale model for estimating the micro residual stresses during the curing process for polymer-matrix composites has been set up. The results show there is a significant difference in the calculation of micro residual stresses by introducing the effect of the multi-scale model. M. Maiarùa et al. proposed a novel computational model to determine the effects of the cure cycle on the mechanical response of polymer matrix composites. Virtual testing is performed using the commercial codes and program ABAQUS have been used as the finite element method solver, supplemented by user written subroutines, that model the curing process and the mechanical response including the onset of damage and subsequent failure [17].

In general, it is necessary to state that the monitoring of microcracks occurrence and evolution in the parts of polymer composite materials under operating loads is the important challenge.

2. Goals and Object of study

The production of parts and structures with pre-predicted properties is a difficult task provided significant anisotropy of physical properties and complex microstructure. That is why the direction of creating smart carbon fiber reinforced polymer has started to develop actively [18 – 20]. It is worth noting the following areas of scientific work:

- Detection of residual technological deformations in the parts from polymer composite materials.
- Detection of deformations and correct temperature in the structures made of polymer composite materials under operating loads.
- Monitoring the occurrence and evolution of microcracks in the parts of polymer composite materials under exploitation loads.

Traditional non-destructive testing techniques, such as eddy-current, magnetic-particle, radiographic, ultrasonic, digital radiography are used to monitor the damage state of CFRP in order to control damage and identify types of defects. That is why strength monitoring, such as that can be achieved using fiber Bragg grating (FBG) sensors, is more in accordance with the need for future development parts and constructions for aerospace applications made from CFRP [21, 22].

The aim of the scientific work is to develop algorithms for prediction and detection of primary microcracks in CFRP under loading using FBG sensors and multiscale mathematical modeling.

IOP Conf. Series: Materials Science and Engineering 1060 (2021) 012014 doi:10.1088/1757-899X/1060/1/012014

The unidirectional CFRP plates consist of 12 layers (reinforcement – carbon fiber UKN–M–6K (UMATEX Co., Russian Federation), matrix – epoxy, volume fraction of carbon fiber was 0.62) that are the object of the current study. Carbon fiber UKN–M–6K properties are presented in table 1.

The carbon fiber prepreg was used to make the plates. The dimensions of the CFRP plates were: length – 200 mm, width – 40 mm, thickness – 1.8 mm. In these plates FBG sensors with wavelengths of 1560, 1570, and 1575 nm were embedded between the six and seventh layers. The FBG sensors were embedded in the longitudinal and transverse directions. The FBG sensors embedded in the CFRP were copper coated or polyimide coated. The diameter of the FBG sensors is 90 μm.

Table 1. Properties of carbon fiber UKN–M–6K

Number of Filaments	Density (g/cm^3)	Tensile Modulus (GPa)	Tensile Strength (MPa)	Elongation (%)
6000	1.75	225	3500	1.5

It is important that FBG sensors were embedded in CFRP with minimal distortion of the microstructural of layers. As seen in figure 1, the FBG sensor is located along the carbon fibers and provides minimal distortion of the microstructural of layers. However, if the FBG sensor is located across the carbon fibers different pores may occur.

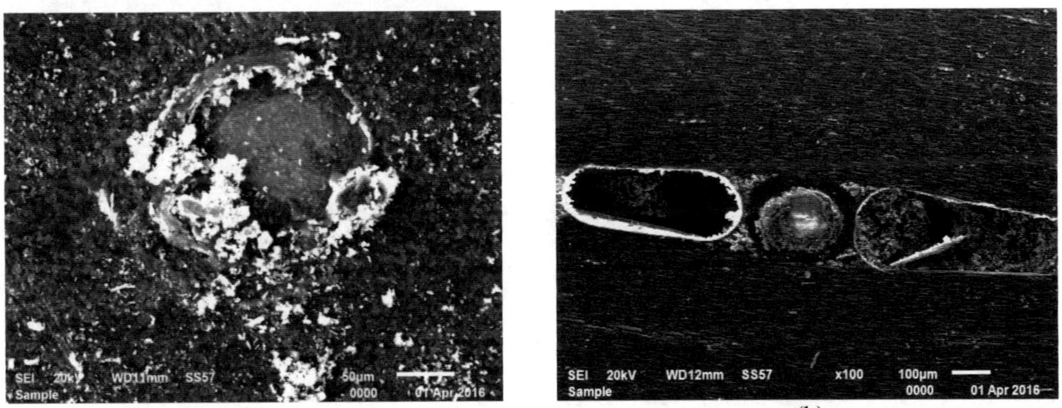

(a) (b)

Figure 1. The microstructure of CFRP in the area of the FBG sensor:
(a) – The FBG sensor is located along the carbon fibers
(b) – The FBG sensor is located across the carbon fibers.

3. Experimental data

First, it is necessary to measure the various properties of the component materials of the CFRP. For elementary carbon fibers, the measured properties were: the elastic modulus, the elongation, tensile strength. For the matrix: the elastic and shear modulus, and the tensile and compressive strength. Also, for glass fibers with FBG sensors: the elastic modulus, the elongation, tensile strength.

In the next step, the FBG sensors were embedded into the green body of plate of the prepreg with carbon fibers and epoxy matrix. After this, the autoclave curing cycle was done. The mold subjected by a vacuum pressure was introduced in autoclave connected at the autoclave vacuum system. Autoclave cycle steps are: heat the mold at 80°C for 30 min; dwell at 80°C for 60 min; heat the mold at 1°C/ min. ramp rate to 150°C;

dwell at 150°C for 180 min.; cool the part at 2°C/min. During polymerization process of the plates, the residual strains and temperatures up to 150°C were measured using embedded FBG sensors. Residual strains have to use for preloading the geometrical model plate of CFRP under stage of the mathematical modelling.

Then, under flexural test of unidirectional CFRP plates was measured indications of longitudinal FBG sensors for each step of loading: 150 N, 220 N, 290 N, 360 N, 430 N (figure 2). The long-beam test was performed by maintaining a span-to thickness ratio of 80:1 and using a loading roller and supporting rollers of 10 mm and 20 mm radii respectively. All CFRP plates were loaded to failure on a machine UTS110-M after setting the velocity of the loading head to 1 mm/min.

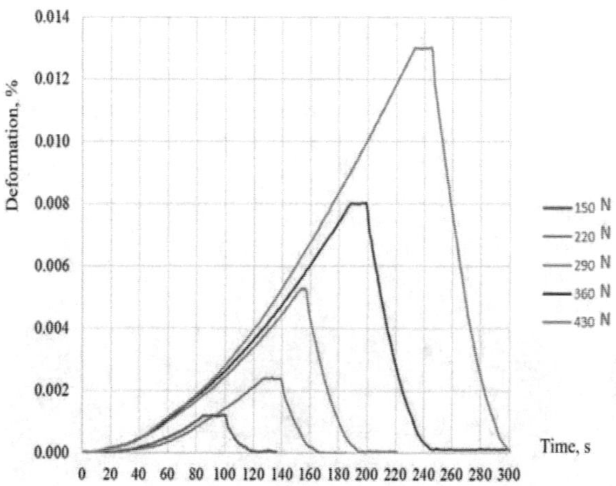

(a)

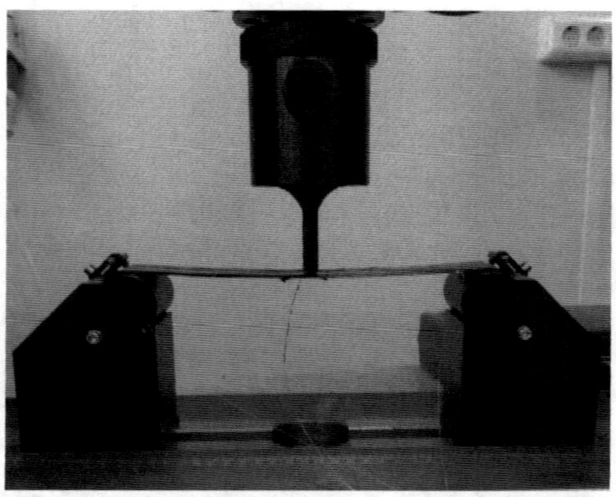

(b)

Figure 2. Indications of a longitudinal FBG sensor (1560 nm):
(a) For each step of loading during flexural test of unidirectional CFRP plate, and (b) Flexural test.

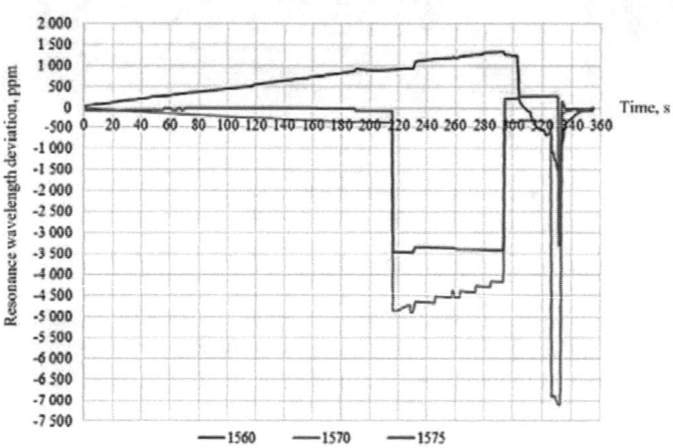

(a)

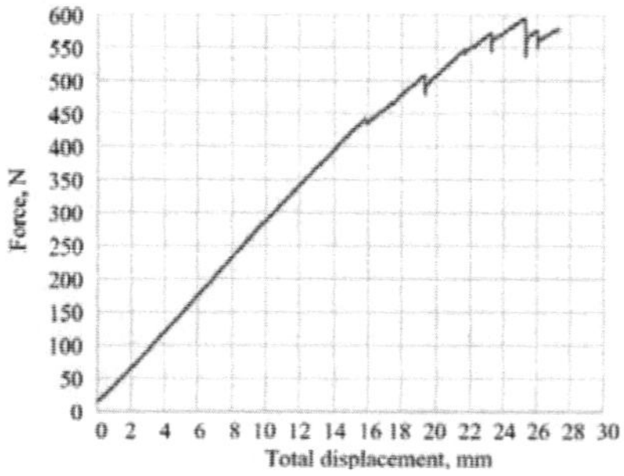

(b)

Figure 3. (a) Indications of FBG sensors, and (b) Diagram of loading up to 600 N
during flexural test of unidirectional CFRP plate.

Figure 3 illustrates the indications of FBG sensors and diagram of loading up to 600 N during flexural test of unidirectional CFRP plate. It is seen how primary microcracks and delamination in CFRP plate under loads were occurred.

Based on the data obtained from the FBG sensors, it is proposed that with an increase in the load, splitting of the peaks of the resonance wavelength, which indicates the occurrence of primary microcracks and delamination.

4. Numerical modelling

The next step to develop algorithms for prediction and detection of primary microcracks in carbon fiber reinforced polymers under loading used multiscale mathematical modeling. For mathematical modelling geometrical model of CFRP plate at flexural tests and representative volume element was built. Materials properties are presented in tables 2 and 3.

Table 2. Epoxy matrix.

Density (g/cm^3)	Tensile Modulus (GPa)	Poisson's Ratio	Tensile Strength (MPa)	Compressive Strength (MPa)
1.02	4	0.3	95	84

Table 3. High-strength steel (loading head and the supports).

Density (g/cm^3)	Tensile Modulus (GPa)	Poisson's Ratio	Tensile Strength (MPa)	Compressive Strength (MPa)
7.8	200	0.3	390	370

The CFRP plate and representative volume element CFRP, the loading head and the supports were simulated using 3D SOLID 186 elements and program ANSYS. The total number of finite elements used was:

- CFRP plate with FBG sensors 328624 (element size for plate – 0.1 mm, element size for FBG sensors – 0.006 mm). Types of element: 10 noded tetrahedron – Tet10, 20 noded hexahedral – Hex 20 and 15 noded prismal – Wed 15;
- Representative volume element CFRP 1415063 (element size for carbon fiber – 0.0008 mm, element size for matrix – 0.001 mm). Types of element: 20 noded hexahedral – Hex 20 and 15 noded prismal – Wed 15;
- The loading head and the supports 24572 (element size – 2 mm). Types of element: 10 noded tetrahedron – Tet10, 20 noded hexahedral – Hex 20 and 15 noded prismal – Wed 15.

INTER 204 elements (element size – 0.0002 mm, 3D 16-node quadratic interface element) can simulate an interface between two surfaces and the subsequent delamination process, when used with SOLID186 elements [23]. The frictionless contact model represented by element types CONTA 174 and TARGE 170 was used to simulate the contact between the loading head and the plate of CFRP and between the plate and supports. In addition, in the geometrical model of CFRP plate was embedded FBG sensors (figure 4). Representative volume element was the area with FBG sensor located in the center of the carbon fibers yarn.

Total deformation in CFRP plate with integrated FBG sensors at flexural test was determined. The results of numerical modeling are in good agreement with the results of experimental data on the values of total displacement.

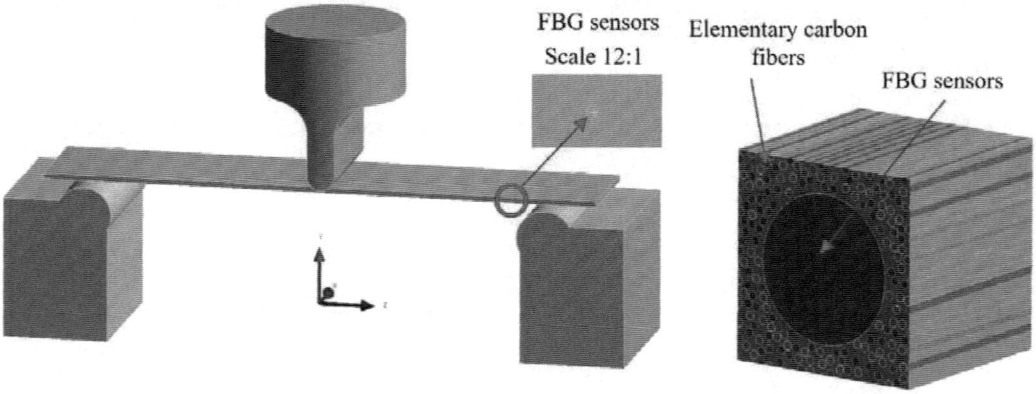

Figure 4. Geometrical model of CFRP plate at flexural tests (left) and representative volume element (right).

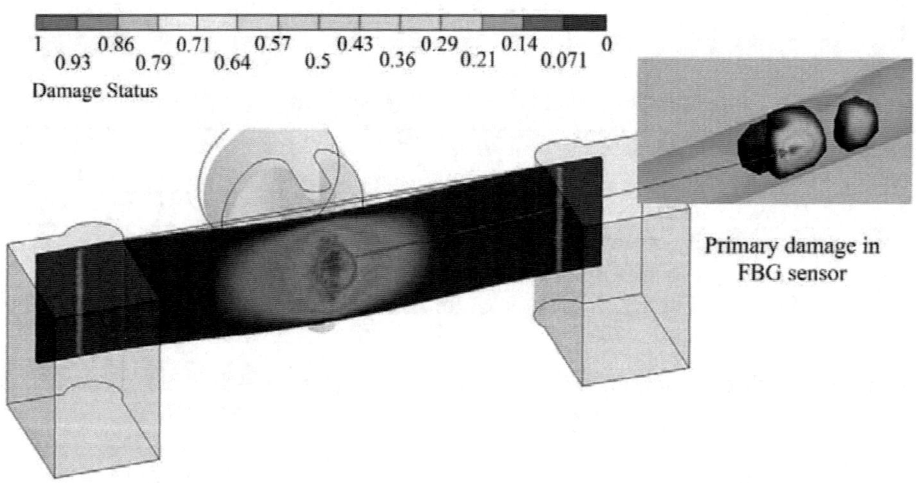

Figure 5. Damage in CFRP plate with integrated FBG sensors at flexural test. Microcracks and delamination.

In this work virtual testing is performed using the commercial codes and program ANSYS have been used as the finite element method solver, supplemented by user written subroutines, that Hoffman`s model including of damage and subsequent failure. Hoffman's model [24] for damage initiation and evolution was used to identify the regions within the CFRP plate or representative volume element at which tensile or compressive damage in the fiber and matrix was initiated. The damage evolution law is known as Continuum Damage Mechanics method that models gradually increasing damage and defects. This is defined in terms of the fracture energy dissipated during the damage process.

The failure in CFRP plate with integrated FBG sensors at flexural test under different loading was presented. It was possible to predict the appearance of primary microcracks and delamination at the level of the CFRP plate. Due to numerical modelling a primary fracture region occurred in the FBG sensor was detected (figure 5).

In additional, a primary fracture region occurred in the matrix near FBG sensor. Multiscale mathematical modeling of a numerical experiment to evaluate the occurrence of areas of primary microcracks in CFRP under loads was performed. Failure criteria in representative volume element of CFRP plate with integrated FBG sensors at flexural test under loading up to 510 N was detected. Based on mathematical modeling, it was found that in representative volume element of CFRP plate with integrated FBG sensors at flexural test under loading microcracks occur in the matrix also in the regions between the carbon fibers and near the sensor (figure 6).

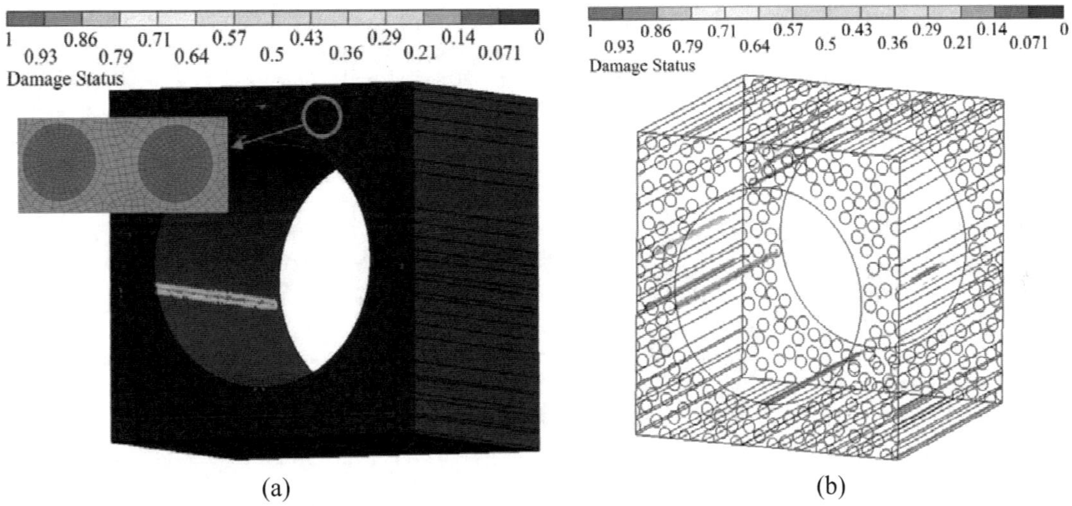

Figure 6. Damage in representative volume element of CFRP plate with integrated FBG sensor: (a) At flexural test, and (b) Microcracks in matrix.

The multiscale mathematical modeling of a numerical experiment was carried out to evaluate the occurrence of areas of primary microcracks region in CFRP under loads. It was shown that the results are in good agreement with the experiment.

5. Discussion

Based on the above, the correctness of the performed mathematical modeling assessment acquires importance. It is necessary to compare experimental data and numerical simulation results in the following parameters: maximum displacement of the plates during the flexural test compared to the modeling, also the primary microcracks region (table 4).

Table 4. Comparison of results.

Density (g/cm^3)	Total displacement (mm)	Force (N)	Detect primary microcracks
Experiment	19.2	507	Primary fracture region occurred near the FBG sensor
Numerical modelling	19.18	510	Primary fracture region occurred near the FBG sensor, also in epoxy matrix: fiber-matrix debonding, intralaminar and interlaminar matrix cracks

As table 4 shows the results of numerical modeling are in good agreement with experimental data but appends it in primary microcracks regions. These results are of particular interest for further research in microcracks prediction and propagation in the volume of composite materials.

6. Conclusions

Algorithms for prediction and detection of primary microcracks in CFRP under loading using FBG sensors and multiscale mathematical modeling were developed.

The results of measuring the primary residual deformations in the plates made of CFRP at the manufacturing stage using embedded FBG sensors, and their testing under loading were presented.

Multiscale mathematical modeling of a numerical experiment to evaluate the occurrence of areas of primary microcracks in CFRP under loads was performed.

It is demonstrated that the occurrence of primary damage and microcracks is indicated by the peaks splitting of the resonant wavelengths of the embedded FBG sensors.

References

[1] Zimin V N, Koloskov I M, Meshkovsky V E and Usyukin V I 2001 *Trans. Mod. and Sim.* **30** 497–504
[2] Reznik S V, Prosuntsov P V, Mikhaylovskiy K V 2018 *IOP Conf. Series: Materials Science and Engineering* **1134** 012048 7
[3] Irving P E, Soutis C 2019 *Polymer composites in the aerospace industry* (Sawston: Woodhead Publishing)
[4] Mikhaylovskiy K V, Baranovski S V 2018 *Vestn. Mosk. Gos. Tekh. Univ. im. N.E. Baumana, Mashinostr.* **5** 15–28
[5] Gorodetsky M A, Mikhaylovsky K V, Reznik S V 2019 *IOP Conf. Series: Materials Science and Engineering* **683** 012075 8
[6] Shokrieh M M *et al.* 2012 *Comput Mater Sci.* **65** 66–73
[7] Farmand-Ashtiani E, Cugnoni J, Botsis J 2015 *Int. J Solids Struct.* **55** 58–65
[8] Dimitrienko Y I, Sborchikov C V and Sokolov A.P. 2013 *Mech. Comp. Mater. Constr.* **19(3)** 365–383
[9] Mikhaylovskiy K V 2010 *Vestn. Computer. and Inform. Technol.* **11 (77)** 17–22
[10] Hosoi A, Kawada H 2018 *Materials* **11** 1182 16
[11] Ogi K, Yashiro S 2009 *J. Jpn. Soc. Compos. Mater* **35** 212–220
[12] Liaojun Y *et al.* 2018 *Composites Part A.* **115** 175–186
[13] Shokrieh M M, Heidari-Rarani M 2011 *Mater. Sci. Eng. A.* **529** 265–269
[14] Stutz S, Cugnoni J, Botsis J 2011 *Compos. Sci. Technol.* **71** 443–449
[15] Farmand-Ashtiani E *et al.* 2015 *Compos. Sci. Technol.* **119** 85–92
[16] Zhenyi Y *et al.* 2018 *Composites Part B.* **155** 49–61
[17] Maiarùa M *et al.* 2018 *Composites Part B.* **149** 285–295
[18] Andreadesa C, Mahmoodib P, Ciampaa F 2018 *Composite Structures* **206** 456–466
[19] Takeda S, Okabe Y, Takeda N 2008 *J. of Int. Mater. Sys. and Struct.* **19** 437–444
[20] Minakuchi S, Takeda N 2013 *Photonic Sensors* **3(4)** 345–354
[21] Mikhaylovskiy K V, Bazanov M A 2016 *Constr. From Comp. Mater.* **2** 54–58
[22] Raffaella D S 2015 *Sensors*, **15**, 18666–18713
[23] Ansys Inc. 2011 *Ansys Mechanical APDL Element Reference* (Canonsburg: Ansys Inc.)
[24] Hoffman O 1967 *J. of Comp. Mater.* **1** 200–206

Temperature state of the housing and shell for an inflatable aerodynamic decelerator

Abramova E.N.

Department SM-13 "Rocket and Space Composite Structures" of the Bauman Moscow State Technical University, 2nd Baumanskaya str., b. 5/1, Moscow, 105005, Russia

E-mail: abramova_en@bmstu.ru

Abstract. Space debris is a growing concern that becomes more urgent with the increase in space exploration. This is the reason for equipping spacecraft with disposal systems. One of the methods for space debris removal is deorbiting a small spacecraft by using a decelerator. This paper presents issue analysis related to a selection of design and technology solutions and temperature analysis for inflatable aerodynamic decelerator of small spacecraft.

1. Introduction

The space debris is a growing concern that becomes more urgent with increase of space application. In near future every satellite should be required to have a certain measure to de-orbit at the end of its lifetime [1-2]. One of the methods for space debris removal is deorbiting a small spacecraft re-entry by using a decelerator.

Operating satellites this is only the proportion of moving objects in near-Earth space. After its mission is complete, a satellite continues its motion orbiting, increasing space debris population. As a result, the near-earth space is becoming increasingly polluted. In near future space junk will make launching spacecraft into orbit more difficult. At the same time, the motion small spacecraft will have to be adjusted. Therefore, a debris removal system for space and rocket equipment, especially for satellites [3], should be considered.

To remove space debris from orbits the different methods can be applied including the satellite deorbiting with subsequent complete and safe destruction. For instance, towing vehicles or extra engine modules can be used for deorbiting. However, such removal methods are not always applicable for small spacecraft due to mass and energy limitations. Inflatable aerodynamic decelerator (IAD) that decreases the orbit height of a redundant satellite by using of a lightweight braking device with a large cross-section area is preferable.

The main element of IAD is ultra-lightweight and thin, inflatable envelope. Shell shape and size must be appropriate for the purpose of the brake device. The shell must withstand the load for the time necessary to sufficiently reduce the orbit. The report presents the results of the design of the body of an IAD. The results of heating the hull and shell in near-Earth orbit are considered. Design deployment options are proposed.

2. The inflatable aerodynamic decelerator design variations

The satellite TNS-0 №2 [4], which was launched on August 17, 2017 has been chosen for the IAD development. Initially, it was planned that the service time of the device will last 5-7 months, but the satellite work time was extended, and its operation is planned until 2021. The main purpose of the nanosatellite was to obtain flight qualification of the devices and sensors installed on it. In particular, the satellite has experimental navigation equipment and a set of solar sensors of different types, as well as a new onboard computer.

The shape of the IAD housing was determined based on the spacecraft for which it has been intended, so the housing intended for use on TNS-0 №2 (figure 1) should have the shape of a hexagonal prism.

Figure 1. TNS-0 №2 [4]

Since the development has been carried out for a nanosatellite, special attention has been paid to the mass perfection of the design. Topological optimization of the structure has been carried out to reduce the housing weight. A thin-walled hexagonal prism has been chosen as the optimized design. Carbon fiber has been chosen as the body material because it combines such properties as high specific strength, stiffness, thermal conductivity and a low coefficient of linear thermal expansion. The goal function of optimization was to minimize flexibility, and the restriction was the mass reduction by more than 50%. The following boundary conditions have been set for analysis: sealing of the bottom edge and compressive axial load for the element of side surface and sealing the perimeter of the element and distributed pressure over the surface for the triangular parts of the base and lid. These loads are caused by housing injection into Earth orbit using a launch vehicle. As a result, the shape of the structural elements has been determined (figure 2). Thus, the resulting design is a topologically optimized power part of the housing, covered with thin plates.

The IAD concept is so that at the moment when the satellite finishes its work at an orbit, the shell opens, increasing the cross-section area of the vehicle, which, in turn, causes an increase of the drag force. Under the influence of this force, the vehicle with an open IAD reduces the height of its orbit and burns up in the dense layers of the atmosphere. A number of research activities have been conducted to determine the shape and characteristics of the shell [5-16]. According to [17], a spherical device with a diameter of 2 m allows to take a satellite weighing 5 kg into the dense layers of the atmosphere in 18 hours. The supercharging system of the IAD shell must ensure the structure opening when the satellite finishes its work in orbit. It is assumed that the supercharging and opening of the housing should occur simultaneously. The use of loop connections allows the housing to be laid out in a flat structure, which minimizes the possibility of shell damage during deployment.

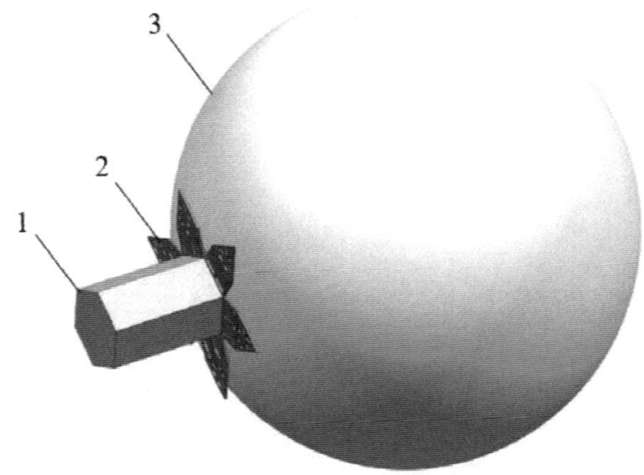

1 – Spacecraft
2 – Housing
3 – Shell.

Figure 2. Spacecraft with IAD

One method of deployment is the shell inflation with compressed gas. This method is often used in structures, including rocket and space ones. For this method of supercharging the main task of an engineer is to determine the necessary minimum of system elements and to calculate the amount of compressed gas, as well as to ensure the safety of the entire structure in space operating conditions. The study of structures deployed in the outer space allows to make a conclusion about the following minimum set of elements of the boost system: attachment to the shell (fitting), flexible pipeline, valve, pressure regulator and compressed gas cylinder.

Another way of deployment is the shell inflation with sublimating chemical. Paper [18] describes an experiment on supercharging a thin-film structure with benzoic acid in a thermal vacuum chamber. The main purpose of the test was to evaluate the dynamics of the process and determine the feasibility of using this substance as a gas generator. In addition, the expansion processes that can occur with residual air after vacuuming and determining their influence on the dynamics of the supercharging process were of interest. Preliminary tests have been conducted in the Cascade TEK high-vacuum furnace located at the Langley Research center, NASA. This high-vacuum furnace is usually used for baking or drying parts, but provides a pressure in the range of 0.1-0.00001 Pa. The test has been carried out in two stages: first for the evacuated shell without a sublimating substance at room temperature (to study the effect of the gas remaining in the shell after vacuuming), then using benzoic acid and heating from 20°C to 127°C. The shell with a diameter of 42 cm was tested, and as a result of the tests, the effectiveness of the sublimation boost system was confirmed.

3. The inflatable aerodynamic decelerator thermal state

When the vehicle moves along the orbit, the main heating of the structure occurs due to radiation flows coming from the Sun and the Earth [19-21]. To reduce the heat load, it is possible to use surface metallization, which makes the surface reflectivity much greater than the radiative one (figure 3). This allows to reduce the temperature ratio during the orbital flight significantly.

A study of the thermal state of the IAD hull was conducted. An orbit with shadow sections was considered to estimate the temperature difference. In one case, the optical properties of carbon fiber plates were used. In another case, the properties of plates covered with a thin layer of aluminum were used.

One of the research tasks is to determine whether it is necessary to use a metallized coating of the shell. It allows to reduce the maximum temperature and temperature difference, which helps to extend the life of the structure. But if the method of deployment with a sublimating substance is used, then the use of metallization does not allow heating the substance inside the shell. In this case, additional heaters are required, which weigh down the final design of the IAD. It is necessary to designate possibility of heating sublimating chemical without the heaters.

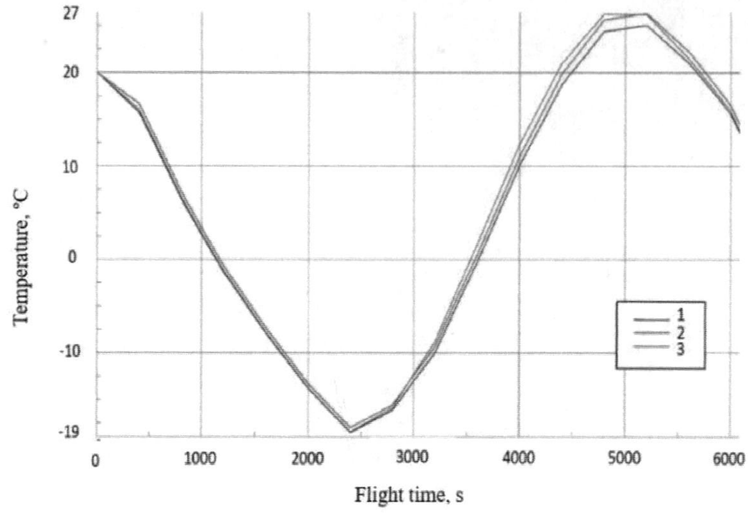

(a) Metallized plates are used

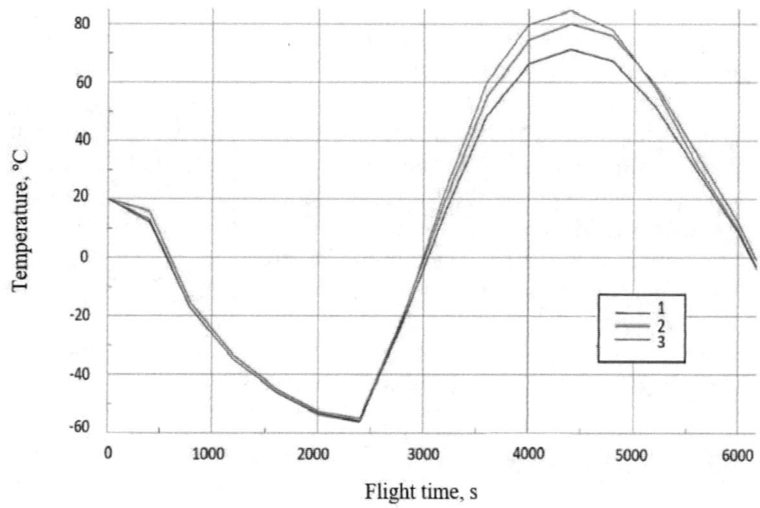

(b) Uncoated plates are used

Figure 3. Comparison of the temperature fall on the surface of the housing at a point on the top of the base: (1) Above the edge; (2) A point on the top of the base; (3) A point on the side surface.

Heating simulation was performed in the Siemens NX software package using the Space System Thermal module. The orbit parameters correspond to the parameters of the TNS-0 No. 2 satellite. Solar heating, own and reflected, Earth radiation, and free molecular heating were taken into account.

The results (table 1) show that the sublimating substance is heated to the temperature necessary to start the sublimation process. However, the shell is heated to a critically high temperature. Therefore, it is more rational to use a shell with a reflective coating. In this case, the chemical reaches the desired temperature using heaters. This will be considered in further research.

4. Discussion

Creating an IAD is a complex interdisciplinary task. It includes the design of the shell, supercharging system, and housing. Also, during the research process, auxiliary systems (for example, protection or heating/cooling) may be necessary.

One of the main assumptions made in this study is that the shell is considered as a sphere. But at the time of opening, the shell is compactly folded into a container and has an arbitrary shape before the chemical is sublimated. In further research, this assumption must be rejected or justified. A deployment shell study is required, which includes the process of opening the shell.

Polyethylene terephthalate was chosen as the shell material. It is assumed that the shell will be folded all the time the satellite is running. This may take several years. It is expected to conduct research that will reduce the adherence of the shell material.

The temperature state of housing and shell was studied separately. In future research, it is necessary to conduct simulations for a model that will include all the main parts and take into account the influence of the satellite.

Table 1. The results of the heat shell

	Shell without a coating	Shell with metallized coating
Maximum shell heating temperature	250°C	70°C
Minimum shell heating temperature	-40°C	-70°C
Maximum temperature of the sublimating chemical	100°C	Does not exceed the initial temperature
Time required for sufficient heating of the chemical	6 hours	–

5. Conclusion

This paper discusses the general issues of creating an inflatable brake device, its appearance and possible methods of supercharging. The temperature state of housing during an orbital flight is investigated. The use of metallized plates reduces the temperature difference.

The temperature state of the shell with the chemical inside is investigated. The use of a metallized film reduces the temperature difference on the surface of the shell, but in this case the chemical does not heat up. Using an uncoated film allows the chemical to be heated to the temperature necessary to start the sublimation process. At the same time, the surface temperature of the shell exceeds the operating temperature of the film. Therefore, it is necessary to use a metallized shell. Heater design will be the subject of further research.

References

[1] Mrusek B M 2019 Satellite maintenance: an opportunity to minimize the Kessler effect. *Int. J. Aviat. Aeronaut. Aerosp.* **6**(2) DOI: 10.15394/ijaaa.2019.1323

[2] Drmola J and Hubik T 2018 Kessler syndrome: system dynamics model *Space Policy* **44-45** 29-39 DOI: 10.1016/j.spacepol.2018.03.003

[3] Straub J, Marsh R and Whalen D J 2017 *Small spacecraft development project-based learning* (Switzerland: Springer International Publishing) pp. 53-63 DOI:10.1007/978-3-319-23645-2

[4] Ivanov D S 2017 TNS-0 No. 2 Nanosatellite Orientation System *Preprinty IPM im. M.V. Keldysha* No. 118 DOI: 10.20948/prepr-2017-118

[5] Voinov S I, Zhelezina G F, Solovyova N A, Yamshikova G A and Timoshina L N 2015 Influence of the external environment on the properties of carbon fiber obtained by pressure impregnation (RTM) *Electronic scientific journal "PROCEEDINGS of VIAM"* No. 2

[6] Antonov F K, Golubev M V, Khaziev A R and Azarov A V 2004 Method of manufacturing products from composite materials by 3D printing and device for its implementation RU2674138C1

[7] Mikhailovskii K V, Reznik S V and Prosuntsov P V 2019 Method for modeling the interaction between transformable shells of spacecrafts and small space debris objects *AIP Conference Proceedings* 2171 030017 DOI: 10.1063/1.5133183

[8] Abramova E N and Reznik S V 2019 Small spacecraft's inflatable aerodynamic decelerator design issues analysis *AIP Conference Proceedings* 2171 040002 DOI: 10.1063/1.5133188

[9] Golomazov M and Ivankov A 2017 Software package for the development of thermal protection systems for space vehicles descended in the atmospheres of the planets *Vestnik NPO im. S.A.*

Lavochkina **3**, 41–53

[10] Nesterin L M, Pichhadze K M, Sysoev V K, Finchenko V S, Firsyuk S O and Yudin A D 2017 Proposal for the creature device to deorbit nanosatellites CubeSat in low earth orbit *Vestnik NPO im. S.A. Lavochkina* **3**, 20–6

[11] Koryanov V V, Oriat F, Griselin H 2019 Use of autonomous inflatable devices for optimal landing of transportation systems *The 2nd Int. Conf. Robot Systems and Applications* DOI:10.1145/3378891.3378893

[12] Koryanov V V, Toropkov A, Kazakovtsev V P and Nedogarok A 2019 Study of the possibility of using the mechanics of inflatable braking devices to remove the spacecraft *J. Phys.: Conf. Ser.* **1215** 012031 DOI: 10.1088/1742-6596/1215/1/012031

[13] Zarubin V S, Zimin V N and Kuvyrkin G N 2017 Temperature distribution in the spherical shell of a gauge-alignment spacecraft *J. Appl. Mech. Tech. Phys.* **58** 1083–90 DOI: 10.1134/S0021894417060141

[14] Zarubin V S and Kuvyrkin G N 2019 Thermal state of a polymer dielectric layer with dielectric characteristics that depend significantly on temperature *J. Eng. Phys. Thermophys.* **92** 1109–16 DOI: 10.1007/s10891-019-02026-1

[15] Hrycak P 1963 Influence of conduction on spacecraft skin temperatures *AIAA J.* 1 11 2619-2621

[16] Horn A C 2017 A low cost inflatable CubeSat drag brake utilizing sublimation *MS, thesis, Old Dominion University* DOI: 10.25777/1xaw-be17

[17] Reznik S V, Smirnov G K and Varlamov S A 2020 On the way to the optimal design of an inflatable aerodynamic decelerator of space debris removal system for CubeSat nanosatellites *IOP Conf. Ser.: Mater. Sci. Eng.* **709** 044105 DOI: 10.1088/1757-899X/709/4/044105

[18] Nakasuka S, Senda K, Watanabe A, Yajima T and Sahara H 2007 Simple and small de-orbiting package for nano-satellites using an inflatable balloon. *Trans. Jpn. Soc. Aeronaut. Space Sci.* **7** 26 31–36 DOI: 10.2322/tstj.7.Tf_31

[19] Reznik S V 2018 Thermal regimes of space composite structures. Part I *MATEC Web of Conferences* 194 01048

[20] Reznik S V 2018 Thermal regimes of space composite structures. Part II *MATEC Web of Conferences* 194 01049

[21] Reznik S V, Prosuntsov P V and Mikhaylovskii K V 2018 Thermal regime of large space structure with transformable elements from hybrid composite *J. Phys.: Conf. Ser.* **1134** 012048

IOP Conf. Series: Materials Science and Engineering 1060 (2021) 012016 doi:10.1088/1757-899X/1060/1/012016

Aerodynamic space tether system as a system with distributed parameters

D V Elenev[1,2], Yu M Zabolotnov[1] and A J McMillan[2]

[1] Samara National Research University, Samara, Russia
[2] Wrexham Glyndwr University, Wrexham, UK

E-mail: elenev@ssau.ru

Abstract. The motion of the aerodynamic tether system includes two stages – the deployment and the motion of the deployed system. The initial condition for the deployment model is that of a spacecraft with a rigidly connected second body in a circular orbit about the planet. The deployment process is based on the aerodynamic forces acting on bodies. After the separation of the rigid connection, further motions of the bodies are then controlled by the tether release mechanism which unreels the tether and controls the rate at which the tether is unreeled. The motion of the system is described by the multi-point model, splitting the tether into a number of parts and using the series of separations of material points from the spacecraft, thus forming a mathematical model of the system with distributed parameters. The aerodynamic forces acting on the tether are taken into consideration. This model makes it possible to calculate numerically elastic deformations and curvature of the tether and provides more accurate modelling if compared with the two-point model.

1. Introduction

The aerodynamic space tether system is a system of two rigid bodies connected by a long thin stretchable tether where the aerodynamic forces has a major influence on the motion of the system. The orientation, or attitude, is the part of the description of the placement that the rigid body occupies in space. The other component of placement is its linear position. The attitude is defined by the angles between the longitudinal axis of the spacecraft and some reference line or plane [1] such as the vector of velocity of the air flow. Therefore, stabilization means the process of maintaining attitude during motion or tending to restore the original attitude after receiving a displacement or perturbation. For a spacecraft, attitude control is the process of achieving and maintaining a desired orientation in space by a corrective action.

Aerodynamic space tether systems can be used for a wide number of applications and purposes. The most obvious application is to provide the appropriate orientation of a spacecraft during descent. This method does not require energy expense, works based on aerodynamic forces, and requires only the presence of the atmosphere. Therefore, the aerodynamic stabilization, using the tether system, can be used not only during descent to Earth, but to stabilize the motion before landing onto the surface of other planets.

The aerodynamic tether system is applicable for returning payloads from outer space to the surface of the planet, especially using small landing modules from the space station. The payload in small modules can include, but is not limited to, materials manufactured in the absence of gravity, devices for service and maintenance, or examples of space objects. In this case, the tether system is helpful in lowering the orbit of the landing module before descent and enabling stable motion during the second stage of descent. Tethered aerodynamic stabilization can be used to target the detachable stages of rocket launchers after their separation from the launcher with greater accuracy towards the desired landing site area. The problem of reducing the search area over which detachable parts of the launchers might land has a great economic and ecological significance, because without aerodynamic stabilization these areas can cover up to thousands of square kilometres.

The most comprehensive investigations of tether systems were associated with National Aeronautics and Space Administration (NASA) [2], where the models of motion and deployment are described, as well as results of experiments on tether systems. Estimates of heating and deformation of the tether under the influence of gravity forces are included.

Tether systems are applicable to investigations of the higher layers of the atmosphere, where the flight of traditional aircraft is not possible [3, 4], to stabilize the motion of spacecraft [5], for removing space debris [6, 7] and for deep space investigations [8]. A tether system with an atmospheric probe is a perspective technology for gathering important information about higher layers of the atmosphere of Earth, at altitudes about 100 to 150 km. The density of the atmosphere at these heights is too low for normal aircraft operation, while on the other hand, traditional spacecraft would meet comparatively high aerodynamic drag. This problem can be solved by using an atmosphere probe, connected by tether to a spacecraft operating at an altitude of about 200 km [9]. It is necessary to mention that direct current can lead to motion instability in the tether system, and that, by changing the amperage in the tether, it is possible to stabilize the motion of the tether system on the electrodynamic basis [10, 11]. A tether system can provide deceleration of a spacecraft before landing [12], but it should be noted that in this case the tether tension force is significantly higher than for upper atmospheric use [13].

2. Mathematical models of the tether system

The tether system is a complex non-linear dynamic system where parameters are distributed. In general, the tether is a flexible mechanical connection which has a distributed mass along its length. Within the tether, there are complex longitudinal, bending and torsion oscillations. The investigations of tether systems focus on two major processes. Firstly, it is the motion of the deployed tether system [14], and secondly, the process of deployment of a tether system [15].

Five types of tether mathematical models were considered here:

1. The *Rigid Rod Model* is a simple model of the tether that ignores its extensibility and flexibility. In other words, the tether has in infinite rigidity. This model was used for an evaluation of the possibility to use tether system for atmosphere investigations [16] and to model the motion of a tether system with a relatively short tether in the atmosphere [13].
2. The *Elastic Rod Model*. Like in the previous model, the tether is a straight rod, but it can be stretched axially. This model ignores the flexibility of the tether, but the extensibility is taken into consideration [17-18].
3. The *Billiard Model* takes into consideration both the axial stretch and bending flexibility of the tether. The tether in this model is a weightless flexible spring. The model allows the bending oscillations to be described and enables limitations on transversal displacements of the tether to be made [19]. This model is inappropriate to describe the dynamics of parts of the tether but makes it possible to research the motion of the tether system in gravity field of the planet.
4. The model of *Rigid Rods with Swivel Joints* was used in [20] to find the parameters for the optimal deployment of the tether. This model has zero torsion stiffness, does not take into consideration the longitudinal oscillations and encounters difficulties in describing the sag of the tether.
5. The *Multi-point Model* describes the tether as a series of material points, also called nodes. The nodes have masses but are connected to neighbouring nodes by weightless springs. The number of points is defined during the mathematical modelling. The more intermediate points included into the mathematical model, the more accurate the model; however, increasing the number of nodes leads to increase in number of degrees of freedom thus greatly increasing the number of

equations and time required for calculations. Therefore, this model commonly used to validate other models [21-23].

Mathematical models of the tether system can be obtained using different but equivalents methods, including the Lagrange Equations, Newton's Laws [24] and Kane's method, originally called Lagrange's form of the D'Alembert principle [25]. The complexity of mathematical models of the tether systems leads to the time-consuming calculations. These time expenses can be unreasonably high during optimization tasks, and can be reduced by using various numerical schemes, such as the artificial increase of periods of fast oscillations in the mathematical model [26].

On modelling a deployed tether system motion in the atmosphere, it can be shown that by changing the parameters of the stabilizer it is possible to obtain a stable tether system motion [14]. In general, the tension of the tether depends most significantly on the geometrical dimensions of the bodies, while length of the tether has only a slight influence on this force. The increase of the tether length during the descent always leads to increased stability of the tether system motion. This leads to a reduction in the amplitudes of the angle of attack oscillations for the bodies and the tether. Therefore, the deployment of a tether during descent can increase the stability of the system.

3. Modelling the motion of the aerodynamic tether system

3.1. *Definitions*

Stable motion is defined as the pre-defined orientation of the tether system such that the angles between the longitudinal axis of bodies and tether do not exceed 90°. The angles of attack of the bodies and the tether should also be less than 90°. The initial condition for the mathematical model is that the tether system is in a circular orbit at an altitude of about 250 km prior to the separation of the *Spacecraft* (first body) and the second body, called here the *Stabilizer*.

The parameters for the body moving through the air of the upper atmosphere include the mass, the reference area and the drag coefficient. It is possible to define the ballistic coefficient of the rigid body as the combinations of these parameters, $\sigma = C_{xv}S/m$, where C_{xv} is the drag coefficient, S is the reference area, and m is mass. Thus, through the design of the tether system, the stabilizer has a significantly higher ballistic coefficient in comparison with the spacecraft. The mathematical model takes into consideration the extensibility of the tether so that the influence of deployment method on the motion of the system can be estimated. The mass and geometrical asymmetries of the spacecraft and the stabilizer, which are modelled as rigid bodies, are also taken into consideration. The aerodynamic deceleration during the descent from planetary orbit is included within the mathematical model.

The separation of bodies starts with a comparatively low relative velocity $\vec{V}_r$. After the separation of the rigid connection between the spacecraft and stabilizer, their further motion is then controlled by the tether release mechanism. The tether release mechanism unreels the tether but is not able to pull the tether back in, therefore the equations of motion of the system include a monotonic deceleration term representing the tether release mechanism. Specific features of regulation system such as discrete work, design tolerances, *etc.*, are not taken into consideration. The tether itself works in tension only. The deployment of the stabilizer body is controlled through the deceleration of the tether. The deceleration control device is situated on the spacecraft.

3.2. *Geometry and Equations of Motion*

The mathematical model describes the process of deployment as well as further motion of the tethered system and includes a perturbation analysis. The aerodynamic resistance of the tether and its mass characteristics are both taken into consideration in the calculation. The mathematical model of the motion during deployment uses the geocentric coordinate system $Oxyz$, which is bound to the plane of the orbit of the centre of mass of the tether system. The plane of the orbit is defined at the moment of the separation of the stabilizer from the spacecraft. The Ox axis is directed to the ascending node of the orbit, the Oz axis is parallel to the vector of total angular momentum of the centre of mass of the system. The spacecraft and stabilizer are rigid bodies of finite length, which are connected by the tether (figure 1). The coordinate systems associated with the bodies $Cx_iy_iz_i$ are used to define the moments of inertia for each of the bodies (figure 2).

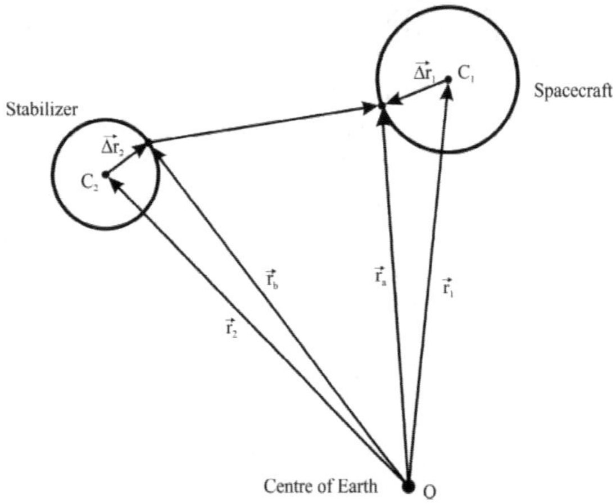

Figure 1. Tether system.

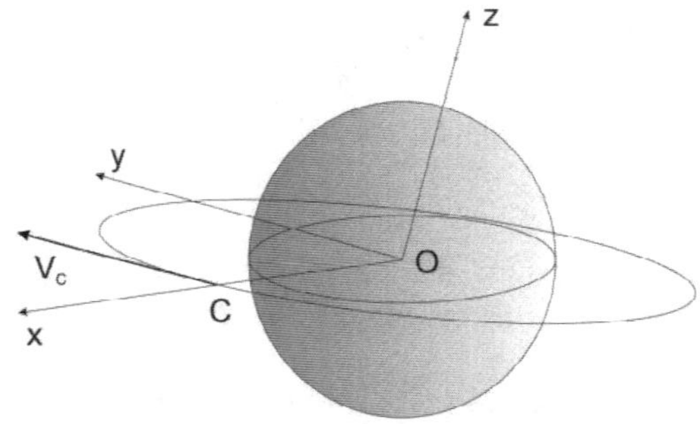

Figure 2. Coordinates system.

The stability is considered lost when the angles ψ_i between longitudinal axis C_ix_i and vector $\vec{r}_{ab}$ or when the angles of attack α_i become higher than $\pi/2$. The ideal orientation of the tether system means that the longitudinal axes C_ix_i and the tether are co-linear with the vector of the velocity of the centre of mass of the tether system. Thus, the angle between the vector of the tether and the vector of velocity of the centre of mass defines the orientation of the spacecraft.

The equations of motion of this system include the equations for the motion of the centres of mass

$$m_i \frac{d^2\vec{r}_i}{dt^2} = \vec{g}_e(r_i) + \vec{T}_i + \vec{R}_i \tag{1}$$

and the equations of rotational motion of the bodies

$$\begin{aligned} I_{xi}\frac{d\omega_{xi}}{dt} + \omega_{yi}\omega_{zi}(I_{zi} - I_{yi}) = l_{xi}, \\ I_{yi}\frac{d\omega_{yi}}{dt} + \omega_{xi}\omega_{zi}(I_{xi} - I_{zi}) = l_{yi}, \\ I_{zi}\frac{d\omega_{zi}}{dt} + \omega_{xi}\omega_{yi}(I_{yi} - I_{xi}) = l_{zi}, \end{aligned} \tag{2}$$

where indexes $i = 1$ and $i = 2$ refer to the spacecraft and the stabilizer respectively, m_i are masses and r_i are radius-vectors of centres of masses for both bodies, $\vec{g}_e(r_i)$ and $\vec{R}_i$ are gravitational and aerodynamic forces, t is the elapsed time during tether deployment starting from the instant of separation, $\vec{T}_i$ is the tension force of the tether, $\vec{T}_1 = -\vec{T}_2$, I_{xi}, I_{yi}, and I_{zi} are the moments of inertia in associated with the coordinate systems $Cx_iy_iz_i$ of the bodies, ω_{xi}, ω_{yi}, ω_{zi} are components of angular velocities of the bodies, and l_{xi}, l_{yi}, l_{zi} are components of the net total angular momentum that act on each of the bodies.

3.3. Forces applied to the Tether System

In the modelling of the motion of the system, the torques from aerodynamic forces and torques arising from the tension forces are taken into consideration. The torques from gravity forces have not been included because they can be shown to be negligibly small. The tension force is defined by Hooke's law for one-side mechanical connection.

$$\vec{T}_2 = T\frac{\vec{r}_{ab}}{r_{ab}},$$

where $\vec{r}_{ab} = \vec{r}_a - \vec{r}_b$ (as shown in figure 1), $\vec{r}_a$ and $\vec{r}_b$ are radius vectors for the tether attachment points. The value of T is equal to zero when r_{ab} is less than the length L_{ND} of non-deformed tether, and $T = k\,(\mathrm{r_{ab}} - L_{\mathrm{ND}})/L_{\mathrm{ND}}$ when $\mathrm{r_{ab}} \geq L_{\mathrm{ND}}$, where k is the spring constant of the tether.

If the tether is not strained, both bodies move freely under the action of aerodynamic forces. For modelling purposes, it is presumed that the motion takes place in low density gas and the hypothesis of diffuse reflection of gas molecules is applicable. Based on this hypothesis, vectors of the aerodynamic forces are co-linear with the vectors of velocities of the bodies defined relative to the atmosphere

$$\vec{R}_i = -\frac{C_i S_i \rho V_i \vec{V}_i}{2}, \tag{3}$$

where C_i are drag coefficients, $\rho = \rho(h)$ is the density of the atmosphere and is function of the altitude h, $S_i = S_i(\alpha_i)$ is the square of projection of the body onto the plane which is perpendicular to the velocity vector $\vec{V}_i$, α_i is the angle of attack during the spatial motion, $i = 1, 2$. If the bodies are spherical, S_i is constant.

The velocities $\vec{V}_i$, calculated relative to the atmosphere, can be found using the equation

$$\vec{V}_i = \frac{d\vec{r}_i}{dt} - \vec{\omega}_{\mathrm{Earth}} \times \vec{r}_i,$$

where $\vec{\omega}_{\mathrm{Earth}}$ is the angular velocity of rotation of the Earth.

The gravity forces are defined by Newton's law

$$\vec{g}_e(r_i) = -G\frac{M_E m_i}{|r_i|^2}\vec{r}_i,$$

where G is the gravity constant, M_E is the mass of the Earth.

The dynamics of the tether regulating mechanism are taken into account in the modelling of the deployment process, as follows

$$m_u\frac{d^2 r_3}{dt^2} = T - F_{\mathrm{control}}(t), \tag{4}$$

where r_3 is the length of the tether, the constant coefficient m_u describes the inertness of the mechanism, $F_{\mathrm{control}}(t)$ is the control force. Then the release of the tether is unguided, the control decelerating force is constant, positive and close to zero. For the guided deployment, the tether release mechanism works on deceleration only, $dr_3/dt \geq 0$. The control force is also greater than zero, $F_{\mathrm{control}}(t) \geq F_{\mathrm{min}} > 0$, where F_{min} is the minimum deceleration force that the tether release mechanism can provide.

3.4. The Kinematic Equations

The equations of motion of the system should be supplemented by kinematic equations. Here, the Euler angles are used for both bodies in the coordinate systems associated with vectors $\vec{V}_i$ and $\vec{r}_i$. The unit vectors for these systems of coordinates are

$$\vec{e}_{1i} = \vec{V}_i/V_i \qquad \vec{e}_{3i} = \vec{V}_i \times \vec{r}_i/|\vec{V}_i \times \vec{r}_i| \qquad \vec{e}_{2i} = \vec{e}_{3i} \times \vec{e}_{1i}. \tag{5}$$

The kinematics equations for Euler angles are

$$\omega_{xi} = \dot{\varphi}_i + \dot{\gamma}_i \cos\alpha_i + \Delta\omega_{xi},$$
$$\omega_{yi} = \dot{\alpha}_i \sin\varphi_i - \dot{\gamma}_i \sin\alpha_i \cos\varphi_i + \Delta\omega_{yi},$$
$$\omega_{zi} = \dot{\alpha}_i \cos\varphi_i + \dot{\gamma}_i \sin\alpha_i \sin\varphi_i + \Delta\omega_{zi},$$

where, $\Delta\omega_{xi}$, $\Delta\omega_{yi}$, and $\Delta\omega_{zi}$ are the corrections to the angular velocities due to the rotation of coordinate systems (5),

$$\begin{pmatrix} \Delta\omega_{xi} \\ \Delta\omega_{yi} \\ \Delta\omega_{zi} \end{pmatrix} = L_i^T \begin{pmatrix} 0 \\ 0 \\ \omega_{vi} \end{pmatrix} \qquad \text{where} \qquad \begin{gathered} \omega_{vi} = V_i/r_i \sin\vartheta_i \\ \vartheta_i = \operatorname{acos}\left(\frac{\vec{r}_i \cdot \vec{V}_i}{r_i V_i}\right), \\ L_i = L_{\varphi i} L_{\alpha i} L_{\gamma i}, \end{gathered}$$

$$L_{\varphi i} = \begin{pmatrix} 1 & 0 & 0 \\ 0 & \cos\varphi_i & \sin\varphi_i \\ 0 & -\sin\varphi_i & \cos\varphi_i \end{pmatrix}, \quad L_{\alpha i} = \begin{pmatrix} \cos\alpha_i & \sin\alpha_i & 0 \\ -\sin\alpha_i & \cos\alpha_i & 0 \\ 0 & 0 & 1 \end{pmatrix}, \quad L_{\gamma i} = \begin{pmatrix} 1 & 0 & 0 \\ 0 & \cos\gamma_i & \sin\gamma_i \\ 0 & -\sin\gamma_i & \cos\gamma_i \end{pmatrix}.$$

After the separation of the stabilizer from the spacecraft, it is necessary to recalculate their velocities using the conservation of linear momentum law

$$\vec{V}_1 = \vec{V}_C - \frac{m_2}{m_1 + m_2}\vec{V}_2, \qquad \vec{V}_2 = \vec{V}_C + \vec{V}_r, \tag{6}$$

where $\vec{V}_C$ is the velocity of the centre of mass of the tether system, $\vec{V}_1$ and $\vec{V}_2$ are velocities of the spacecraft and the stabilizer after the separation in geocentric coordinate system.

3.5. Deployments Models

The implementations of space tether systems require a method for the control of the deployment process. Tether systems with non-electrical tethers use rocket engines for centre of mass motion correction, and special devices to control the unreeling the tether [27]. The control of the unreeling can be with or without feedback. The simplest non-feedback method of the deployment is release of the tether with a minimal decelerating force until the tether is completely unreeled. The deployment can be realized with a decreasing rate of change of tether length basing on the continuous deceleration or basing on the Length rate of change control. A dynamic deployment method with constant decelerating force can be improved by mitigating the limitations of the final tether length and by using the program dynamic method, which includes regions of both acceleration and deceleration.

A dynamic method with constant decelerating force and a programmable change of control force can be improved by making use of the feedback principle. In this instance, the following form of regulating force can be applied

$$F_{\text{control}} = F_n(t) + p_L\left[L - L_p(t)\right] + p_V\left[\dot{L} - \dot{L}_p(t)\right] \tag{7}$$

where $L_p(t)$ and $\dot{L}_p(t)$ are pre-defined program dependencies of tether length and the rate of change of length with respect to time; p_L, p_V are feedback coefficients; L and $\dot{L}$ are perturbed length and the rate of change of length of the tether, which meets the condition (4); $F_n(t)$ is the decelerating force.

This principle of regulating the length and the rate of its change (7) was used during the real orbital tether experiment "YES2" and in other research [18, 22, 28]. Here and below the modelling uses more simple principle based on kinematic control method, which is defined as

$$\dot{L}_p(\tau) = V_{\max} \cos^2(\omega\tau + \nu) \tag{8}$$

where V_{max}, ω, ν are parameters. Boundary conditions are imposed by the solution of the system of non-linear equations

$$\dot{L}_p(t_{fin}) = 0, \quad \frac{d\dot{L}_p}{d\tau}(t_{\text{fin}}) = 0, \quad \dot{L}_p(0) = V_r, \quad \int_0^{t_{fin}} \dot{L}_p(t)dt = L_{\text{fin}}. \tag{9}$$

Parameters t_{fin}, $V_{\max}$, ω, ν are found by solving the system of equations (9). Due to the periodicity of equation (8), the system of equations (9) has more than one solution. Consider the solution where $\dot{L}_p(\tau)$ illustrates the method of deployment with areas of both acceleration and deceleration. The comparison of different methods of deployment has shown the effectiveness of the kinematic control method [29].

3.6. Numerical Calculation Results

The model of the tether system described above uses the model of elastic rod for the tether. This model provides with the basic knowledge of the motion of the aerodynamic tether system during its deployment. The following results are obtained for modelling a deployment of a tether system where the mass of the spacecraft is equal to 200 kg, the mass of the stabilizer is 12 kg, the initial altitude $H = 250$ km, Young's modulus for the tether material is $2.5 \cdot 10^{10}$ Pa, and the diameter of the tether is $0.6 \cdot 10^{-3}$ m. The characteristics of the tether refer to the Dyneema® fibre [30]. The initial velocity of the separation $V_r = 2$ m/s. For the tether system with final length of the tether equal to 15 km, the kinematic deployment method parameters are $V_{\max} = 3.5$ m/s, $\omega = 0.000325$, $\nu = 2.43$, $t_{\text{fin}} = 7040$ s.

The tension force oscillates during the first seconds of deployment due to shock loads caused by the control force. These oscillations are shown in figure 3. The value of tension force reached here (7.09 N) is the maximum tension during further deployment process for the systems with specified length of the tether and for the systems with a shorter tether as well.

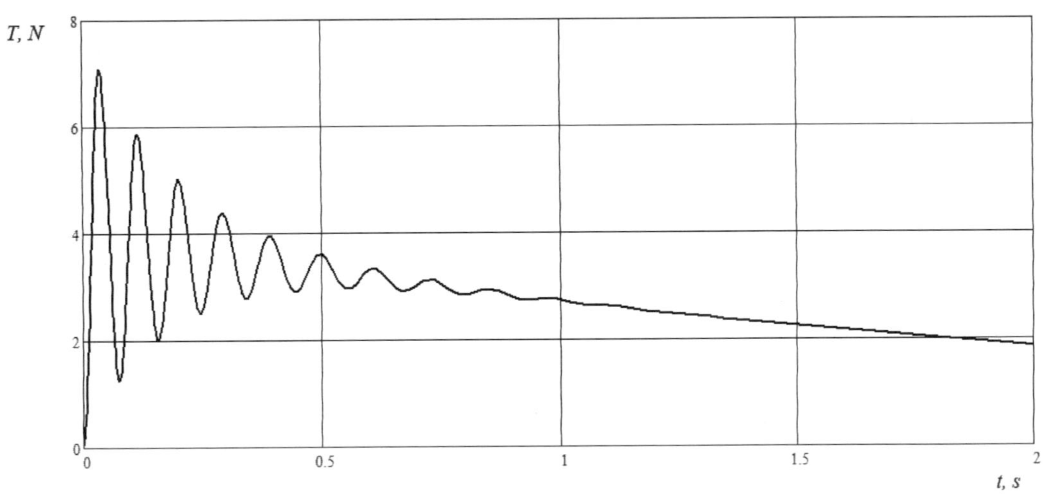

Figure 3. Tension force during the beginning of the deployment.

Figure 4 illustrates the trajectories of relative motion of spacecraft and the stabilizer during the deployment. The values of feedback coefficient are chosen under the condition of non-periodic process while the task of finding optimal values for the coefficients was not considered here.

Both bodies of the tether system may have static and dynamic asymmetry. The static asymmetry is the distance from the centre of mass of the body to the longitudinal axis of the body. The dynamic asymmetry is the difference between the moments of inertia I_{yi} and I_{zi}. The relative value of static asymmetry is defined using the diameters of the bodies and the relative value of dynamic asymmetry is the arithmetic mean of I_{yi} and I_{zi}. It was found that small static or dynamic asymmetry of the bodies does not have a significant influence on the motion of the system but there is a dependence of the angle of attack of the tether on the ballistic coefficient of the stabilizer. The higher the ballistic coefficient of the stabilizer, the lower the maximal values of angle of attack of the tether.

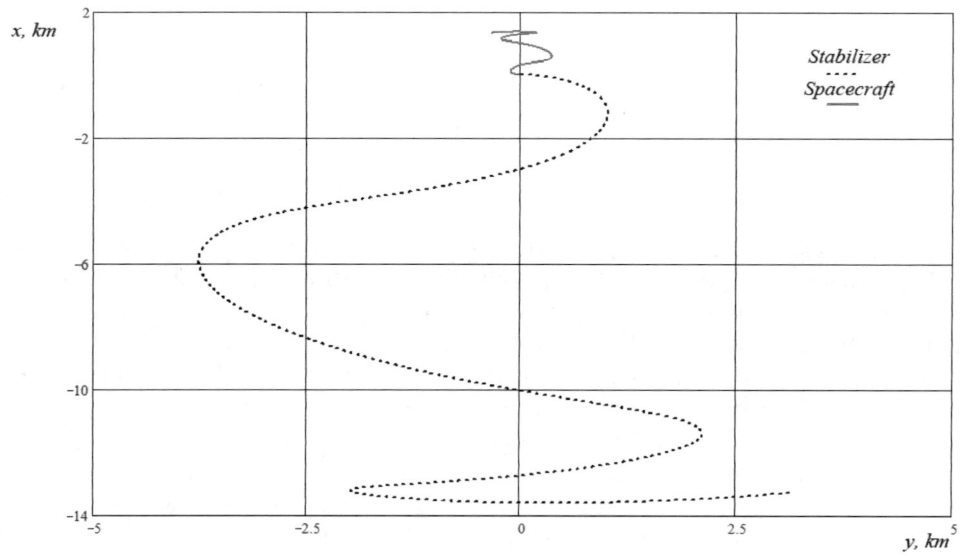

Figure 4. Trajectories of the bodies.

The model described above represents the tether as an inflexible rod – a one-dimensional stretchable mechanical connection without mass. This model does not take the mass of the tether into consideration, and since the tether is assumed to be straight, the elastic deformation and flexure of the tether itself cannot be taken into consideration. For more accurate modelling, it is necessary to use more complex multi-point model.

4. Modelling the tether as a multi-point system

4.1. Model Construction

To model the flexure of the tether it is necessary to split the tether into a number of parts. The higher the number of tether parts, the better the resolution of the model solution. Such a result would show the flexure of the tether system as a set of material points, or nodes, along the length of the tether. The anticipated non-straight line of nodes would indicate the tether response to constant or slow-time varying forces such as the gravity field and the aerodynamic forces. Therefore, the multi-point representation of the deployed tether system consists of n nodes, including two nodes for description of rigid bodies, the spacecraft and the stabilizer, and $n-2$ intermediate nodes defining the tether.

For modelling purposes, the full length of the tether is divided into parts according to the number of intermediate nodes on the tether. Starting from the beginning of deployment, as the tether unreels and the length of the tether reaches $l_{\text{full}}/(n-1)$, where l_{full} is the full length of the non-stretched tether, the new intermediate material point (node) is inserted into the mathematical model of the system (figure 5). The 10-node mathematical model of the tether system includes the following nodes: the spacecraft, 8 intermediate nodes and the stabilizer. The tether for this example is represented by 9 parts, and the lengths of the parts of the tether are equal under the condition that they are not stretched elastically. In other words, the tether is represented numerically, specifying a discretization of the tether into a series of material points (nodes) with elastic connections.

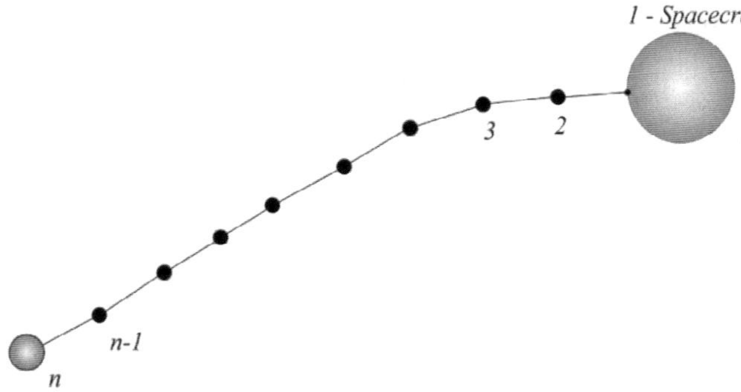

Figure 5. The multi-point model of the tether system.

For every intermediate node, the value of the tether tension is equal on both sides, and at the moment during the deployment that the node was created this value should have been equal to the tension of the tether in this point just before the node was inserted. The vector of velocity of the new node is defined based on the velocities of the neighbouring nodes using proportions. The velocity of the spacecraft after inserting the new node is corrected using of the law of conservation of momentum. It is assumed that the tether does not have internal dissipative forces like friction.

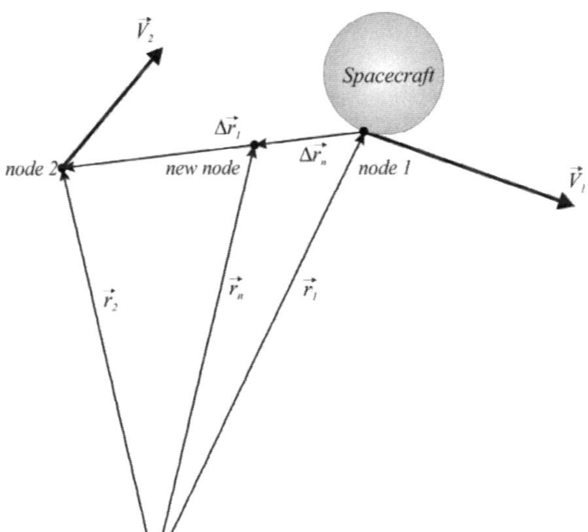

Figure 6. New node on the tether.

The multi-point model of tether describes a series of separations of material points from the spacecraft. These material points describe the stabilizer and the nodes on the tether. As the length of the non-stretched tether coming from the unreeling device reaches a pre-defined value equal to the distance between nodes on the tether, the next material point separates from the spacecraft. This new material point is added between spacecraft and previous node on the tether, or, if this were the first node on the tether, between spacecraft and stabilizer. The scheme of adding each new node is shown in figure 6. It is necessary to take into consideration that before the beginning of the deployment the tether is situated on the spacecraft, and the mass of the spacecraft decreases during the deployment as the tether unreels. The mass of the second body remains the same as in the beginning of the deployment process. When a new node is inserted into the mathematical model of the tether system, the mass of the spacecraft is reduced by the mass of the section of the tether.

4.2. Numerical Calculation Results

The advantage of the multi-point model is that it allows the elastic deformations and curvature of the tether to be calculated numerically. The following calculations are made for the tether system with the following parameters: mass of the spacecraft is equal to 200 kg, mass of the stabilizer is equal to 2 kg, radius of the spacecraft is 0.5 m, radius of the stabilizer is 2 m. drag coefficient for both bodies is equal to 2.4, Young's modulus for the tether material is equal to $2.5 \cdot 10^{10}$ Pa, diameter of the tether is $0.6 \cdot 10^{-3}$ m, the initial altitude is 250 km, and the full length of the tether is equal to 30 km. The deployment starts with initial velocity of separation $V_r = 6\ m/s$. The parameters of the deployment process are defined by numerical solution of (9), and deployment ends at $t_k = 11555$ seconds. The multi-point modelling predicts angles having slightly smaller values than calculated using the two-point model because the aerodynamic forces acting on the tether are now included in the model. Therefore, the aerodynamic forces acting on the tether also have a stabilizing effect. Figure 7 depicts the change of the tension of the tether during deployment. The shock loads seen in the first 200 seconds are non-physical shocks and related only to numerical effects in the node adding algorithm. Ideally, the node adding algorithm should be modified to avoid this.

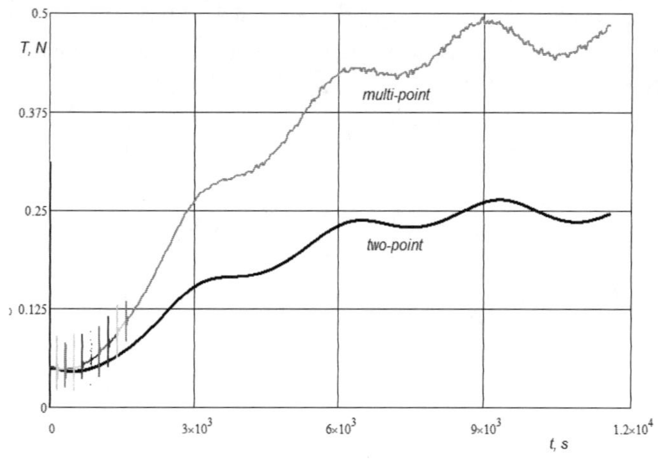

Figure 7. Tension of the tether, N.

The trajectory of the stabilizer relative to the spacecraft is shown in Figure 8. The solid line depicts the motion calculated using the two-point model, and the dashed line represents the motion predicted by the multi-point model. The spacecraft is placed in the upper right corner of the graph, and the stabilizer, affected by the aerodynamic and gravitational forces, moves away from the spacecraft as the tether is being deployed. The difference compared with the relative trajectory shown in figure 4 is based on the big difference in ballistic coefficients of the bodies in the second case.

The shape of the tether during the motion is of great interest. Figure 9 illustrates how the shape of the tether changes during deployment. Each circle on the tether depicts an intermediate node, and different lines show the state of the tether system at the moment of inserting each new node. The tether is almost straight at every moment of deployment, and only when the tether is almost fully deployed and is, therefore, long, is there is a slight curvature near the stabilizer end. A similar analysis was also made for a simpler model, with fewer intermediate points. By increasing the number of intermediate points from 5 to 8 it was found that, for this shape of the tether and specified deployment model, there is no significant difference in results of modelling parameters of the motion of the system. From this it

IOP Conf. Series: Materials Science and Engineering 1060 (2021) 012016 doi:10.1088/1757-899X/1060/1/012016

can be concluded that further increase in the number of nodes is not advisable. Figure 9 also confirms that the altitude of the spacecraft decreases as a result of the influence of the atmosphere. The spacecraft is shown by the dot at the right-hand end of each line describing the tether, and its position goes down as the tether length is increased.

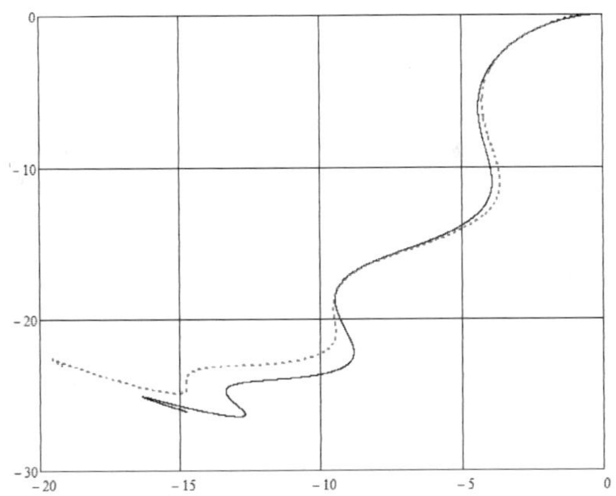

Figure 8. The trajectory of the stabilizer relative to spacecraft, length is indicated in kilometres.

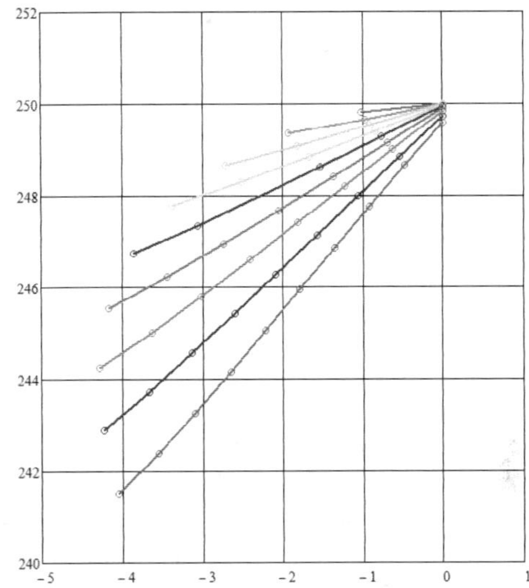

Figure 9. Shape of the tether, length is indicated in kilometres.

5. Peculiarities of multi-point modelling

A comparison of the results of modelling tether system deployment, for different tether lengths, using both the previous two-point model and the multi-point model was made. There is a difference in the motion parameters obtained by these two models. The tension of the tether calculated using the multi-point model is approximately twice that calculated using the two-point model, and the angles defining orientation of the tether system are also lower in this case. This is explained by the fact that the multi-point model represents the aerodynamic forces as acting not only on the stabilizer, but also on the tether, thus predicting a higher tension force than predicted by the two-point model, where only the stabilizer aerodynamic forces were modelled. According to the properties of material from which the tether is made, the value of tension force is admissible in both cases.

The multi-point model is more computationally expensive than the two-point model. As the results obtained by the more complex method agree with those obtained by two-point model, it can be concluded that the two-point model can be relied on, not only for the particular modelling of the deployment of the tether system, but also for automated calculations and selection of parameters of the tether system at the design stage. On the other hand, the multi-point model does show the motion of the tether and the tether system as the system with distributed parameters. When the shape of the tether is close to the straight line, it is possible to reduce the number of intermediate nodes used.

During the real process of the deployment, there can be emergency situations such as a breakage of the tether or jamming in the unreeling device. In these cases, the multi-point model would allow the parameters for the resulting motion of the system to be calculated: the two-point model would not be capable of providing this information about the flight.

6. Discussion

The descent of a tether system from orbit includes two stages. During the first stage the tether system is deployed, and the second stage describes the motion of the deployed system in the atmosphere. There are similar problems in the modelling of the tether motion in both of these stages. The first group of problems refer to the manufacture process, the second group to the perturbations arising from the effects of the surrounding medium, and the third to perturbations not taken into consideration in the mathematical model of the motion. In the manufacture of any product it must be assumed that there can be design deviations arising from inaccuracy, lack of quality control or comparatively high manufacturing tolerances. Examples of such deviations include mass and shape asymmetry of the spacecraft, flaws in the unreeling mechanism or non-uniformity of the tether material. These deviations lead to the perturbations in parameters of the tether system. Simplifications made in the modelling of the atmosphere account for the surrounding medium effects. Existing models of the atmosphere do not include the chaotic character of local flows. The most vivid example of atmosphere perturbations is wind. Wind is a natural phenomenon and takes place in every location on Earth, but it is highly problematic to take it into consideration. There are also differences in the properties of the atmosphere resulting from, for example, time of day, season, and solar activity.

These influences can lead to perturbations during the deployment process or in the motion of the deployed system. For example, there can be errors in the absolute value or direction of the velocity of separation at the beginning of the deployment, and so the intention is that the model of deployment overcome such perturbations, by making use of stabilization methods. Methods such as spinning, or the use of a tether system provide the means of achieving the conditions of stable motion.

The magnetic field of the Earth was not taken into consideration in this research but depending on the materials the spacecraft is made from there can be an influence from appropriate forces. There are other forces acting on the spacecraft during its motion in the atmosphere, including the pressure from the sunlight, gravity forces from space objects and so on. The magnitudes of these forces are small and are usually not included in the mathematical models. It is also necessary to consider the mutual influence of the spacecraft and stabilizer bodies on each other. The perturbations in airflow caused by the presence of one body changes the regime of airflow on the body downstream, thus changing the values and directions of the aerodynamic forces.

Every real system has dissipative forces acting on or within the system, and damping is of great importance during the motion. Damping reduces the amplitudes of oscillations thus tending to make motion more stable. The dissipative forces cannot destroy the stability of motion of the system, but it is important to include damping to get the mathematical model closer to the real system. The problem here is the damping is one of the most complex processes for mathematical modelling, and the appropriate choice of damping model for use in the modelling of space tether systems is the subject for ongoing research [8, 31].

Another important consideration in the tether system design is the cross-sectional shape of the tether. The Dyneema® fibre considered here has a circular cross-section; however, the choice of a tape cross-section would make the tether less vulnerable to micrometeorites attack [8]. Other advantages of a non-circular cross-section could include aerodynamic characteristics that are dependent on the angle between the tether surface and oncoming airflow. In such circumstances it would be necessary to include the localized rotation of the tether with respect to its longitudinal axis.

7. Conclusions

The multi-point model of the tether system allows the examination of the deformation and flexure of the tether during and after the deployment. Compared to the two-point model, the multi-point model takes into consideration the mass of the tether and the aerodynamic forces acting on the tether. For modelling purposes, these forces were treated above as acting on the nodes.

The tether can be used as an aerodynamic stabilizer itself, and the small spacecraft can be stabilized by using a tether represented for modelling purposes by number of nodes during its motion in the atmosphere. Compared to the two-point model, the multi-point model predicts more accurate values for the parameters of the tether system during the motion through the atmosphere. The number of intermediate nodes during the modelling was set to eight: increasing the number of nodes leads to significantly higher model complexity without giving any noticeable improvement in the result accuracy compared to the model with fewer intermediate nodes.

The comparison of results obtained from the two-point and multi-point models show that the change in mass after adding each node and the influence of aerodynamic forces acting on the tether leads to differences in values of the tension force and angles describing the orientation of the bodies. Therefore, the multi-point model has noticeably higher accuracy compared to the models of the rigid and elastic rods. In addition to modelling the motion of the deployed system, the multi-point model also enables the modelling of the deployment process taking into consideration the curvature of the tether.

References

[1] *Space primer* 2009 Air Command and Staff College 356

[2] Cosmo M L and Lorenzini E C 1997 *Tethers in Space Handbook. Third edition.* Smithsonian Astrophysical Observatory. http://www.tethers.com/papers/TethersInSpace.pdf

[3] Kumar K 2006 *Review of dynamics and control of nonelectrodynamic tethered satellite systems*. Journal of Spacecraft and Rockets **43**(4) 705–720.

[4] Alpatov A P, Beletsky V V, Dranovskii V I, Khoroshilov V S, Pirozhenko A V, Troger H and Zakrzhevskii A E 2010 *Dynamics of tethered space systems.* Boca Raton, FL: CRC Press. http://mech.spbstu.ru/images/c/c3/Dynamics_of_Tethered_Space_Systems.pdf [Accessed 20 Sep. 2020]

[5] Levin E 2007 Dynamic analysis of space tether missions. San Diego, Calif: Univelt.

[6] Lim J and Chung J 2019 *Removal of captured space debris using a tethered satellite system* Journal of Mechanical Science and Technology **33**(3) 1131-1140

[7] Sabatini M, Gasbarri P and Palmerini G B 2016 *Elastic issues and vibration reduction in a tethered deorbiting mission.* Advances in Space Research. **57** 1951-1964

[8] Mantellato R, Valmorbida A and Lorenzini E. 2015 *Thust-aided librating deployment of tape tethers*. Journal of Guidance, Control, and Dynamics **52(**5) 1395-1406

[9] Johnson L, Gilchrist B, Estes R D and Lorenzini E 1999 *Overview of future NASA tether applications.* Adv. Space Research **24**(8) 1055–1063.

[10] Ohkawa Y, Kawamoto S, Okumura T, Iki K, Okamoto H, Inoue K, Uchiyama T and Tsujita D 2020 *Review of KITE – Electrodynamic tether experiment on the H-II Transfer Vehicle* Acta Astronautica **177** 750-758

[11] Voevodin P S and Zabolotnov Y M Analysis of the Dynamics and Choice of Parameters of an Electrodynamic Space Tether System in the Thrust Generation Mode Cosmic Research 2020 **58**(1) 42-52

[12] Aslanov V S and Ledkov A S 2017 *Tether-assisted re-entry capsule deorbiting from an elliptical orbit* Acta Astronautica **130** 180-186

[13] Puig-Suari J and Longuski J M 1991 *Modeling and Analysis of Tethers in an Atmosphere.* Acta Astronautica **25**(11) 679–686

[14] Zabolotnov Y M and Elenev D V 2013 *Stability of motion of two rigid bodies connected by a cable in the atmosphere*. Mechanics of Solids **48**(2) 156-164

[15] Yu B S, Wen H and Jin D P 2018 *Review of deployment technology for tethered satellite systems* Acta Mechanica Sinica/Lixue Xuebao **34** 754-768

[16] Pradeep S and Kumar K 2003 *Extension of tethered satellites in the atmosphere.* Acta Astronautica **52**(1) 1–10

[17] Jin D P and Hu H Y 2006 *Optimal control of a tethered subsatellite of three degrees of freedom.* Nonlinear Dynamics **46**(1–2) 161–178

[18] Williams P. 2006 *In-plane payload capture with an elastic tether.* Journal of Guidance, Control, and Dynamics **29**(4) 810–821

[19] Beletsky V V, Kasatkin G V and Starostin E L (1996) *The pendulum as a dynamical billiard.* Chaos, Solutions & Fractals **7**(8) 1145-1178

[20] Williams P 2008 Deployment/retrieval optimization for flexible tethered satellite systems. Nonlinear Dynamics **52**(1–2) 159–179

[21] Zhong R and Zhu Z 2011 *Dynamic analysis of deployment and retrieval of tethered satellites*

using a hybrid hinged–rod tether model. International Journal of Aerospace and Lightweight Structures **1**(2) 239–259

[22] Zabolotnov Y M 2015 *Control of the deployment of a tethered orbital system with a small load into a vertical position*. Journal of Applied Mathematics and Mechanics **79**(1) 28-34

[23] Dong Z, Zabolotnov Y M and Wang C 2017 *Modeling and analysis of deployment dynamics of a distributed orbital tether system with an atmospheric sounder*. 3rd IAA Conference on Dynamics and Control of Space Systems 1053-1063

[24] Mankala K K and Agrawal S K 2008 *Dynamic modeling of satellite tether systems using Newton's laws and Hamilton's principle*. Journal of Vibration and Acoustics **130**(1) 014501-1–014501-6

[25] Kane T R and Levinson D A 1980 *Formulation of Equations of Motion for Complex Spacecraft*. Journal of Guidance and Control **3** 99–112

[26] Zabolotnov Y M and Elenev D V 2014 *A way for accelerated calculation of spacecraft motion in an atmosphere with a tethered aerodynamic stabilizer*. Mathematical Models and Computer Simulations **6**(1) 38-45

[27] Wen H, Jin D P and Hu H Y 2008 *Advances in dynamics and control of tethered satellite systems*. Acta Mechanica Sinica. **24**(3) 229–241.

[28] Kruijff M 2011 *Tethers in Space*. Netherlands: Delta-Utec Space Research

[29] Elenev D V, Zabolotnov Y M and McMillan A J 2017 *Deployment and stabilization of an aerodynamic space tether system* Advances in the Astronautical Sciences **161** 1023-1030

[30] Van Der Heide E J, Kruijff M, Avanzini A, Liedtke V and Karlovsky A 2003 *Thermal protection testing of the inflatable capsule for YES2*. 54th International Astronautical Congress of the International Astronautical Federation (IAF), the International Academy of Astronautics and the International Institute of Space Law. **1** 1719-1726

[31] Aslanov V S and Ledkov A S 2015 *Dynamics of Tethered Satellite Systems*. Beijing: National Defense Industry Press. 179

IOP Conf. Series: Materials Science and Engineering 1060 (2021) 012017 doi:10.1088/1757-899X/1060/1/012017

Modelling the work of the transport complex

V S Kuzmin, D V Elenev

Samara National Research University, Samara, Russia

E-mail: vskuzmin28@gmail.com

Abstract. The modelling of the transport complex is made using an automated system and an example of the enterprise that uses the dangerous materials. The example of such enterprise is a nuclear power station. The automated system used for modelling is a web-based intranet application that realizes monitoring and control functions. The automated system uses a system of objects included into the model of the enterprise, namely: gates, roads, transportation devices and controllers, airtight zone, describing the reactor building, gateway which allows to move cargo into and out from the building, non-airtight zone, which describes the outside area, and the zone of loading and unloading the cargo. The system receives input signals of discrete or analogue origin. Basing on the results of data collection, the algorithms of forecasting the crucial changes in states of main elements of the transport complex were developed. The automated system models the workflow of elements and the whole transport complex.

1. Introduction

The transportation of cargo between different sites of the territory of an organization should meet the requirements of logistics process and safety of the cargo. Some cargos being transported can be dangerous for the environment and staff. In order to obtain safe transportation, special conditions may be used. For example, the transport complex of the nuclear plant or chemical station provides transportation of containers with fuel and radio- or chemical-active elements through the sealed enclosure of the building [1].

The aim of the work is to create cross-platform software (designer) allowing an engineer without special programming skills to develop a complete software tool for operator panels that meets requirements for the functionality and safety of technological processes in production. In order to reduce the repair time when breakage occurs, it is necessary to minimize the time spent on diagnostics of individual elements at their installation sites. This can be done by standard software procedures aimed to monitoring and forecasting procedures for the operations of the transport complex.

In contrast to commonly used software like TIA Portal by Siemens AG and ABB Panel Builder, the automated system described below offers a novel fast user interface based on web application that can be used on different desktop and mobile devices.

2. Literature review

The transportation of cargo is a complex process that has peculiar properties basing on the nature of the cargo. The automation of the transport is necessary to reduce the decision-making time by providing operators and other staff with actual information about current state of the transportation process. The transportation process can be described by a number of process variables including temperature, pressure, distance, liquid levels, etc., and these variables can be received with delays or simultaneously [2]. The connection between people, services and things can be made using the Internet of things (IoT) concept [3] instead of traditional automaton of the manufacturing process [4,5].

Some researchers use their own software modules for simulation testing systems, for example, testing for train-borne controller [6]. Usually these applications are desktop, but for better mobility, it is possible to use web-based technologies [7].

The transportation where dangerous cargo is being transferred, like atomic energetics, should always have the risk management [8-10]. The transportation of the cargo can be consistent, aimed to transport one cargo at a time, while wide number of transport complexes use multi-vehicle transportations in both civil and military areas [11,12]. The cargo itself can be solid, liquid, bulk [13] or may have special properties.

The automation of the transportation process allows to reduce the reaction time and to forecast possible failures in the elements of the complex basing on the history of usage.

3. Architecture

The model of the transport complex is based on a number of typical objects used in software designer. These objects include airtight and non-airtight zones, gateway between them, zones of loading and unloading etc. The automated system receives input signals of discrete and/or analogue origin.

The automated process control system (APCS) during its workflow realizes the data transfer from the field level sensors that performs discrete and analogue signals, to the panel PC level where the web application works (figure 1).

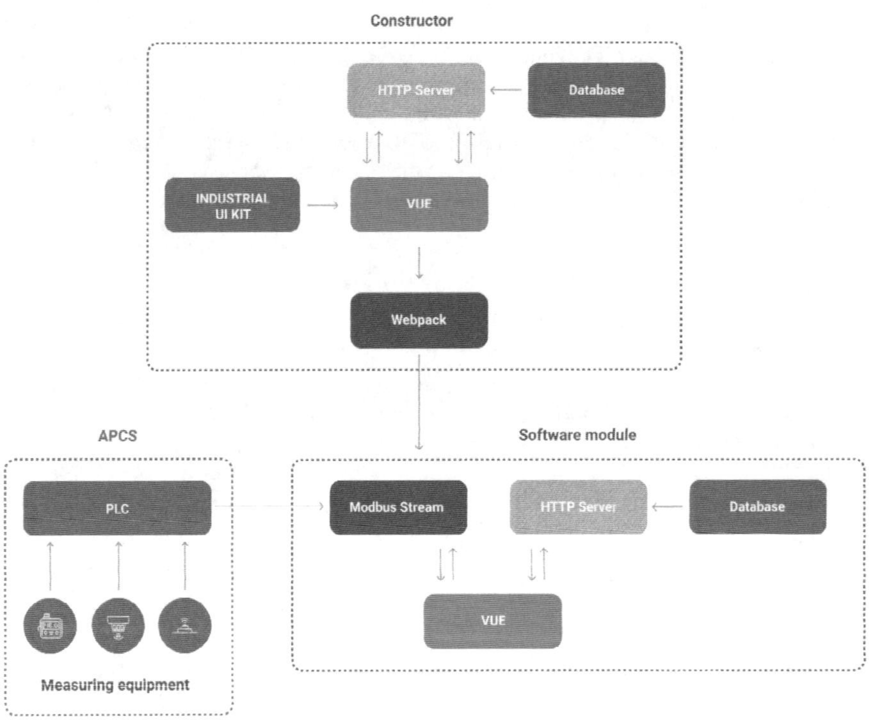

Figure 1. Model of the automated system

As a result of exchange of states between the server and client sides at the Panel PC level, the subsequent visualization of the data in made on the operator panel. The event log with warning and alarm conditions is sent to the Database block.

The software project for modelling the work of the transport complex uses the technological stack where the components work in a single eco-system that does not consume excessive hardware resources. This technological stack consists of the following components:

- Vue.js - a framework for implementing the client part of the constructor

IOP Conf. Series: Materials Science and Engineering 1060 (2021) 012017 doi:10.1088/1757-899X/1060/1/012017

- Vuex.js - a library with a centralized state repository for the Vue.js framework
- Chart.js – a library for converting data into charts
- HTTP server for collecting and transmitting data between the PLC and the operator panel
- Modbus TCP js library – a library for working with the ModbusTCP protocol
- MongoDB – database
- IndustrialUIKIT – author's library of components for implementing user interfaces

This software solution is cross-platform, allowing to use any desktop or mobile device.

4. Modelling and forecasting

The software made for allows to the model the work of the transport complex as a web-based application. This application uses the database of states of the element of the transport complex and helps to predict possible faults of elements. As a result, the application automates the work process of the transport complex.

An example of a constructor web application user interface is shown in figures 2 and 3. The constructor provides the following functions: creating a parameter, collecting statistics from an object, import/export data, filtering, sorting parameters on the screen, warning dialogues.

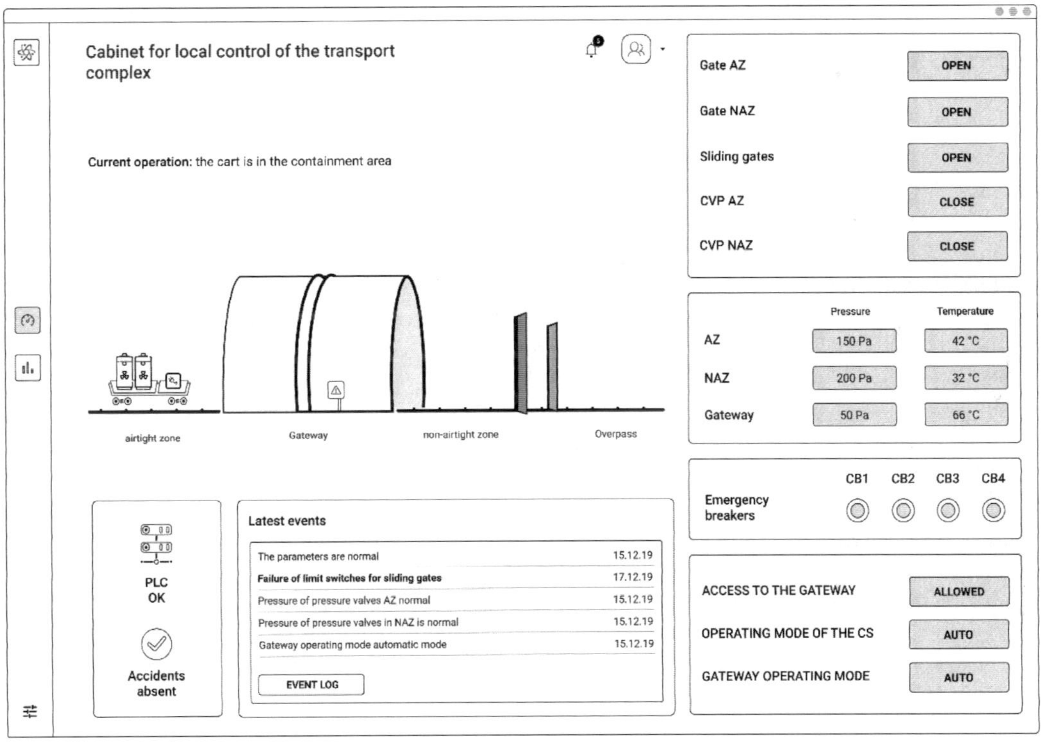

Figure 2. User interface

The software system collects data about elements of the transport complex, and this database allows to forecast the crucial changes in states of main elements. The forecasting algorithms allows to provide emergency messages for operators. The base of the algorithm is the statistics on duration and intensity of usage of every element of the transport complex. The algorithm uses the information about the properties of the elements given by their manufacturers. The software allows to complement this information by user-defined parameters.

Therefore, the upcoming end of lifetime of the elements can be forecasted using the data stored in the database of the system, thus reducing the frequency of failures in the lifetime of the transport complex.

IOP Conf. Series: Materials Science and Engineering 1060 (2021) 012017 doi:10.1088/1757-899X/1060/1/012017

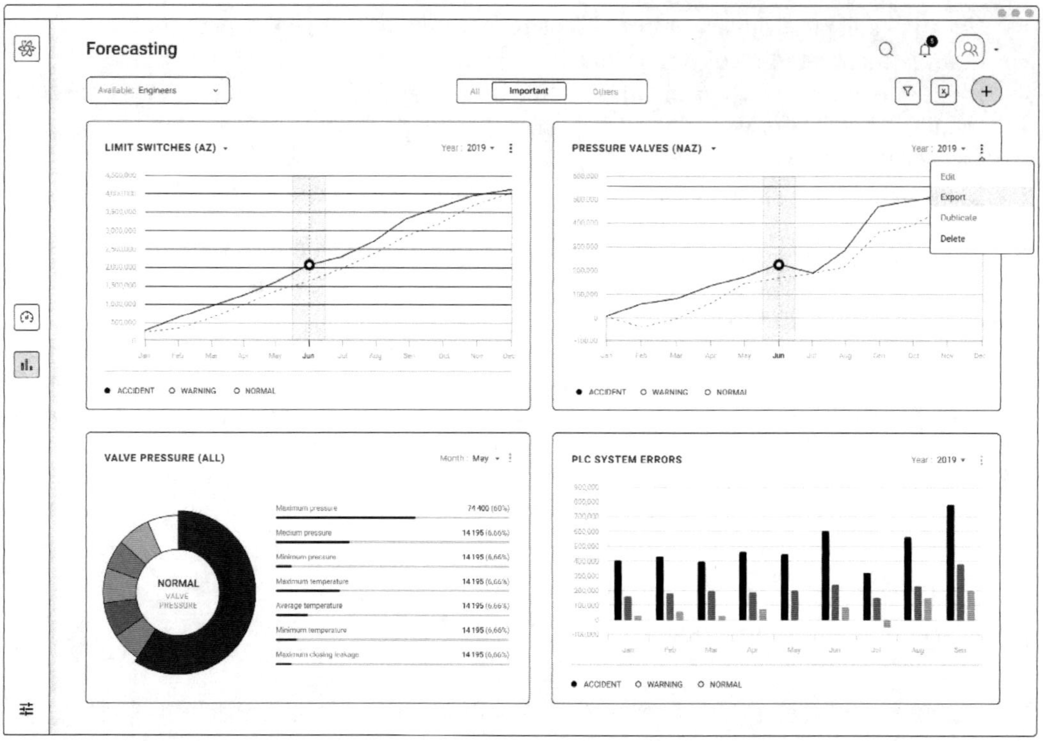

Figure 3. Constructor

5. Discussion

The algorithms of forecasting the important changes use the base of knowledge of lifetime of single elements. This knowledge and logging all events allows to predict the failure of elements due to end of their life cycle thus reducing time while replacement of element. At the same time, the conditions of the surrounding medium like overheating, high humidity, electromagnetic radiation *etc.* lead to the reduce in lifecycle of element, and this factors should be taken into consideration to minimize the events of replacement of single elements thus reducing the cost of ownership.

6. Conclusions

The automated system that models of the work of the transport complex provides operators with actual information about state of the cargo and elements of the complex. The sources of this information are field level sensors, and the analysis of the collected data allows to reduce the number of unplanned repairs.

The software made for the automation uses fast user interface base on web application, thus offering platform independency for the end-user device.

References

[1] Kuzmin V S and Sadova K V 2018 Development of a software module for managing the Belarusian NPP transport complex, in *Youth science: challenges and prospects: Materials of the 1st Russian scientific and practical conference* 4–6

[2] Vara Lakshmi G, Ganesh U, Sai Venkata Sahadev C and Sobha Rani P 2020 Industrial automation with raspberry Pi and IoT *Int. J. Adv. Sci. Technol.* **29**(5) 3534–9

[3] Breivold H P and Sandstrom K 2015 Internet of Things for Industrial Automation-Challenges and Technical Solutions *IEEE International Conference on Data Science and Data Intensive Systems* (8th IEEE International Conference Cyber, Physical and Social Computing; 11th IEEE International Conference on Green Computing and Communications and 8th IEEE International Conference on Internet of Things, DSDIS/CPSCom/GreenCom/iThings) 532–9

[4] Panteleev V N and Proshin V M 2012 *Osnovy avtomatizatcii proizvodstva* [Fundamentals of manufacturing automation] (Moscow: Academy)
[5] Selevtsov L I and Selevtsov A L 2012 *Avtomatizatciya tehnologicheskih processov* [Automation of technological processes] (Moscow: Academy)
[6] Qi X-B, Liu D-D and Chen X-X 2017 Design and realization of simulation testing system *2nd IEEE Int. Conf. Intelligent Transportation Engineering ICITE* 8056889 99–102
[7] Anil R, Satyakumar M and Jayakumar J 2018 Travel Time Estimation and Routing for Emergency Vehicles under Indian Conditions *3rd IEEE Int. Conf. Intelligent Transportation Engineering ICITE* 8492696 247–52
[8] Guseva A, Koptelov M and Kovtun D 2019 Decision Support for Project Risk Management Using the Information-Semantic Field *1st Int. Conf. Control Systems, Mathematical Modelling, Automation and Energy Efficiency SUMMA* 250–4
[9] Byvaykov M E 2019 Interface software channels of large-scale systems in APCS of NPP *Proc. 12th Int. Conf. Management of Large-Scale System Development MLSD* 8911074
[10] Pokrovskaya O, Reshetko N, Kirpicheva M, Lipatov A and Mustafin D 2019 The study of logistics risks in optimizing the company's transportation process *IOP Conf. Ser.: Mater. Sci. Eng.* **698**(6) 066060
[11] Hu Z-H and Wei C 2018 Synchronizing vehicles for multi-vehicle and one-cargo transportation Computers and Industrial Engineering **119** 36–49
[12] Milewski R and Smal T 2018 Decision making scenarios in military transport processes *Archives of Transport* **45**(1) 65–81
[13] Astapenko N V, Koshekov K T and Kolesnikov A N 2018 *Computer Optics* **42**(3) 510–20

IOP Conf. Series: Materials Science and Engineering 1060 (2021) 012018 doi:10.1088/1757-899X/1060/1/012018

Opportunities and prejudices in synthetic experiments

A J McMillan

Wrexham Glyndwr University, Mold Road, Wrexham, LL11 2AW, UK.

E-mail: alison.mcmillan@physics.org

Abstract. Synthetic experiments, meaning experiments performed using models of reality, are used in a variety of contexts. In recent years, advances in computing capability have led to the development of modelling capability that replicates reality to such an extent that in some circumstances it is difficult to distinguish the model from reality. From a science or engineering perspective, computational analysis models are used to replicate or predict phenomena such as fluid flow or stress in a machine component, but the capability has also crossed over into the arts enabling the computer generated imagery (CGI) used in films, and the augmented reality (AR) of video games. This paper reviews the application of synthetic experiments to replace physical experiments and discusses some of the advantages and pitfalls of such an approach in the context of education. The trend for democratizing analysis is considered, and the dangers of misinterpreting results are discussed. A discussion of the issue of model over–simplification is given, and finally the use of synthetic experiments as part of a systematic scientific investigation is also considered.

1. Introduction

In many scientific contexts it is either impossible, or rather infeasible, to perform a physical experiment, or sufficient physical experiments, to be able to bring an investigation to a firm conclusion. In place of physical experiments, it is increasingly likely that computational models will replace or extend the investigation to provide access to insights that would not otherwise be available. Likewise, in engineering product design practice, it is becoming increasingly common to replace verification testing with modelling and rely on a smaller number of physical product validation tests. In both science and engineering the principle is the same: the model used is believed to contain all the pertinent physics principles that apply in the real world example: all the *good practice* principles that used for that type of modelling are scrupulously applied, for example, mesh size control, time step size, and other such methods for ensuring numerical stability and accuracy; and therefore the outcome of the model analysis is then believed to be *correct* within some precision tolerance band. Let us call such experiments *synthetic experiments.*

Clearly, the computer age provides a great opportunity for experimental study and engineering verification using models. Nevertheless, it is inevitable that even the most diligent researchers will approach modelling with a degree of prejudice, and it is important to recognise this and avoid complacency. For example: one might be diligent enough to select a suitable mesh, which is a mathematical analysis stability or convergence requirement; but fail to recognise some other special features, such as a requirement to represent some pertinent physics within the model.

In the context of education, there is a growing trend towards teaching through simulations, and a reduction in laboratory practical classes. In the post–COVID–19 era, the move away from the laboratory will be accelerated. The fallacy is this: it is generally believed that the laboratories merely confirm theory and develop the students' laboratory skills. This is simply not the case: the fundamental role of laboratory experiment is to show that even the most meticulous experimental work will not replicate theory perfectly, because there will inevitably be other physics principles involved. It is in recognising this that enables a good researcher to avoid prejudice in modelling work.

2. The avoidance of prejudice

The principle of avoidance of prejudice is nothing new. Even as long ago as in Ancient Greece, Aristotle [1] listed thirteen *fallacies*, of which the eighth, *Secundum quid et simpliciter* concerns the confusion between that which is true within a particular context with that which is absolutely true. This is precisely to the point of modelling: models must be recognised as being simplifications of reality, and understood to be applicable only within a particular regime.

Somewhat more recently, in 1267, Roger Bacon [2] listed four *hindrances*, the second of which relates to entrenched custom and practice, while the fourth warns against hiding personal ignorance and giving an outward display of knowledge. In today's highly disciplined and controlled engineering industry, companies that design and manufacture safety critical machines and structures have detailed written procedures and processes for every aspect of the design, validation, manufacture and inspection of their products. The computational modelling work has become an increasingly significant part of the validation process, and is rightly tightly controlled by the procedural documentation; however, observing custom and practice alone is insufficient as a guard against a failure to capture all the relevant physics in the model. Furthermore, the beauty and apparent certitude of a rainbow coloured stress analysis model image provides misleading evidence for those with insufficient knowledge or experience.

In 1620, Francis Bacon [3] illustrated the problem of prejudice in science by means of his *Idola mentis* – false images of the mind. The first of these, *Idola tribus*, relates to prejudices shared by all humankind, and arising from our shared limitations. The metaphor Bacon used to explain this is that what we understand of reality is distorted or discoloured, by details not apparent to us, in the same way that an imperfect mirror can provide a distorted or discoloured image. Later in the second book of his *Novum Organum*, he describes the process of listing instances of a phenomena, and from this identifying *causes*: this seems to be the first scientific publication in which it is recognised that for any (experimental) observation the result might comprise effects from more than one cause, and that one can gain value from designing experiments in such a way as to be able to separate those effects.

In modern times, we are becoming increasingly aware of *human factors* [4] in science and engineering. Even diligent researchers will approach modelling with prejudice: we remember to make a suitable mesh, but do we fail to recognise some other special features? It is important to recognise this and avoid complacency.

3. Definitions and motivations

In the context of the present paper, the focus is on **synthetic experiments**, *i.e.* experiments made on the basis of computational (or possibly other) models. Where modelling can be considered a useful and faithful representation of reality, the use of models as proxies for experimental test specimens can be advantageous.

This is precisely the role of models in component design *verification* and *validation* [5]: modelling enables design *verification* at lower cost and more quickly than used to be possible with the complex test programmes that were necessary before the advent of the sophisticated Finite Element Analysis (FEA) software available today. Nevertheless, the process of *validation* still requires a final physical test proof to confirm the model predictions. The scepticism implied by the physical test process is justified: it is understood that a model is based on assumptions, and while those assumptions might be justifiable on the basis of prior experience within a similar regime (a process known as *read–across*), it is possible that for the particular regime of the new design, the model is less than adequate. Product development is an expensive process, the modelling work can be very useful in reducing the risk that the design would need to be changed, and this is a matter of importance to the finances of a company engaged in product design and manufacture. Failure of a product in–service, particularly if that failure were to be catastrophic and lead to loss of life, is a matter of far more importance, and has severe implications for reputation, as well as stringent financial penalties. It is for this reason that the final physical test programme is, and will continue to be, required.

In terms of science exploration, one might think that there is nothing more of novelty to do after the modelling capability is developed. Such a model might represent the requisite physical phenomena, and perform in a mathematically and computationally stable way. Nevertheless, from the point of view of research opportunities, this is only the beginning. Unless a model is extremely simplistic it cannot be explored in its entirety using mathematical analysis techniques alone, and thus must be explored by computational analysis with a myriad of choices of variable parameters and applied conditions (*e.g.* loads, temperatures and boundary conditions). Surprising results may ensue, and these might be described as *emergent phenomena*. If the model is an appropriate representation of reality, then the same phenomena might then be discoverable in reality. In this way, we might facilitate the process of real science discovery using the computational model as a tool.

3.1. Why do we make models?
Model–making is probably one of the very first tools of humanity. It allows us to stand outside our world and theorise about it. Understanding our world is the first step in being able to control it. Originally, more control meant a life more free from predators and reduced uncertainty about food supply. Over time this enabled the increasing ability to devise more complex physical tools, which has become a virtuous circle leading all the way through to modern civilisation.

But we are not gods: we know that models are limited. For example, the cycle of the day and the year are fixed with the precision of celestial mechanics and provide a basis for expectations about weather in particular seasons, but predictions of what the actual weather would be on a future day is limited, even with the powerful forecasting capability that is available in modern times. In the same way, engineering product design validation must make some assumptions about the loads and conditions that a product might be expected to withstand during a reasonable working life, but the individual product might be used in a wide variety of circumstances.

3.2. Confusing models with reality
Unfortunately, a time is approaching where many people become so inured with living with models, that the distinction between reality and the capacity of a model to represent reality becomes difficult to gauge. In the past, models were basic and therefore obviously approximate. In the model results of early FEA and computational fluid dynamics (CFD), the approximation was made obvious by the crudeness of the visualisation imagery. More recently, and to an ever–greater degree, FEA and CFD models look like the real problem. For example, in a 1D beam finite element model, many FEA results viewers will render the image using the geometrical information provided for the calculation beam bending stiffness. Similarly in CFD, the representation of particles entrained in flow can mislead understanding about how those particles interact with the flow, each other, or other physics not in the analysis.

In modern films, computer–generated imagery (CGI) monsters look and move like real monsters, and the image manipulation algorithms which provide realistic textures and shadows demonstrate incredible capability. The cross–over between games and real applications using combined computational analysis and augmented reality (AG) [6] has already begun. These image presentation capabilities are gradually being applied in computational models, as a means to drawn in the non–expert, and *democratize* analysis [7]. While it can be helpful for the modelling specialist to convey to company senior management the significant aspects of the modelling work undertaken, it can also instil the erroneous expectation that such models provide an entirely *faithful* representation of reality, not merely a *useful* one.

3.3. Typical uses for computational modelling
Computational modelling methods are typically used when it is difficult to perform a physical experiment and when the can model provide insights that would not otherwise be available. Judicial use of modelling can lead to a reduction of the costs in engineering product design by reducing the number of physical tests needed for product validation. The same principle can be used in science discovery, where modelling can anticipate the physical test result, and help in the design of an appropriately sensitive physical test.

In using such a model there is a belief that it will contain all the pertinent physics principles. This belief can be tested by assessing candidate physical principles and estimating the extent of their influence. This is not a fool–proof method, but it is a common process in model–building to test a model before and after adding such an enhancement to establish whether the change is negligible or provides a modification to the result of the type and magnitude expected.

In modelling, the recognised and established *good practice* procedures are applied. In terms of computational mechanics modelling, this will include adequate checks on parameters such as mesh size, time step size, and other such methods for ensuring numerical stability and accuracy. After this then the model analysis can be considered *repeatable* meaning that there should be no errors arising from mathematical or computational inadequacy or instability, but this provides no assurances as to the adequacy of the physics represented in the model. Thus, it must be understood that however much we might increase reliance on computational modelling, and reduce the need for testing, there will remain a need for *some* testing.

4. Models in the context of STEM education and training
In the UK, most STEM degrees are accredited by professional bodies. An accredited degree programme provides prospective students an assurance that the content of the programme meets the specification set by the professional body that accredited it. The degree then forms part of education and training route toward professional qualifications. In almost all cases, laboratory based practical work forms a part of the specification [8, 9] required for accreditation.

It is generally believed that student laboratory work serves to confirm theory, consolidate understanding, and to provide the student with laboratory skills. For a campus–based student, it is certainly true that time spent in the laboratory allows for personal interaction with fellow students and laboratory demonstrators, thereby providing the student with an opportunity to ask questions and improve understanding. The performance of prescribed laboratory experiments is somewhat artificial: the apparatus and test pieces are generally pre–designed, so that the student follows a standard recipe. Generally, the laboratory experiments are designed in such a way that the student is unlikely to break anything significant, and if an accident were to happen then the consequences would be minimal. Familiarity with handling apparatus, leads to confidence, develops hand skills, and might also serve to inspire the student to think beyond the particular experiment set.

The Open University, and other universities which offer a distance learning option, manage the laboratory experience requirement by holding some residential practical sessions. For many of those students, travel can be a hindrance, and allocating blocks of time away from home might be difficult. For the simpler laboratory experiments, laboratory kits can be sent by post to perform at home, and more recently, the Open University has developed remotely operated physical laboratory facilities, which the students can access and drive over the internet [10]. In addition to these practical solutions, there is a growing trend of teaching practical skills using simulation tools and CGI models. This is driven in part by cost reduction, in part by convenience, and in recent months by COVID–19 and the need to reduce social interaction. After the pandemic, it is likely that this trend will be accelerated.

The advantages of such simulation–based models is that they are clean, safe, unbreakable, require minimal space to store or to operate, and that they behave consistently. Such advantages are also disadvantages in that they conceal reality: one important role of the laboratory practical is to show that, even with the most meticulous experimental practice, there will be occasions when experimental results will not replicate theory. There will always be some other physics principle involved, and sometimes that will be significant. A good researcher learns to recognise a when an experiment is a good differentiator of the significant effects and the negligible ones. Thus, although there are many advantages to teaching through simulated laboratory experiments, it is clear that students will require some real, physical laboratory experiences as part of their professional formation.

5. Models in the context of STEM research and product development

We live in a macroscopic world, which informs our *common sense*. Furthermore, our education conditions us to believe books not our own eyes, and we have become like rats running around in the intellectual model mazes that we have learned to navigate. The discipline boundaries have become *comfort blankets* protecting our minds from model regimes that clash, and while we are being exhorted to embrace interdisciplinary studies, there is little recognition of the additional intellectual burden of navigating through multiple discipline areas [11]. Here are some examples.

5.1. Material is "Bulk"

Classical physics and most of modern–day engineering is built on the notion of *bulk* materials. Almost everything that we interactive with daily involves Avogadro numbers of atoms, molecules or electrons. Most of the time most material (solid or liquid) behaves more like a bulk material than a surface, line or individual particle, and this reinforces our natural prejudice for classical thinking. While this might be expected to be the case for engineers, since they have relatively little exposure to modern physics and chemistry concepts, physical scientists can be just as trapped by classical or common sense thinking. The construction of mathematical models such as quantum mechanics and condensed matter provides prediction capability but does relatively little to shape human insight.

Two examples, spanning the last century, of such an intellectual struggle, include: in physics, the ever–continuing debate over what quantum mechanics really means;

> *Physicists do not believe quantum mechanics because it explains the world, but because it predicts the outcome of experiments with almost miraculous accuracy*
>
> John Horgan [12]

and in engineering, the development of the theory of fracture mechanics from Griffiths crack surface energy argument through to the misunderstanding and mis–application of the theory which is still common and embodied in engineering standards even now, 100 years later [13].

5.2. The "Perfect" geometry

In performing engineering component stress analysis, the initial starting point is the specification of the component geometry. Generally, this is a Computer Aided Design (CAD) model which

IOP Conf. Series: Materials Science and Engineering 1060 (2021) 012018 doi:10.1088/1757-899X/1060/1/012018

specifies the geometry through the definition of a number of reference points fixed in space, from which solid geometry is defined using splines and parametric curves. In practice, for some complex geometric shapes, the initial CAD models created by the design engineers are inadequate for use in stress analysis because of geometrical flaws such as mis–match of the edges of adjoining surfaces leading to tiny steps or curvature that are not intended features in the real component. Such geometry flaws must be inspected and *healed*.

The healed geometry can now be meshed: the term *meshing* really means subdividing the entire modelled component into sub–volumes, called *elements*, and these elements are defined by geometric points called *nodes*. The surface geometry of the component is thus defined by the surfaces of the elements which are in the outer layers of the mesh, and hence by the geometric points of the nodes and the interpolated element surface, which is usually linear or quadratic with respect to the nodes. Thus, the analysis geometry is similar to, but not the same as, the CAD geometry.

Even if we are prepared to accept these subtle differences, we should observe that this is still an idealised mathematical model of an object with a perfect geometry, comprised of perfect bulk material, and subjected to idealised loads. Such a component has no polycrystalline substructure, no unintended porosity or foreign material inclusions, nor does it have a surface texture. In such an idealised component, the stress state within the material is perfectly varying with no localised extremes, and is an object through which only bulk stress waves propagate.

One particular problem in mechanical analysis concerns the assumption of geometric linearity. For small loads and small deflections, the assumption is reasonable, but problems can arise when a structure is slender. Within the linearised model for a beam or shell, the mathematics does not allow for the eventuality of buckling under compressive loading. In order to model for buckling, one first has to recognise that buckling is possible, and then apply and appropriate mathematical model. In solving using FEA, it is usually necessary to introduce some artificial asymmetry to the geometry to precipitate the analysis to model the buckling mode. Since it is generally possible that a structure may have more than one buckling mode, the modification to the geometry has to be considered carefully, as otherwise it is possible that the lowest buckling mode would not be excited, and the load predicted for buckling initiation would be too high. In today's modern light–weight structures, there is increasing reliance on designs including slender features and the use of features like corrugations to inhibit buckling. Small geometrical variations, or in–service damage, can lead to substantial variation in the loads that can be supported by such structures.

A similar issue arises in modelling fluid flow. A fluid flowing along the idealised smooth surface of a component encounters nothing to trigger turbulence. In order to help the model represent observed reality, additional mathematical artefacts are added: hence the turbulence models used in CFD [14]. In both cases, the existing model is simplified, essential physical effects have been excluded, and then put back in using a different, and possibly rather arbitrary, mathematical form.

5.3. Dimensional reduction of the mathematical model — The Elastic Beam

The beam is a particular case of a model simplification process in which the real–world component is first modelled in mathematical form using a governing equation, and then the physical dimension of the problem is then reduced, based on some aspect of the geometry or symmetry of the component.

In such cases, there might be a ranking of plausible equations, based on how well they match a particular class of problem, and the mathematical elegance of the equation. For example, equation (1) shows the general equations for linear elasticity in 3D, equation (2) gives the Euler-Bernoulli representation for wave propagation through a 1D beam, and equation (3) shows the Timoshenko beam with a transverse shear correction term. Each equation is *correct*, but only in the particular context of its applicability.

$$\nabla \cdot \sigma + \mathbf{F} = \rho\, \ddot{\mathbf{u}} \tag{1}$$

$$EI\frac{\partial^4 w}{\partial x^4} = \rho A\frac{\partial^2 w}{\partial t^2} + q \tag{2}$$

$$EI\frac{\partial^4 w}{\partial x^4} = q(x) - \frac{EI}{\kappa AG}\frac{\partial^2 q}{\partial x^2} \tag{3}$$

5.4. Physically inadequate mathematical model — Navier-Stokes Equations

Until recently, most CFD packages were based on numerical solution of the Navier–Stokes equations for fluid dynamics. In the derivation of the Navier–Stokes equations [15], the following process took place: first all the classical physical effects were written down and assembled into an equation, then all the mathematically difficult terms were removed, and this was justified by a reduction of the context of applicability. In CFD packages, some of those difficult terms have been re–introduced, or they have been modified or replaced by empirical terms to make them more numerically amenable.

An alternative approach is to develop from the fundamental physics principles of the mean free path of a particle travelling and interacting with other particles. This method is based on the Boltzmann equation [16]. Equivalence between this formulation and Navier–Stokes has been shown, for the context at which the latter is applicable. By basing numerical solution on the Boltzmann equation, a different numerical scheme is required — Lattice Boltzmann CFD, and this circumvents the various limitations of model building using CFD based on Navier–Stokes.

6. Results demonstrating model prejudice

The purpose of this section is to provide some illustrations of some large–scale effects arising from physical details that are usually excluded from analysis models. It is not immediately apparent that these details are significant, and their exclusion is frequently left unremarked. On the basis that these are significant effects arising from seemingly insignificant details, we might call these effects *emergent*.

6.1. Model of a test specimen with imperfect geometry included

The Saint–Venant principle is a guiding principle in stress analysis, which indicates that small geometric or stress distribution variations are a local phenomenon only, and their effect diminishes to insignificant at a sufficient distance from the disturbance.

The rationale for this study was the observation that for a perfect material, with uniform properties and perfectly smooth geometry, and when loaded to a level below the elastic limit can then be unloaded and return to the original state without damage or introducing a residual stress–strain state. On the other hand, it is well known that a real component, when cyclically loaded to a moderate fraction of the elastic limit, will eventually fail though *fatigue*. Generally the process of fatigue is characterised by test or empirical formulæ derived from test data. In these studies [17, 18, 19], it is argued that fatigue must be a physical process, and as such should be possible to model directly.

Figure 1 shows a simple test specimen, and sequence of section details leading to a local model of the gauge section. A cyclic load is applied to 90 % of the elastic limit at the gauge section, and the material properties include an isotropic strain–hardening plasticity model. Figure 2 shows von Mises stresses for three models: where the gauge section has surface roughness only; where the gauge section has sub–surface porosity only; and where the gauge section has both roughness and porosity. The interface between pale and dark blue represents the perfect model "nominal stress" at 90 % of the elastic limit stress. The green represents stresses above the

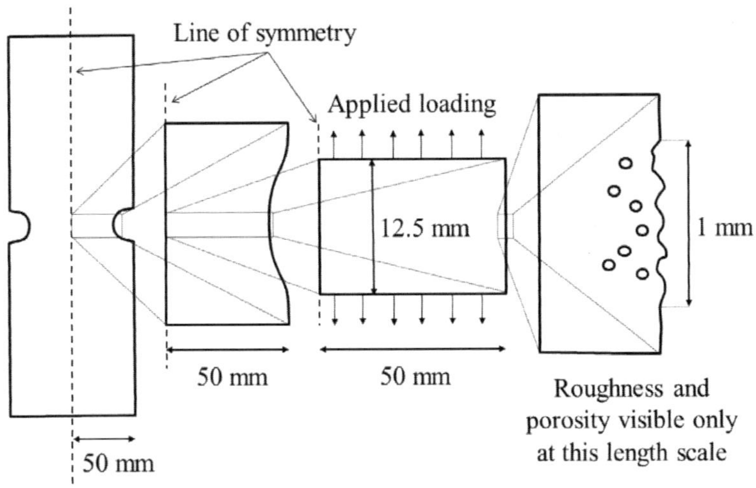

Figure 1. Schematic of a test specimen illustrating a local representation of the gauge section.

elastic limit leading to plastic deformation. Note that for a model that does not include surface roughness nor sub–surface porosity, the nominal stress over the gauge section would be uniform, and always below the elastic limit. In such a model, the cyclic loading would be unable to generate damage accumulation, potentially leading to fatigue.

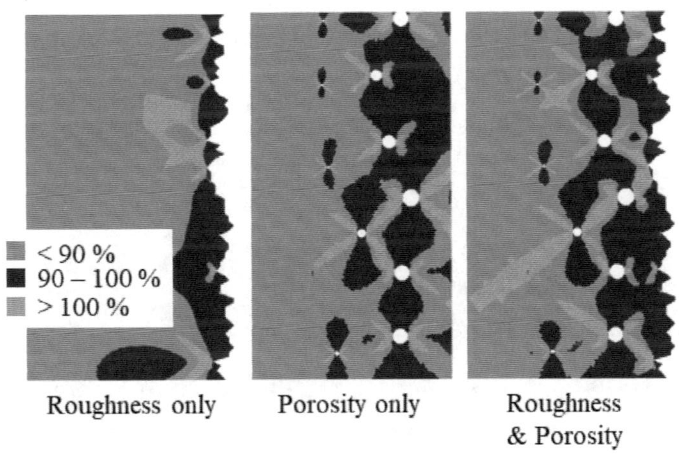

Figure 2. Von Mises stress field in three models of the gauge section (the contour colour bands indicate ranges of percentages of the elastic limit stress: the nominal stress is 90 %).

Notice that the effects of surface roughness and porosity are different. Surface roughness introduces high stresses at deeper roughness troughs and tends to link troughs. The high stress region penetrates into the surface by about 1.5 times the peak to trough height in the roughness. The effect of porosity is to form diagonal bands extending from the surface and penetrating quite deeply into the sub–surface. The combined effect of roughness and porosity is significantly greater than either effect considered individually.

6.2. Low Reynolds number and steady laminar flow

In this example, the purpose is to illustrate that the simplistic view that low Reynolds number alone implies that the flow is steady and laminar can be misleading. Take the simple example of plane Poiseuille flow [15] between two parallel walls, as illustrated in figure 3. In this case, the walls are perfectly smooth, and the analytical solution predicts a velocity distribution that is of parabolic form.

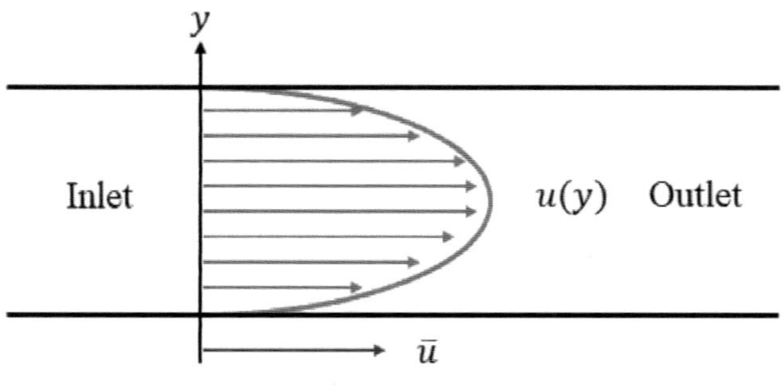

Figure 3. Plane Poiseuille flow between two infinite walls.

This problem can be replicated as a simple 2D CFD model [20], the results of which are shown in figure 4(a). In this case, the walls were considered perfectly smooth with a no–slip condition. The inlet condition at the left–hand side is for a uniform velocity distribution equal to the mean flow velocity. The outlet condition at the right–hand side is that of zero pressure difference. Applying the usual due diligence one would evaluate the Reynolds number, and on confirming that it is sufficiently low for the fully developed flow to be laminar, one would then consider the size requirement for the *lead–in* region. A lead–in region is required because the inlet conditions represent the mean flow, and not the velocity profile of the fully developed flow. The velocity profile image shows that the fully developed flow develops within the first 5 % of the model geometry.

Figures 4(b–f) show similar models, but where the lower wall has some irregularity. In the case of figure 4(b) the gap narrows, but away from the gap step changes, the velocity profile

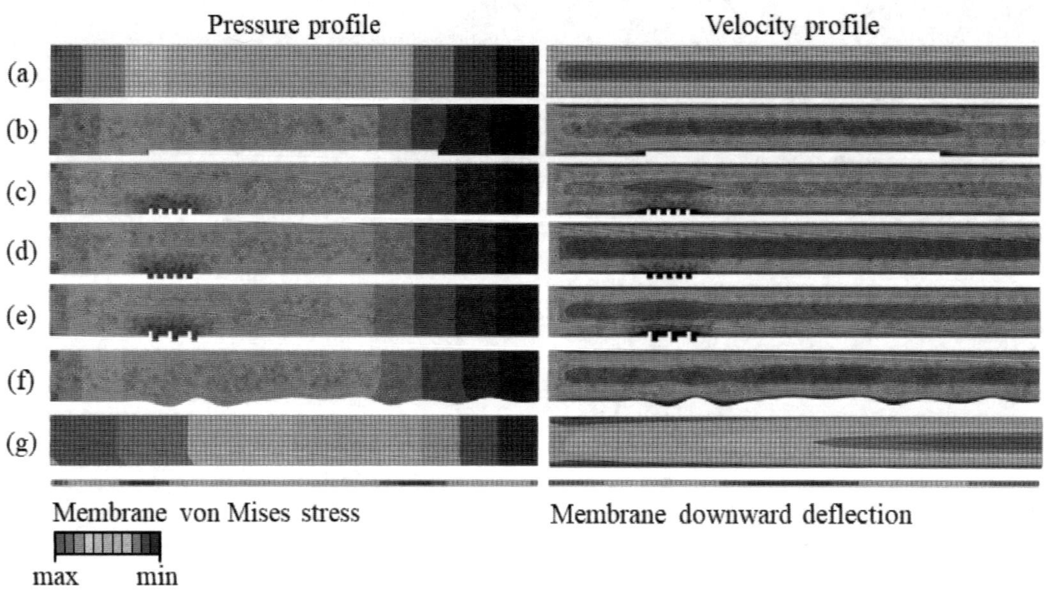

Figure 4. CFD models of plane Poiseuille flow between walls with imperfections (the pressure and velocity are non–dimensionalized and presented on a max–min scale).

and pressure profile are very similar to the perfectly smooth wall case. In figures 4(c–e), the gap varies over a shorter region, and is comprised of several inward or outward repeated steps which are closer together than the lead–in length scale. The flow treats these as if they were a single step, and the presence of the outward steps seems almost unimportant to the velocity profile. Figure 4(f) shows a continuously varying gap profile, with features that a comparable or somewhat larger than the lead–in length scale. In this case, the velocity profile is also continuously varying, but consistent with what might reasonably be expected.

In all the above cases, the walls were modelled as rigid, but in figure 4(g) the lower wall was modelled as a flexible elastic structure that can flex or vibrate at low amplitude as a response to the fluid pressure. The analysis made use of co–simulation methods, with the fluid flow modelled in CFD, in the same way as previously, and the wall modelled in FEA. The structure model is of a 1D shell and represented in the lower two figures as a strip of one element in height. Numerical stability of the combined analysis system was difficult to achieve, except for an elastic structure with high stiffness. In the result presented here, the upper two figures show the pressure and velocity profiles in the fluid.

The higher pressure at the inlet end extends over a larger distance. The velocity profile suggests that on average, across the whole stream, the flow is faster towards the right–hand end of the figure. This can only be explained by a pumping effect driven by the expansion and contraction of the fluid volume and resulting from the lower wall flexure. Thus, when the elasticity of the wall is taken into account, the fluid flow becomes pulsed. This is again, an *emergent* phenomenon, since the set–up of the model did not include any explicitly time varying parameters, yet a time varying result was obtained.

7. Discussion

The observations and analysis results presented in the previous sections have highlighted the importance of understanding the limitations of models, while recognising that the pertinent limitation is removed, models can provide new insights not previously observed, or recognised for their significance.

In the context of education and training, the use of simulations in place of real experiments can save time and cost, as well as greater convenience for both the institution and the student. The two significant disadvantages are, *(i)* that the student misses the interaction with other students or laboratory demonstrators, and *(ii)* that the results of the simulated experiment can be too consistent, and as such do not provide the student with the problem–solving experience of working out why his or her results might not have been the same as those of other students.

In the context of industry, the use of models as a *verification* strategy is highly useful and underpins and de-risks the product development process. At the end of the development process, the product is then *certified* for operational use based on real proof *validation* tests.

Finally, in the case of science research, the use of models is not just restricted to the development of new modelling methods and the validation of those against previously validated models or experiments. A fundamental science opportunity exists in using the models in unexpected ways, to challenge the accepted assumptions that were made in earlier times, when it was not possible to run more complex computational models.

8. Conclusions

Although models are not real, and are only representations of reality, they can be very effective in many usage scenarios. Models should be useful representations reality: to investigate that reality in new ways, the models can be used for *synthetic* experiments. When making use of the excellent modelling capabilities that are available today, one should also challenge the accepted ideas about what features or physics should be included in the model. This is an especially important lesson for students and should be explored through a combination of simulation and physical laboratory practical classes. Finally, one should be aware that models might not always be sufficiently trustworthy, and validation through physical testing will continue to play an important role in engineering product development.

References

[1] Aristotle *Organon* [Trans. Owen O F 1853 (Kindle Edition)]

[2] Bacon R 1267 *Opus Majus* [Ed. Bridges J H 1900 Williams & Norgate]

[3] Bacon F 1620 *Novum Organum* [Trans. & Ed. Urbach P and Gibson J 1994 (Open Court)]

[4] Calero Valdez A and Ziefle M 2018 Human Factors in the Age of Algorithms. Understanding the Human-in-the-loop Using Agent-Based Modeling. In: Meiselwitz G. (eds) *Social Computing and Social Media. Technologies and Analytics. SCSM 2018* Lecture Notes in Computer Science, vol 10914. Springer, Cham.

[5] McMillan A J 2006 Mechanical validation of organic matrix composite aero-engine components *SWERDA Forum on Advanced Composites for Aerospace Applications (Tokyo)*

[6] `www.3ds.com/heart`

[7] Hanna R K and Weinhold I 2020 The Democratization of CFD *Mentor Graphics Corp.* Available from `https://revolutioninsimulation.org/resources/democratizing-simulation/`

[8] UK-SPEC Third Edition 2013 `https://www.engc.org.uk/ukspec.aspx`

[9] The Physics Degree 2011 `https://www.iop.org/sites/default/files/2019-10/the-physics-degree.pdf` `https://www.iop.org/education/support-work-higher-education/degree-accreditation-recognition`

[10] `http://stem.open.ac.uk/study/openstem-labs`

[11] Lyall C, Bruce A, Tait J and Meagher L 2011 *Interdisciplinary Research Journeys: Practical Strategies for Capturing Creativity* (Bloomsbury Open Access).

[12] Horgan J 1997 *The end of science: facing the limits of knowledge in the twilight of the scientific age* (Little, Brown).

[13] Kendell K 2020 Remove fracture mechanics errors *A cracking approach to inventing tough new materials: fracture stranger than friction* Royal Society, London.

[14] CFD Online `www.cfd-online.com/Wiki/Turbulence_modeling`

[15] Batchelor G K 1967 *An Introduction of Fluid Mechanics* (Cambridge University Press)

[16] Chen S and Doolen G D 1998 Lattice Boltzmann method for fluid flows *Annu. Rev. Fluid Mech.* **30** 329–64.

[17] McMillan A J 2012 Material strength knock–down resulting from multiple randomly positioned voids *Journal of Reinforced Plastics & Composites* **31**(1) 13–28.

[18] McMillan A J, Jones R, Peng D and Chechkin G A 2018 A computational study of the influence of surface roughness on material strength *Meccanica* **53** 2411–2436.

[19] McMillan A J and Jones R 2020 Combined effect of both surface finish and sub-surface porosity on component strength under repeated load conditions *Engineering Reports* **2**(9) e12248.

[20] McMillan A J 2017 Computational tribology: modelling the lubrication of rough surface elastic bodies *2017 SIMULIA UK Regional User Meeting, Warrington*

IOP Conf. Series: Materials Science and Engineering 1060 (2021) 012019 doi:10.1088/1757-899X/1060/1/012019

Parametric identification of thermophysical characteristics of heat-protective decaying materials

N S Aldebenev, S Yu Ganigin, D A Demoretsky, A N Diligenskaya and M Yu Livshits*
Samara State Technical University, Molodogvardeyskaya street, 244, Samara, 443100, Russian Federation

* mikhaillivshits@gmail.com

Abstract. The paper presents a method of identifying the thermophysical properties of heat-protective materials of complex multi-component composition used in structures to meet the specified requirements for their resistance to external temperature influence. The research is dedicated to the study of a thermal barrier coating material containing chemically active components that enter into a chemical decomposition process, accompanied by the heat absorption. The equivalent specific heat capacity of the material reflects its thermophysical properties, taking into account the endothermic reaction. The equivalent specific heat capacity is identified by solving the inverse heat conduction problem using the results of temperature measurement in a multi-layer structure element during the thermophysical experiment. Regularization of this incorrect inverse problem is performed when it is transformed into a mini-max parametric problem of semi-infinite optimization. The results of differential thermal analysis and known qualitative laws of endothermic reactions make it possible to narrow the set of feasible solutions to the level of a compact set of piecewise continuous functions of a special structure, which provides a reduction to the parametric optimization problem. The minimax problem is solved by an alternance optimization method. The results obtained confirm the effectiveness of the proposed method for solving applied problems.

1. Introduction

The problem of creating effective thermal protection is relevant for a wide range of applications – from the design of reusable spacecraft carriers and emergency recorders to the creation of specialized containers and safes protected from the high-temperature effects of fire [1–5]. Thermal protection of structures based on degraded heat-shielding coatings (HSC) containing components that enter into a chemical endothermic reaction when the temperature rises above the critical level has become widespread. The heat absorption during the reaction provides the effective protection of the structure inner surface from the effect of high ambient temperatures on its outer surface. The design of such heat-shielding structures is based on mathematical modeling, which in turn requires the determination of the thermophysical characteristics (TPC) of the construction materials, including the TPC of the heat-shielding coating. The paper is dedicated to the problem solution of TPC identification of heat-shielding degraded coatings. Let us briefly consider the theoretical foundations for the subject under consideration.

2. The theoretical foundation of the methodology for the identification problem solution

The Temperature field $T(x,t)$ of a wide class of objects of technological thermal physics within the open part D of the spatial area $\overline{D}, x \in \overline{D}, \overline{D} \in R^1$ or $\overline{D} \in R^2$ with boundary S is described by the inhomogeneous heat conduction equation:

$$L[T(x,t)] = f_1\left(x,t,h_{D_1}(x),h_{D2}(t),h_\beta(x,t,T)\right), x \in D, t \in [0,t^{end}] \quad (1)$$

with boundary

$$\Gamma[T(x,t)] = f_2(h_S(x,t)), x \in S, t > 0 \quad (2)$$

and initial conditions

$$N[T(x,t)] = h_{\overline{D}_0}(x), x \in \overline{D}, t = 0. \quad (3)$$

Here t is the time, $h_{D_1}(x), x \in D$ and $h_{D2}(t)$ are the internal heat sources (sinks), $h_S(x,t), x \in S$ is the space-time boundary impact on an area D from the side of its environment, $h_{\overline{D}_0}(x), t = 0, x \in \overline{D}$ is the initial condition, $h_\beta(x,t,T)$ is the characteristics or parameters expressing thermophysical properties and generally depending on the spatial coordinate x, time t and/or temperature $T(x,t)$; L, Γ, N are given linear or nonlinear (in general) differential operators; f_1, f_2 are known functions of their arguments that meet the ordinary smoothness requirements. The solution to the inverse heat conduction problem (IHCP) involves the determination of the vector characteristic $h = \left\{h_{D_1}(x),h_{D2}(t),h_S(x,t),h_{\overline{D}_0}(x)\right\}$ in the boundary value problem (1)-(3) based on *a priori* or experimental information on the temperature distribution $T^*(\psi)$ by a temporal $t \in [0,t^{end}]$ argument in some measurable area. As an area of temperature measurement in the basic formulations of typical IHCP, a finite number (usually small, including equal to one) of special coordinates $x^* \in \overline{D} \in \Omega_1$, corresponding to the installation sites of temperature sensors is considered [6, 7].

Hereafter we restrict ourselves to the condition that the desired characteristics belong to the corresponding sets

$$h_{D_1}(x) \in H_{D_1}(x), h_{D2}(t) \in H_{D2}(t), h_S(x,t) \in H_S(x,t), h_{\overline{D}_0}(x) \in H_{\overline{D}_0}(x) \quad (4)$$

of piecewise smooth functions. Then an optimal (variational) problem is formulated with an optimality criterion with respect to the temperature distribution $T\left(x,t,h_{D_1}(x),h_{D2}(t),h_S(x,t),h_{\overline{D}_0}(x)\right)$, [8,9]:

$$\psi = t: I_1 = \max_t|T(x^*,t,h) - T^*(x^*,t)| \to \min_h. \quad (5)$$

Problem (5) is included in a wide range of typical statements of IHCP: boundary, internal, initial and coefficient inverse heat conduction problems, when unknown boundary conditions, the space-time distribution function of internal heat sources, the initial state or thermophysical characteristics of the object are subject to identification.

Taking into account the specific features of a particular applied problem and *a priori* information about an object, as a rule, allows one to reduce the identified characteristics to a parametric form in a selected set$H_{\overline{D}_0}(x)$, $H_{D2}(t)$ of a given class functions. The most typical classes are constant functions, functions that are continuous with their first derivative, polynomial functions, or functions of special structure. More universal parameterization methods are possible [8-12].

Parametric representation of the identified characteristic $h = \left\{h_{D_1}(x),h_{D2}(t),h_S(x,t),h_{\overline{D}_0}(x)\right\} = h(\Delta)$ allows to obtain a parametric form of the temperature distribution $T(x,t,h_{D_1}(x),h_{D2}(t),h_S(x,t)) = T(\Delta)$. Thus, the desired characteristic and the corresponding temperature distribution are uniquely determined by the corresponding vector of parameters $\Delta = (\Delta_1, \Delta_2 ... \Delta_n)$.

The found parametric representation of the temperature field $T(\Delta)$ leads to the original inverse problem of a special nonsmooth mathematical programming problem with respect to the desired vector of parameters.

Temperature measurement results are used $T^*(x^*, t)$ on the identification time interval $t \in [0, t^{end}]$ at a certain fixed point x^*, corresponding to the spatial coordinate of the temperature sensor.

Mathematical model (1) – (3) of the temperature field, taking into account the parametrized shape of the identified influences, makes it possible to obtain an expression for the temperature on the identification interval $T(x^*, t, \Delta)$, characterized by the vector $\Delta = (\Delta_n), n = \overline{1, N}$ on a set of parameters G_N. Specific type of parametric description $T(x^*, t, \Delta)$ and vector dimension value N of parameters are determined by the representation form of the general solution to the boundary value problem and the selected class of solutions in the form of a given compact set.

Thus, the preliminary parameterization of the desired characteristic $h \in P$ on a compact set P of piecewise smooth functions of time, spatial coordinate, or temperature in the class P and the parametric representation of the temperature field $T(\Delta)$ provide a reduction of the original problem (5) to a special nonsmooth mathematical programming problem with respect to the desired vector of parameters Δ

$$\psi = t: I_2(\Delta) = \max_t |T(x^*, t, \Delta) - T^*(x^*, t)| \to \min_\Delta. \tag{6}$$

3. The alternance method in inverse heat conduction problem

The problem of mathematical programming (6) can be solved with the use of the general scheme of the alternance method based on the special properties of optimal solutions to the problem, similar to the properties of nonlinear Chebyshev approximations. On the basis of these properties a closed system of ratios with respect to the parameters $\Delta = \Delta_0$ of the optimal problem solution is composed, which is further transformed into a closed system of calculation equations of desired values and unknown functional values $I_4(\Delta_0)$

$$\begin{gathered} T(x^*, t_q^0, \Delta_0) - T^*(x^*, t_q^0) = \pm(-1)^{q+1} I_4(\Delta_0), q = \overline{1, N} \\ \frac{\partial}{\partial t}\left(T(x^*, t_q^0, \Delta_0) - T^*(x^*, t_q^0)\right) = 0. \end{gathered} \tag{7}$$

According to the system of calculated ratios (14), threshold criteria $|T(x^*, t_q^0, \Delta_0) - T^*(x^*, t_q^0)|$, equal to $I_4(\Delta_0)$, are reached only at individual points of the alternance t_q^0.

As a result of solving the equation system (7), the desired vector of parameters is found Δ_0, which allows to implement the parametric identification of unknown quantities h in a parametrically specified class of functions. Found number values N and vector of parameters Δ implement the approximation of the desired characteristic with the required accuracy at the identification interval.

As the number increases $N = 1,2, ...$ a sequence of functional values is made I_2^N (6), converging, under ideal conditions, to the minimum possible value $I_2^\infty = I_2^{inf}$. In this case, for any value N it is ensured that the solutions belong to the selected compact set and, thus, the correct statement of problem (6) is preserved.

In practical applications, the sequence of parametric optimization problems (6) is solved for increasing values $N = 1,2, ... N^*$ up to a certain value N^*, that ensures the accuracy requirements, according to a given value of the functional $I_2^0(\Delta^0)$.

4. The problem of heat-shielding materials TPC identification

The stated general method for solving IHCP is used to identify the thermophysical properties of a heat-shielding material having a complex multicomponent composition.

The design and optimization of the weight and size characteristics of a heat-shielding structure is based on the thermophysical characteristics of materials [1–5].

Modern approaches to the creation of effective thermal protection coatings provide for the use of chemically unstable components in the HSC, the heating of which causes their interaction accompanied by an endothermic reaction.

In this case, the amount of heat to be absorbed by the HSC during the endothermic reaction and providing the specified requirements for the fire resistance of the structure is determined by the decision of the corresponding IHCP [6,7]. To formalize this problem, let us introduce into consideration the equivalent time-varying heat capacity of the HSC $C^{eq}(T)$, providing the equivalent real heat content of the HSC, when it is heated, accompanied by an endothermic reaction and a change in the phase composition of the components.

The article reflects the study of the thermophysical characteristics of HSC, which has a complex multicomponent composition, which includes, in addition to the basic components, two chemically active fillers that degrade with the effect of heat absorption at temperatures120-150 °C and 300-320 °C respectively.

HSC, due to its efficiency, is not a thermally thin body; therefore, it is difficult to identify its TPC, because the process of decomposition of unstable components begins in the HSC layer not simultaneously, but gradually as the critical temperature for each filler is reached during the heating of the layer. Due to the low thermal conductivity of the HSC composition, large temperature gradients appear in this layer, which leads to the fact that, during heating, the HSC material has a different phase composition and, as a consequence, different thermophysical characteristics.

The inverse heat conduction problem for determining the TPC of the investigated HSC was solved on the basis of experimental information on the temperature distribution in the prototypes obtained as a result of a series of experiments.

5. Differential thermal analysis

A materials laboratory testing [13,14] of the HSC fillers for each of the two stages of the endothermic reaction was carried out separately.

As endothermic fillers by differential thermal analysis (DTA) on a DTA GSM instrument, which determines the temperature difference between two test components that have different decomposition temperatures. For the analysis, weighed portions of 50 mg of the test substances were formed. The determination of the endothermic effect was carried out using thermocouples of the DTA instrument placed in test tubes inside the test substances. Temperature deviation is monitored during tube heating T_{sub} inside the substance from the substance T_{tb} of the outer tubes. The experimental setup is shown in figure 1. The experiment stops when the temperature inside the substance is equal to the temperature of the surface of the tube, which means the end of the endothermic reaction. The results of the experiment were recorded using the software of DTA GSM and are presented in figure 2 for each of the reaction stages. The axis of ordinates shows the deviation of the temperature$\Delta T = T_{\text{sub}} - T_{\text{tb}}$ inside the sample T_{sub} from the temperature of the tube wall T_{tb}. The axis of abscissa shows the values of the tube wall temperature T_{tb}, which increases at a rate 5 ^{0} C /min (0,083 K/sec).

The temperature curve changes of the studied composition of the first reaction stage are shown in figure 2a. It shows that the endothermic reaction in the substance begins 22 minutes after the heating start and proceeds in the temperature range ΔT =112-133 ^{0}C during 4,2 minutes.

The maximum endothermic effect (deviation of the temperature in the substance from the heating temperature) is observed at T_{tb}=120 ^{0}C (ΔT = 6,8 ^{0}C).

The specific endothermic effect of the first stage chemical reaction is: $\Delta H^0_{\text{reaction}} = 97{,}89$ kJ (for the decomposition of 1 mole of a substance, 97.89 kJ of heat is spent).

The temperature curve changes in the medium of the studied composition of the second reaction stage are shown in figure 2b. It shows that the endothermic reaction in the test substance begins 60 minutes after the heating start and proceeds in the temperature range T_{tb}=300…382 ^{0}C during 16,4 minutes.

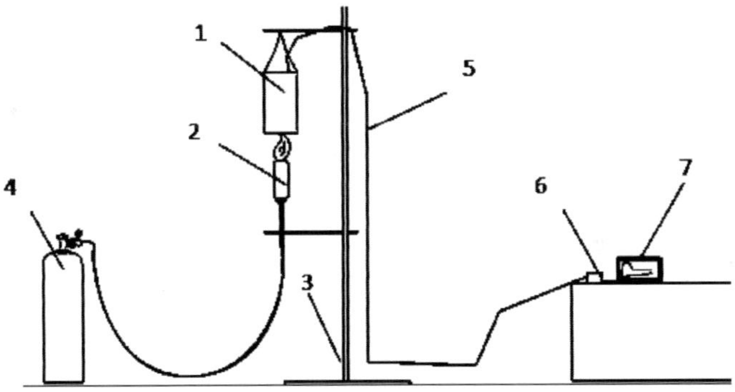

1 – test prototype
2 – gas-burner
3 – stand rod
4 – gas cylinder
5 – connecting leads
6 – analog-digital converter
7 – computer

Figure 1. – Laboratory setup diagram:

The maximum endothermic effect (deviation of the temperature in the substance from the heating temperature) is observed at T_{tb}=365 ^{0}C (ΔT = 19,8 °C).

The specific endothermic effect of the second stage chemical reaction is: $\Delta H^0_{reaction}$ = 175,9 kJ (for the decomposition of 1 mole of a substance 175,9 kJ of heat is spent).

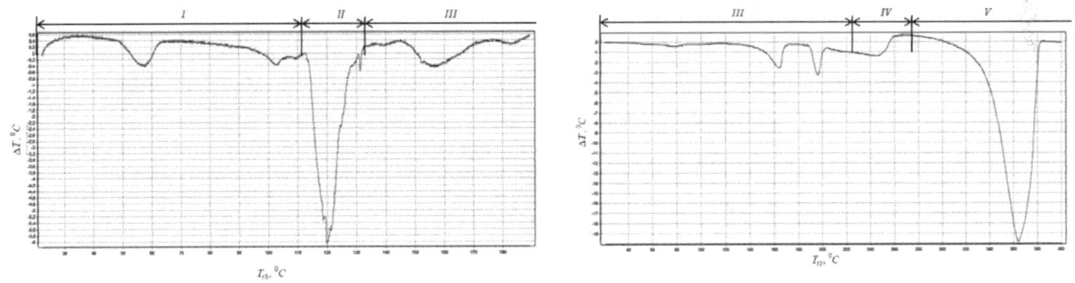

a b

Figure 2. DTA results for: (a) the first, and, (b) the second chemically active components of the HSC

6. Thermophysical experiment

For a preliminary assessment of the effectiveness of the use of HSC, a series of thermophysical experiments was carried out in the thermal protection structure using three variants of the model design. In all cases, the test models were multi-layer structures of steel plates and a thin air gap and differed in the filling of the HSC layer or its absence.

As a model, an experimental element of a real multi-layer structure with dimensions is studied: $2H \times L$, $L = 0.263\ m$, $H = 0.075\ m$, $x_1 = 0.185\ m$, $x_2 = 0.23\ m$, $x_3 = 0.239\ m$, $x_4 = 0.242\ m$, $x_5 = 0.251\ m$, $r = 0.038\ m$, $s = 0.02\ m$, (figure 3 a, b). The outer part of the structure consists of two steel plates separated by a thin air gap and connected by a longitudinal stiffener. The prototype from the lateral surfaces is insulated with a layer of asbestos, and from the outer surface it is exposed to the action of an open flame. A multicomponent heat-shielding coating is applied to the inner shell and must provide the specified requirements for resistance to external temperature effects. The maximum permissible temperature $T(x, y, t)$ on the outer surface of the containment should not exceed 180 °C for reasons of structural reliability, fire and explosion safety of the contents protected by the structure. To control the temperature conditions, the temperature of the outer surface is measured (point A in figure 3), as well as the temperature at point B of the inner wall of the shell using thermocouples (accuracy class 0.2).

The first design option does not contain a heat-shielding coating (figure 3a), the second and third design options contain HSC of different composition (figure 3b). In the second variant, an inert material is used as HSC, which does not contain endothermic components, and in the third, a

IOP Conf. Series: Materials Science and Engineering 1060 (2021) 012019 doi:10.1088/1757-899X/1060/1/012019

multicomponent composition of HSC is used, containing chemically active fillers that enter into an endothermic decomposition reaction.

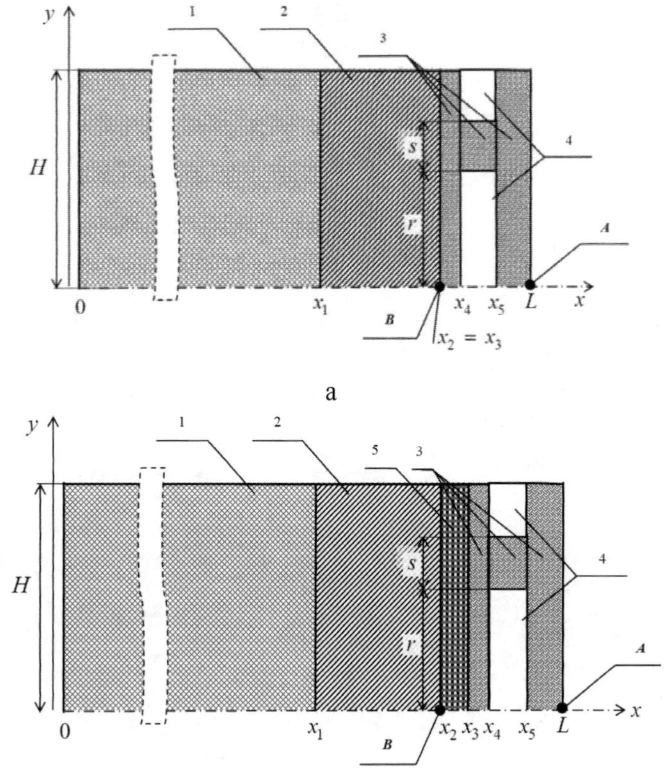

b

1 – contents of the thermal protection structure
2 – inner shell
3 – outer shell
4 – air gap
5 – HSC, *A* and *B* – thermocouple locations.

Figure 3. Test layouts in a preliminary thermophysical experiment:
a – without HSC
b – with HSC

In the course of the experiments, the outer surface of the model was under the conditions of an open flame, the temperature of which at the initial stage was maintained at about 900 °C, after which the experimental conditions were tightened, and the flame temperature exceeded 1200 °C (figure 4).

Heating continued until reaching the maximum permissible temperature of 180 °C at point *B*). The experimental results are shown in figure 4. The results of the experiment make it possible to visually assess the effectiveness of the HSC. The introduction of decomposing components (curve 5) into the composition, which is accompanied by heat absorption, makes it possible to increase the time to reach the critical temperature of 180 °C at the control point. Thus, the results of a series of thermophysical experiments have confirmed the effectiveness of using chemically active materials as fillers for HSC and substantiate the use of a chemically active HSC with two decomposing components in a heat-shielding structure.

On the basis of DTA (see Figure 2) and the results of thermophysical experiments (see figure 4), the possible stages of the HSC state are identified:

- I - initial state characterized by constant specific density $\rho_5 = \rho_5^{(I)} = \text{const}$, heat-conduction coefficient $\lambda_5 = \lambda_5^{(I)} = \text{const}$ and specific heat capacity $C_5 = C_5^{(I)} = \text{const}$ of thermal protective composition, which determine the regular temperature condition of the sample;
- II - the first stage of the endothermic reaction, at which there is a sharp increase in the specific heat capacity $C_5 = C_5^{(II)}(t)$, followed by a sharp decline;
- III - an intermediate steady state of HSC after the end of the decomposition stage of the first endothermic filler and the beginning of the decomposition of the second active component.

Here, due to a change in its phase composition, the component under study is characterized by other, different from the previous, constant values of the specific heat capacity $C_5 = C_5^{(III)} =$ const, density $\rho_5 = \rho_5^{(III)} =$ const and heat-conduction coefficient $\lambda_5 = \lambda_5^{(III)} =$ const;

- IV – the process of decomposition of the second chemically active component, accompanied first by an increase and then a decrease in the value of the equivalent specific heat $C_5 = C_5^{(IV)}(t)$;
- V – final state described by constant values of thermophysical characteristics $C_5 = C_5^{(V)} =$ const, $\rho_5 = \rho_5^{(V)} =$ const, $\lambda_5 = \lambda_5^{(V)} =$ const and established upon completion of the endothermic reaction.

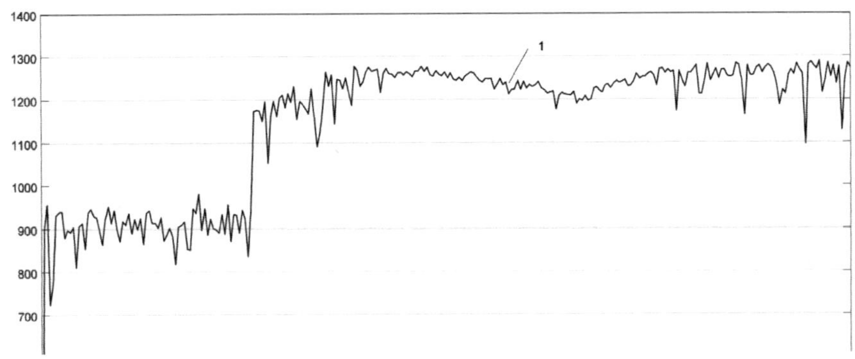

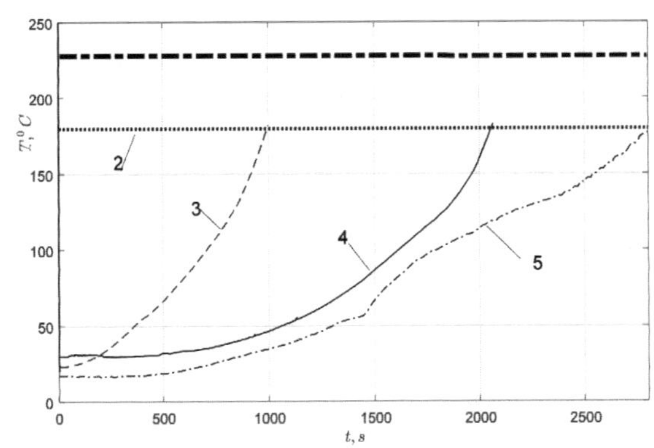

1 – Temperature at point A (flame area)
2 – Maximum permissible temperature
3 – Temperature at test point B without HSC
4 – Temperature at test point B when using non-degradable HSC
5 – Temperature at control point B when using a reactive HSC.

Figure 4. Results of thermophysical experiment

7. Mathematical model of the HSC thermal conductivity process

To identify the TPC of HSC, it is necessary to concretize the general mathematical model of the boundary value problem (1) - (3). For the temperature distribution in the test sample, the boundary value problem (1) – (3) is considered, which takes the form of a two-dimensional boundary value problem of unsteady heat conduction, containing a system of two-dimensional heat conduction equations for temperature $T_i(x, y, t), i = \overline{1,5}$ in every i structural element (figure 3b), where the third layer (steel shell) contains three elements $(x, y) \in \Omega_3 \div \{3a, 3b, 3c\}$, and the fourth layer consists of two sections of the air gap $(x, y) \in \Omega_4 \div \{4a, 4b\}$. Every i element ($i = \overline{1,5}$) is characterized by specific heat capacity C_i, material density ρ_i, heat-conduction coefficient λ_i and the corresponding boundary conditions. Thus, the mathematical model for temperature $T_i(x, y, t)$ is described by the system of Fourier equations:

$$C_i\rho_i\frac{\partial Ti}{\partial t}=\lambda_i\left(\frac{\partial^2 T_i}{\partial x^2}+\frac{\partial^2 T_i}{\partial y^2}\right), i=\overline{1,5}: 0<t\le t^{end} \quad (8)$$

$i=1: (x,y)\in[0,x_1]\times[0,H]$ $\quad i=2: (x,y)\in[x_1,x_2]\times[0,H]$ $\quad i=5: (x,y)\in[x_2,x_3]\times[0,H]$
$i=3: (x,y)\in[x_3,x_4]\times[0,H]$ $\quad (x,y)\in[x_4,x_5]\times[r,r+s]$ $\quad (x,y)\in[x_5,L]\times[0,H]$
$i=4: (x,y)\in[x_4,x_5]\times[0,r]$ $\quad (x,y)\in[x_4,x_5]\times[r+s,H]$

Initial conditions (3) for all structural elements are taken uniform and constant:

$$T(x,y,0)=T_0=const; 0\le x\le L, 0\le y\le H. \quad (9)$$

Boundary conditions are formulated. Axial symmetry condition is specified on the boundary $y=0$,

$$\frac{\partial T_i(x,0,t)}{\partial y}=0 \quad (10)$$

$i=1;2;3: (x,y)\in[x_3,x_4]\times[0,H]$ $\quad [x_5,L]\times[0,H]$
$i=4: (x,y)\in[x_4,x_5]\times[0,r]$ $\quad i=5$

The high-temperature effect of an open flame on the outer surface is given in the form of a complex heat exchange containing radiation and convective components

$$\lambda_3\frac{\partial T_3(L,y,t)}{\partial x}=\varepsilon\sigma\left({T_3}^4(L,y,t)-T_{am}^4(t)\right)+\alpha\big(T_3(L,y,t)-T_{am}(t)\big) \qquad 0<y<H \quad (11)$$

where ε – emissivity factor, σ – Stefan-Boltzmann constant, α – heat-exchange coefficient, $T_{am}(t)$ – environment temperature.

The boundary conditions on the remaining external surfaces correspond to the thermal insulation conditions:

$$\frac{\partial T_1(0,y,t)}{\partial x}=0 \qquad 0<y<H, \frac{\partial T_i(x,H,t)}{\partial y}=0 \quad (12)$$

$i=1;2;3: (x,y)\in[x_3,x_4]\times[0,H]$ $\quad [x_5,L]\times[0,H]$
$i=4: (x,y)\in[x_4,x_5]\times[r+s,H]$ $\quad i=5$

On the inner contact surfaces of the layers, the conjugation conditions formulated for spatial variable x are satisfied:

$$T_i(x_k,y,t)=T_j(x_k,y,t) \qquad -\lambda_i\frac{\partial T_i(x_k,y,t)}{\partial x}=-\lambda_j\frac{\partial T_j(x_k,y,t)}{\partial x} \quad (13)$$

$i=1, j=2, k=1, 0\le y\le H$ $\quad i=2, j=5, k=2, 0\le y\le H$
$i=5, j=3, k=3, 0\le y\le H$
$i=3, j=4, k=4, 0\le y\le r$ $\quad i=3, j=4, k=4, r+s\le y\le H$
$i=3, j=4, k=5, 0\le y\le r$ $\quad i=3, j=4, k=5, r+s\le y\le H$

and, similarly, for variable y:

$$T_3(x,r,t) = T_4(x,r,t) \qquad -\lambda_3 \frac{\partial T_3(x,r,t)}{\partial y} = -\lambda_4 \frac{\partial T_4(x,r,t)}{\partial y} \tag{14}$$

$$x_4 \le x \le x_5 \qquad \begin{matrix} i = 3: (x,y) \in [x_4, x_5] \times [r, r+s] \\ i = 4: (x,y) \in [x_4, x_5] \times [0, r] \end{matrix}$$

$$T_3(x,r+s,t) = T_4(x,r+s,t) \qquad -\lambda_3 \frac{\partial T_3(x,r+s,t)}{\partial y} = -\lambda_4 \frac{\partial T_4(x,r+s,t)}{\partial y}$$

$$x_4 \le x \le x_5 \qquad \begin{matrix} i = 3: (x,y) \in [x_4, x_5] \times [r, r+s] \\ i = 4: (x,y) \in [x_4, x_5] \times [r+s, H] \end{matrix}$$

The nonlinear direct boundary value problem of heat conduction (8) – (14) can be solved by known numerical methods if all the thermophysical characteristics included in it are known. The solution of this boundary value problem allows, using the appropriate optimization procedures, to determine the design parameters that provide effective thermal protection.

8. Thermophysical characteristics identification based on the IHCP solution

On the basis of the mathematical model (8) – (14), the task of designing a structure with given heat-shielding properties is posed, the solution of which provides for the stage of identifying the thermophysical parameters of the HSC. To solve the identification problem, IHCP is considered on the basis of model (8) - (14), where the equivalent specific heat capacity is to be determined $C_5^{eq}(T)$ (i=5) HSC (element 5 in figure 3b), reflecting its heat content, taking into account endothermic reactions, based on the experimentally obtained temperature $T(x^*, y^*, t)$ at point B $x = x^*, y = y^*$. TPC of other structural elements is considered to be known with a sufficient degree of reliability. The problem was solved according to the general methodology described above. The inverse problem is formulated in a variational setting as an optimization problem and provides for the search for the function $C_5^{eq}(T)$, that minimizes the discrepancy between the experimental data $T^*(t)$ (obtained at point B) and the numerical solution of the boundary value problem (8) – (14) obtained at the same point B [8,9].

A set V of physically realizable functions is considered under constraints

$$C_5^{eq}(T) \in V, T \in [t^{beg}, t^{end}] \tag{15}$$

and the problem is posed of finding an equivalent specific heat capacity $C_5 = C_5^{eq}(T)$, that ensures the fulfillment $(0, t^{end})$ of the minimax relation on the identification interval (6), which in the considered case will take the form:

$$I(C_5^{eq}) = \max_{t \in [0, t^{end}]} \left| T^*(x^*, y^*, t) - T(x^*, y^*, C_5^{eq}(T), t) \right| \to \min_{C_5^{eq}(T) \in V}. \tag{16}$$

The search for solutions is carried out on the set V of piecewise continuous measurable functions with the following data: heating end time $t^{end} = 2820$sec; operating temperature range of HSC $[T^{beg}, T^{end}] \approx [20^0 C, 800^0\ C]$), where $T^*(x^*, y^*, t)$ — experimental temperature values at the sample point (x^*, y^*); $T(x^*, y^*, C_5^{eq}(T), t)$ — the values obtained by numerical solution to the boundary value problem (8) – (14) at the same point.

At the stages of decomposition of the components, it is advisable to approximate the desired characteristic $C_5^{eq}(T)$ (see figure 2) by a shifted Gaussian function $c + k\, exp - (T - T^*)^2 \alpha^{-1}$, on the basis of which a set of piecewise continuous functions of the following form is considered:

$$C_5^{eq}(T) = \begin{cases} c_{50}, T \in \left[T^{beg}, T^{(1)}\right] \\ c_{51} + k_1\, exp(-(T - T_1{}^*)^2)\, \alpha_1^{-1}, T \in \left[T^{(1)}, T^{(2)}\right] \\ c_{52} + k_2\, exp(-(T - T_2{}^*)^2)\, \alpha_2^{-1}, T \in \left[T^{(2)}, T^{end}\right]. \end{cases} \quad (17)$$

Such an approximation makes it possible, during formalization, to combine the stage of decomposition of the filler, accompanied by a significant change $C_5^{eq}(T)$, with subsequent stages of permanence $C_5^{eq}(T) = c_{5j} = \text{const}, \mathrm{j} = \overline{0,2}$.

Thus, for an effective solution of the IHCP, a parametrization is performed, which, in accordance with the general methodology, narrows the area V of the existence of the sought functions. The values of temperature $T^{(1)}$ and $T^{(2)}$ are determined on the basis of a general description of the desired function as the moments of the onset of a sharp increase $C_5^{eq}(T)$ in the active stages of the endothermic reaction. The value T^{end} is specified according to the calculated temperature field for a point $(x_3, 0)$ on the outer boundary of the HSC. Thus, the desired characteristic $C_5^{eq}(T)$ is uniquely determined by the vector of unknown parameters $\Delta = \Delta^{(N)} = [c_{50}, c_{51}, c_{52}, k_1, k_2, T_1, T_2, \alpha_1, \alpha_2]$ with the dimension $N = 9$.

Dependency approximation (17) $C_5^{eq}(T)$ in the selected class $V(T)$ of functions accurate to the vector of parameters Δ $\boldsymbol{\Delta}$ provides the parameterization of the temperature field $T_i(x, y, \Delta, t), i = \overline{1,5}$ as a function of the argument $C_5^{eq}(\Delta)$ (17) in the numerical solution of the problem (8) – (14), by determining the temperature as a function of the vector of parameters Δ at any fixed point(x^*, y^*). Parametric representation $T_i(x, y, \Delta, t)$ allows you to go from the original IHCP setting (16) to an equivalent mathematical programming problem of the form (6) relative to the vector Δ, which in this case will take the form

$$I^*(\Delta) = \max_{t \in [0, t^{end}]} |T^*(x^*, y^* t) - T(x^*, y^* \Delta, t)| \rightarrow \min_{\Delta} \quad (18)$$

The solution of the mathematical programming problem (15), (17), (18) is performed by the alternance optimization method.

The main basic relations of the alternance method (7) [10] put in correspondence the number R of alternance points, in which the limiting relations are fulfilled (18), which have alternating character, with the number of unknown parameters, which include parameters Δ of the desired characteristic and the value of the maximum deviation $I(\Delta)$. On this basis, a closed system of algebraic equations relative to the unknown components of the vector is compiled $\Delta = \Delta^{(N)} = [c_{50}, c_{51}, c_{52}, k_1, k_2, T_1, T_2, \alpha_1, \alpha_2]$ as a result of the solution of which the required parameters are found that ensure the minimization of the functional $I^*(\Delta)$ (18). Often, when solving applied problems, the requirement to achieve a global minimum $I^*(\Delta)$ turns out to be excessive and can be replaced by the condition:

$$\max_{t \in [0, t^{end}]} |T^*(x^*, y^*, t) - T(x^*, y^*, \Delta, t)| \leq \varepsilon_{dop}, \quad (19)$$

where ε_{dop} – predetermined permissible temperature discrepancy at the control point (x^*, y^*).

As a result of solving a mathematical programming problem (15), (17), (19) based on gradient numerical methods using special software the vector of parameters was determined Δ^0, that unambiguously identifies the equivalent specific heat capacity of the HSC, the results of which are presented in figures 5, 6.

The specific heat capacity function of the HSC reconstructed on the basis of the values of the parameters vector $\Delta^0 = \Delta^{(9)} = [c_{50}, c_{51}, c_{52}, k_1, k_2, T_1, T_2, \alpha_1, \alpha_2]$ found as a result of solving the problem (15), (17), (19) is shown in figure 5.

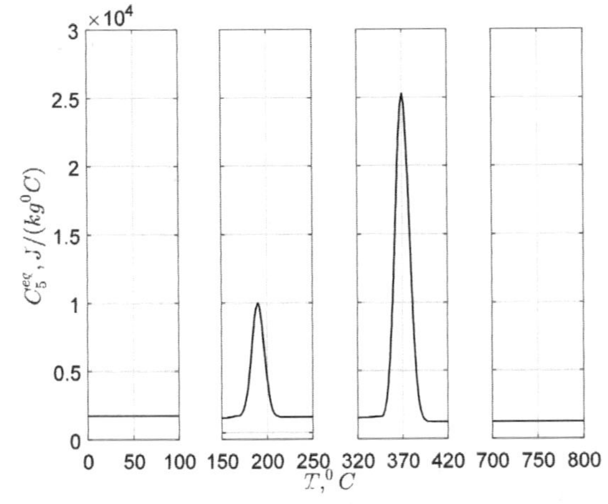

Figure 5. Dependence of specific heat capacity on temperature $C_5^{eq}(T)$

The obtained parametric representation of the desired function $C_5^{eq}(T)$ allows to carry out thermophysical studies on the studied identification interval, in which the absolute error in the deviation of the temperature distribution does not exceed 7,5 °C and is 4,16 %, which is quite satisfactory for engineering calculations (figure 6a). The permissible deviation of the calculated temperature from the experimental one obtained at the control point B is set from the conditions of technical benefit and is within 10 %.

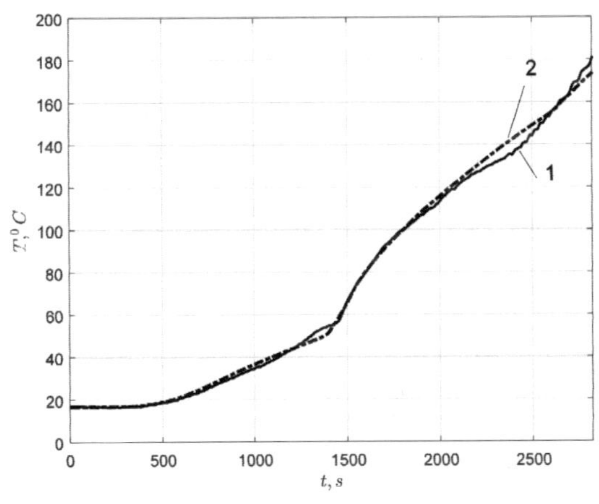

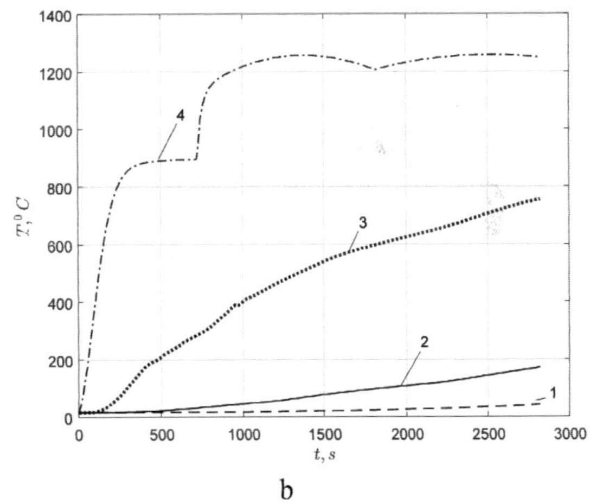

Figure 6. IHCP solution results
a: Experimental (1) and calculated (2) temperature distribution at point B $(x_2, 0)$
b: 1, 2, 3, 4 – the results of temperature calculations at points with coordinates $(x_1, 0), (x_2, 0), (x_3, 0), (L, 0)$ respectively

The calculated temperature curves in the test sample are shown in Figure 6b and confirm the efficiency of the HSC.

9. Conclusion

A solution to the problem of identifying the thermophysical characteristics of heat-shielding coatings with decomposing components is obtained. The use of the alternance method for minimax optimization of the error in the deviation of the calculated temperature from that obtained experimentally at the control points of the heat-protective coating, carried out in a uniform estimation metric in a given class of functions, provides high identification accuracy.

The obtained results can be used to solve urgent problems of designing thermal protection containing chemically active components.

Acknowledgements
This work was supported by the Russian Foundation for Basic Research (projects No. 20-08-00240, No. 18-08-00565)

References

[1] Okhapkin A S 1985 *J. Eng. Phys.* **49**, 6, 1469–1473
[2] Alifanov O M, Cherepanov V V 2010 *J. Eng. Phys.* **83**, 4, 770–782
[3] Alifanov, O M, Budnik S A, Nenarokomov A V, Netelev A V, Titov D M 2010 *The 61th International Astronautical Congress* (Prague, Czech Republic), 10
[4] Dec J A, Braun R D, Lamb B 2012 *J Thermophys Heat Tr.* **26**, 2, 201–212
[5] Reznik S V, Prosuntsov P V, Mikhailovskii K V 2015 *J. Eng. Phys.* **88**, 3, 594–601
[6] Alifanov, O M 1988 *Inverse Heat Transfer Problems* (Moscow: Mashinostroenie)
[7] Beck J.V, Blackwell B and St. Clair C R 1985 *Inverse Heat Conduction: Ill- posed Problems* (New York, John Wiley and Sons)
[8] Rapoport E Ya and Pleshivtseva Yu E 2009 *RF Avtometriya* **45**, 5, 103–112
[9] Diligenskaya A N, Rapoport E Y 2016 *J. Eng. Phys.* **89**, 4, 1008–1013
[10] Rapoport E Ya 2000 *Alternance Method in Optimization Applications* (Moscow: Nauka)
[11] Livshits M Y, Nenashev A V, Borodulin B B 2020 *Studies in Systems, Decision and Control*, **260**, 79–88
[12] Livshits M Y, Nenashev A V 2019 *RF. J. Samara State Tech. Univ., Ser. Phys. Math. Sci.* **23**, 2, 361–377
[13] Fatima Zohra El Wardi , Abou-bakr Cherki , Soumia Mounir , Abdelhamid Khabbazi , Youssef Maaloufa 2019 *Energy Procedia* **157**, 480–491
[14] Ying Li, Lu Zhang, Rujie He, Yongbin Ma, Keqiang Zhang, Xuejian Bai, Baosheng Xu, Yanfei Chen 2019 *Aerospace Science and Technology* **91**, 607-616

Effects of Surface Reactions during Friction on Tribological Characteristics of PTFE-Muscovite Composites

S A Sleptsova[1], A A Okhlopkova[1,2], Iu V Kapitonova[1], N N Lazareva[1], A P Vasiliev[1], L A Nikiforov[1], P N Tarasova[1], S Laukkanen[1]

[1]Chemistry Department, Institute of Natural Sciences, North-Eastern Federal University, 58, Belinsky Str., Yakutsk 677000, Russian Federation

[2]Material Science Laboratory, Institute of Oil and Gas Problems of the Siberian Branch of the RAS, 1, Oktyabrskaya str., Yakutsk, Russian Federation

E-mail: lazareva-nadia92@mail.ru

Abstract. This work presents the results of tribological and structural investigations of polymer composites based on polytetrafluoroethylene and muscovite. It is shown that the wear resistance of the composites increased up to 675 times in comparison with the initial polymer. The increase of wear resistance is attributed to the occurrence of tribo-oxidation processes on the friction surface, which have been confirmed by various investigations. Thus, IR spectroscopy showed that carboxylate anions are formed on the friction surface of composites during friction processes. An increase in oxygen and a decrease in carbon and fluorine concentrations on the composite surface have been detected after friction using the EDS analysis. The results of XPS have confirmed the destruction of C-F bonds and the formation of new oxygen-containing compounds. These research methods indicated the occurrence of tribo-oxidative processes during friction.

1. Introduction

Nowadays polymers are widely used in various branches of human life, including the engineering industry, due to their strength, low density, weather, and corrosion resistance. One of the most widely used polymers is polytetrafluoroethylene (PTFE). PTFE is usually used as a solid lubricant and for manufacturing parts of friction units due to the low friction coefficient, corrosion, and thermal resistance. However, initial PTFE has some disadvantages such as poor wear resistance and creep. To improve its properties, fillers are added to the initial PTFE. Fillers reduce the slippage between the polymer lamellae, thus reduce the wear of materials [1]. Also, it is considered that the transfer film between the surface of the PTFE composite and counter body surface contributes to wear reduction.

The previously mentioned hypothesis is the main theory of wear reduction, but recently the new assumption has been suggested. According to this, the tribo-performance of PTFE composites is closely related to the chemical reactions between polymer surface and counter-body, the ambient air, the formation of metal fluoride bonding. Many researchers have found pieces of evidence in their studies [2-4]. At the same time, there is not enough discussion about the effect of the tribo-chemical reactions on the tribo-performance of the PTFE based composites.

Layered silicates as fillers have been used in our previous studies to investigate the role of filler on tribo-performance and tribo-chemical reactions [5]. Hence, the PTFE/layered silicate composites are used in this work as the continuation of studies. The influence of tribo-chemical reactions on the tribo-performance will be analyzed and discussed. The objectives of this study are to investigate the effects of muscovite on the tribological behavior and wear mechanism of PTFE.

2. Experimental

The polymer matrix was polytetrafluoroethylene (PTFE) from Perm manufacture (Russia), trademark PN-90 (GOST 10007-80) with an average molecular weight 500000, an average particle size 90 µm, density 2150-2260 kg/m^3. The analogue of PTFE PN-90 is Tarflen® SM-2 PTFE. PTFE was filled with natural layered silicate-muscovite $KAl_2[AlSi_3O_{10}](OH)_2$ was from the Kovdor deposit (Murmansk) with a Mohs hardness 2.5 – 3 and density 2760 – 3100 kg/m^3. Magnesium spinel (MS) $MgAl_2O_4$ was synthesized at the Institute of Solid-State Chemistry and Mechanochemistry of the Siberian Branch of the Russian Academy of Science (Russia) using a mechanochemical method. MS has an average particle size of 70-80 nm and a specific surface of 170 m^2/g and was used as a functional additive.

The preparation of muscovite included its drying at a temperature of 120 °C for 4 h. Then it was mechanically activated in the "Activator 2S" planetary mill (Russia) for 2 min. PTFE powder was provided with a particle size of approximately 90 mesh number. PTFE, muscovite, and magnesium spinel powders were mixed using a paddle mixer at 3000 rpm to make the mixture uniform and dispersive. After mixing, the uniform mixture was molded at 50MPa at room temperature for 2 min to prepare a sample. Then, samples were sintered in the laboratory oven at temperature 300-375 °C and cooled with rate of 0.03 °C / s. The amount of muscovite was varied in the interval 1 – 5 wt. %, and the amount of magnesium spinel was varied from 0.5 to 1 wt. %.

3. Results

Tribological properties of polymer composites were determined with the UMT-3 tribometer (CETR) at room temperature using a pin-on-disc friction scheme. Cylindrical specimens (10 mm diameter and 20 mm height) and a steel disk (brand 40X, HRC hardness 45–50) were used for the test. The applied load was 2 MPa, sliding speed 0.2 m/s, test duration 3 hours. Figure 1 shows the schematic representation of the pin-on-disc friction scheme. The rate of mass wear was evaluated by the weight loss of samples per time unit.

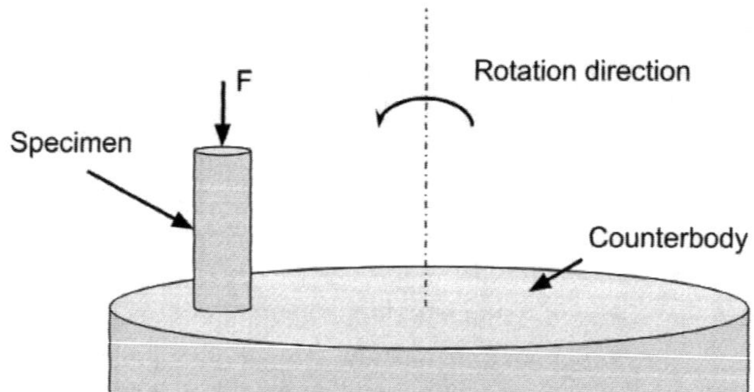

Figure1. Pin-on-disc friction scheme

Energy dispersive spectroscopy (EDS) combined with scanning electron microscopy (SEM), X-ray photoelectron spectroscopy (XPS), and infrared spectroscopy (IR) analysis was used to determine the chemical composition of friction surfaces. A combination of XPS, EDS, and IR analysis can reveal plenty of useful information on processes and mechanisms which occur on a composite surface at friction. EDS test was conducted using the EDS X-MAX20 (Oxford) attachment of SEM microscope JEOL JSM-7800F. XPS data were obtained using K-Alpha (Thermo scientific) spectrometer, Al Kα emission with 100 W capacity, and 12 kV voltages. The diameter of the analysis field was 30 ~

400 μm. Fourier transform infrared spectroscopy (FTIR) analysis of composites surface was carried out using FTS 7000 Varian IR-spectrometer before and after friction.

4. Discussion

Table 1 shows the effect of the mica and MS concentration on the wear rate of PTFE composites. The results of the investigation showed that the introduction of MS increases wears resistance of the composites. As was confirmed by Okhlopkova *et al.* [6] the higher MS concentration corresponds to the better wear resistance. The mass wear rate of composites containing muscovite reduced by 600 times, composites with muscovite and MS – by 675 times.

Table 1. The results of tribological tests of PTFE-muscovite composites

Muscovite content, wt. %	MS content, wt. %	I, mg / h	*f*
0	0	120±5.50	0.20±0.01
0.5	0	4.18±0.20	0.24±0.01
	0.1	1.76±0.08	0.26±0.01
	0.5	9.50±0.45	0.18±0.01
	1.0	1.69±0.05	0.22±0.01
1.0	0	1.20±0.03	0.25±0.01
	0.1	1.84±0.05	0.23±0.01
	0.5	2.57±0.07	0.22±0.01
	1.0	3.79±0.10	0.18±0.01
2.0	0	1.65±0.04	0.29±0.01
	0.1	1.43±0.03	0.26±0.01
	0.5	1.05±0.03	0.23±0.01
	1.0	0.78±0.05	0.23±0.01
5.0	0	0.45±0.01	0.29±0.01
	0.1	0.59±0.01	0.20±0.01
	0.5	0.43±0.01	0.23±0.01
	1.0	0.40±0.01	0.23±0.01

Note: MS – magnesium spinel, I – mass wear rate, f - friction coefficient.

It can be seen from table 1 that the friction coefficient increases with the increase of muscovite content. The migration of the mineral particles to the friction surface can explain the increase of the friction coefficient. The increase of the friction coefficient of the composite is negligible. According to the author's previous investigations [7], the filler particles crop out to friction surface during the friction test and form a protective screen. These filler particles prevent destruction and wear.

The most significant influence on the mechanism of lubricating processes was given by the chemical interaction of composites with metal counter-body at severe conditions. As a result, films with lubricating properties are formed on both metal and composite friction surfaces [8].

To assess the effect of fillers on friction and wear processes composites surfaces were analysed by IR spectroscopy. The IR spectra of the composites, which have been recorded after friction in the range of 3600 – 2600 cm $^{-1}$, contain a wide absorption band. In accordance with the data from [9], such band in this range appears due to the formation of associated hydroxyl groups, moreover, the band corresponded to peroxides is also in this range. The presence of these bands in the spectra indicates the formation of the intermediate products of carbon chain oxidation. The appearance of these oxygen-containing groups serves as evidence of the participation of peroxide radicals in destruction reactions. The further oxidation of these products inevitably leads to the formation of carboxylates with the participation of cations of the metals contained in the fillers or the counter-body material. This is confirmed by the presence of two characteristic peaks in the range of 1680 – 1400

IOP Conf. Series: Materials Science and Engineering 1060 (2021) 012020 doi:10.1088/1757-899X/1060/1/012020

cm^{-1} in the IR spectra of all composites (figure 2). It is possible that the formation of carboxylates favours the inhibition of the oxidation processes since carboxylic acids are the end product of oxidation. The probability of the simultaneous occurrence of the processes of structure formation and destruction of polymer macromolecules is high because radical particles will react with each other to form a new bond or, in this case, a cross-linking and thus will create a secondary wear-resistant layer on the surface of friction.

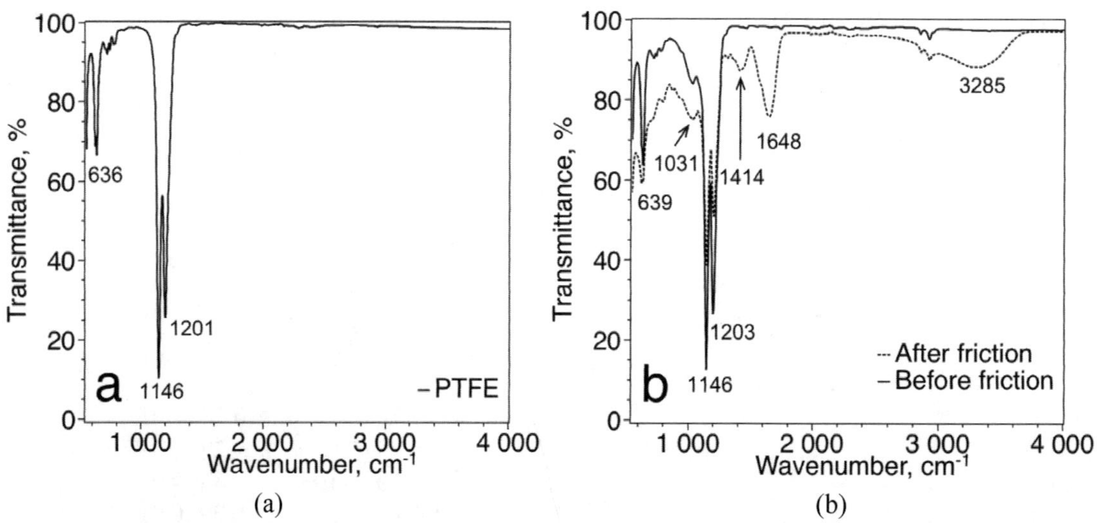

Figure 2. FTIR spectra of: (a) initial PTFE, (b) muscovite-PTFE composites before and after friction.

As shown in figure 2 (a), the initial PTFE exhibits transmittance bands at 1201, 1146, and 636 cm^{-1} corresponding to C-F stretching in –CF_2- groups of PTFE. The bands corresponding to wagging vibrations γω (CF_2) are transmittance at 639 cm^{-1}. Besides, new broad peaks at 3285, 1648, and 1414 cm^{-1} were observed in the spectrum after friction, see figure 2 (b). There are not these new peaks in the spectra of not worn surfaces of both PTFE and composite. The broadband at 3400 cm^{-1} corresponds to the stretching vibration of the hydroxyl bonds of associated with the carboxylate salts water [9]. The bands near 1648 and 1414 cm^{-1} were assigned to the antisymmetric and symmetric oscillations of two equivalent C=O bonds [10]. Pitenis *et al.* have revealed these carbonyl groups are generated tribo-chemically and are part of carboxyl groups of perfluorinated carboxylic acids chelate salts [9]. This can occur due to the destruction of PTFE by different mechanisms. Because PTFE has a high molecular weight (~500000), it can be assumed that the whole chain of PTFE cannot be transferred to the metal counter-body during friction, therefore the PTFE chain can be split [11]. Also, it should be taken into account that one of the specimen preparation process stages usually is conducted in the presence of oxygen and it gives reason to assume that the PTFE could be oxidized during friction. Similar hypothesis has been reported by Pitenis *et al.* in [9], where they have supposed that terminal carboxylate groups are generated by the tribo-chemical breaking of C-C bonds in the presence of a favourable environment with water vapor. Harris, Kajdas *et al.* [11-12] attribute the broad peaks at 1650, 1432 cm^{-1} observed in the IR spectrum of PTFE / Alumina composites to carboxylic acid end groups. The appearance of these peaks coincides with carboxylate peaks in the XPS spectra.

Furthermore, heat generation contributes to the change of sample structure. All the above-listed factors lead to the cross-linking of separate fragments of macromolecules thermo-oxidative decay and forming cross-linked structure with high wear resistance.

As can be observed in figure 3, the spectrum of worn composite surface contains peaks of chromium and iron which most likely were transferred from counter-body. Also, oxygen content increases while carbon and fluorine concentrations decrease (Figure 3, Table 2).

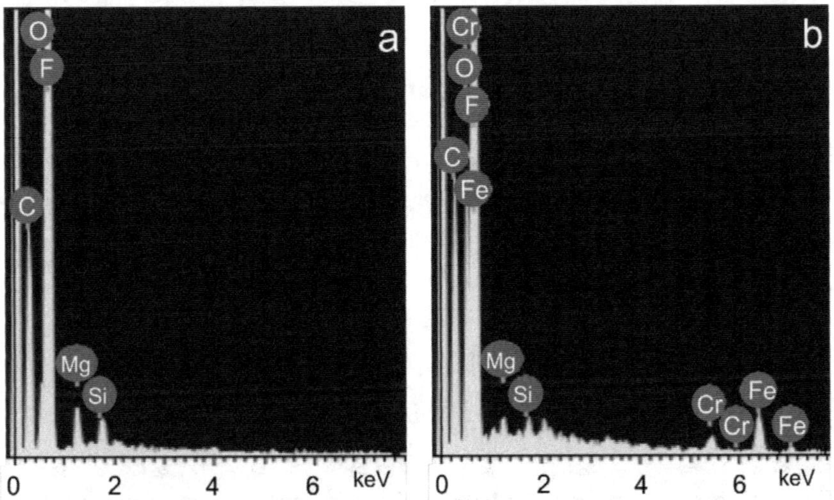

(a) (b)

Figure 3. EDS analysis of muscovite-PTFE composite:
(a) initial surface (b) worn surface

Table 2. EDS analysis of muscovite-PTFE composites before and after friction

Element	Initial surface		Worn surface	
	Weight, %	Atomic, %	Weight, %	Atomic, %
C	17.09	24.57	16.96	25.01
O	3.06	3.31	7.53	8.33
F	77.87	70.79	68.78	64.13
Mg	1.21	0.86	0.57	0.42
Si	0.76	0.47	0.40	0.25
Cr	-	-	0.97	0.33
Fe	-	-	4.79	1.52
Total	100.00		100.00	

XPS spectra of the lines F1s and C1s recorded for the PTFE and PTFE with 1 wt. % muscovite composite after friction in figure 4. The initial PTFE has the position and shape of the F1s line at 688.3 eV and C1s at 291.8 eV that correspond to groups CF_2-CF_2 [13]. After the friction process, the new peak of O1s line at 531.4 eV was found. The presence of oxygen after friction indicates the occurrence of the destruction of the polymer during friction in oxygen-containing environment. Peaks assigned to C-F (291.7), -COO- (~288.7) and C or C-C (284.6 eV) are detected on the worn composite surface [14]. The peak at ~289 eV complies with the R-COOH group [9]. These facts confirm the destruction of the polymer, the decrease in fluorine content on the surface, and hence, the formation of C–C bonds and carboxylate anion. According to Zuo experiment [3] the activation energy of PTFE defluorination is only about 1.0 kcal/mol, and it means defluorination is high-probability process during friction. Besides the peak at 687.9 eV, the new peak at ~683.5 eV was detected in the XPS spectra of the composite. It corresponds to the existence of metal–F bonds according to the researchers [15, 16]. This assumption about the formation of Al-F and Fe-F bonds during friction was reviewed in [17, 18]. In Gao et al., studies [4] a possible mechanism of chemical reactions that proceed on the surface of PTFE and metal counter-body was described. Gao suggested the formation of metal–F bonds via radical reactions.

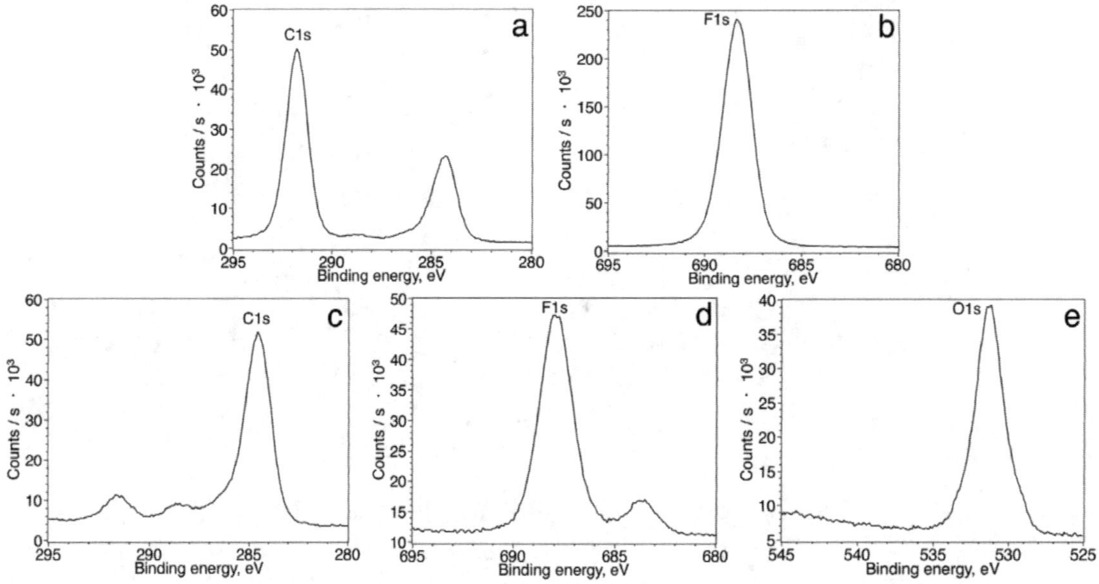

Figure 4. XPS spectra of specimens of (a)–(b) PTFE and (c)–(e) PTFE + 1 wt % of muscovite composite subjected to friction: (a) C1s; (b) F1s; (c) C1s; (d) F1s; and (e) O1s

5. Conclusion

The tribological tests of composites based on PTFE have shown that the PTFE destructs during friction and new terminal carboxyl groups formed at the presence of ambient air. It means that the tribo-chemical reactions take place on the friction surface of PTFE and its composites and these carboxyl groups are formed by breaking of C-C bonds. The IR spectroscopy and XPS have confirmed PTFE macromolecules break during friction. Also, XPS detected Me-F chemical bonds on surfaces of PTFE and its composites after friction.

Thus, the formation of the wear resistant surface of composites is a complex process that involves chemical reactions between components of composite, metals of counter-body, filler particles and the presence of atmospheric oxygen and water vapor.

6. Acknowledgement

This work was supported by the Ministry of Education and Science of the Russian Federation, state assignment No. FSRG-2020-0017.

References

[1] Friedrich K 1986 *Friction and Wear of Polymer Composites* vol 1 (Amsterdam: Elsevier Science Publishers) p 478
[2] Gong D L, Zhang B, Xue Q J, Wang H L 1990 *Wear* **137** pp 267–273
[3] Zuo Z, Yang Y, Qi X, Su W, Yang X 2014 *Wear* **320** pp 87-93
[4] Gao J T 2000 *Wear* **245** pp 100-106
[5] Sleptsova S A , Okhlopkova A A, Kapitonova Iu V, Lazareva N N, Makarov M M, Nikiforov L

A 2016 *J. Frict. Wear* **37** pp 129-135
[6] Okhlopkova A A., Popov S N, Sleptsova S A, Petrova P N, Avvakumov E G 2004 *J. Struct. Chem.* **45** pp 172-177
[7] Kirillina I V, Nikiforov L A, Okhlopkova A A, Sleptsova S A, Yoon C, Cho J H 2014 *Bull. Korean Chem. Soc.* **35** pp 3411-3420
[8] Zhang Z, Xue Q, Liu W, Shen W 1997 *Wear* **210** pp 151-156
[9] Pitenis A A, Harris K L, Junk C P, Blackman G S, Sawyer W G, Krick B A 2015 *Tribol. Lett.* **57** p 4
[10] Nakanishi K 1962 *Infrared absorption spectroscopy: practical* (San Francisco: Holden-Day) p 243
[11] Harris K L, Pitenis A A, Sawyer W G, Krick B A, Blackman G S, Kasprzak D J, Junk C P 2015 *Macromolecules* **48** pp 3739–3745
[12] Kajdas C K 2005 Tribol. Int. **38** pp 337-353
[13] Wheeler D 1980 *NASA Tech. Pap.* **1728**
[14] Shen J T, Top M, Ivashenko O, Rudolf P, Pei Y T, De Hosson J Th M 2015 *Appl. Surf. Sci.* **331** pp 482-489
[15] Cadman P, Gossedge G M 1979 *Wear* **54** pp 211–215
[16] 1979 *Handbook of X-ray photoelectron spectroscopy*, ed by Wagner C D, Riggs W M, Davis L E, Moulder J F, Muilenberg G E (Minnesota: Perkin-Elmer Corporation) p 190
[17] Gong D L, Xue Q J, Wang H L 1991 *Wear* **148** pp 161–169
[18] Suzuki M, Prat P 1999 *Wear* **225** pp 995–1003

IOP Conf. Series: Materials Science and Engineering 1060 (2021) 012021 doi:10.1088/1757-899X/1060/1/012021

Inter-Disciplinary Approach to Suborbital Reusable Spaceplane Composite Wing Design

E R Ashikhmina*, P V Prosuntsov and S V Reznik

Bauman Moscow State Technical University, 2nd Baumanskaya str., b. 5, Moscow, 105005, Russia

*corresponding author: ashikhmina.ekaterin@gmail.com

Abstract. The development of the wing of suborbital reusable spaceplane is a complex task due to conflicting requirements that are sufficient for its structure. They include high mass and cost efficiency, sustainability, low g-forces, effective heat protection under the consideration of technological restraints. To satisfy outlined requirements in the design of the composite wing the inter-disciplinary approach is presented relying on advanced methods of mathematical modelling. The approach starts with wing shape optimization enabling to get good aerodynamic performance. Subsequent optimal re-entry trajectory planning which use both angle-of-attack and angle-of-bank changes to control the re-entry provides a set level of g-forces and heat flux. Then based on the obtained trajectory the aerodynamic loads for the wing are gained. To acquire composite materials properties of the wing the analysis/test approach is applied that comprises multilevel modelling of mechanical and thermal material properties by commercial software. The results of modelling are validated by experiment. In addition, to enhancing mass efficiency of the wing structure the topology optimization of its wing box and structural optimization of the wing skin is conducted which uses both angles of orientations and thicknesses of composite monolayers as the variables with constraints on the strength and stiffness of wing skin. Furthermore, the approach includes thermal design due to high aerodynamic heating taking place during re-entry. Thermal loads are analysed by modelling the thermal state of the structure. The analysis showed that the wing skin needs thermal protection. As the thermal protection coating, the material based on glass microspheres was chosen. Its thickness through wing is obtained in the result of parameter optimization. As the result, the interdisciplinary approach to the design of the spaceplane composite wing is developed that allows feeding back mass, economic, aerodynamic, heat and strength impact into the spaceplane wing design synthesis.

1. Introduction

Human spaceflight and space exploration has traditionally been a well-funded domain of government [1]. Nevertheless, over the past decades, the interest to the business-driven space transportation systems has increased considerably [2]. Nowadays robust technological development by certain private companies, entrepreneurs, and research groups to make private spaceflight an everyday reality has surged. SpaceX Crew Dragon launch to the ISS successfully demonstrated private sector capability to design, test and operate a spacecraft for human spaceflight [3]. This event marked the start of a new era of commercial access to space.

The prominent branch of the space industry that has advantageous commercial application is space tourism [4-7]. Current market trends show a huge potential demand for space tourism especially for suborbital one [5]. In contrast to orbital space tourism, suborbital travel represents a brief spaceflight that fails to complete one full orbit of the Earth. The suborbital spaceflight bases on using reusable launch vehicles that can achieve an altitude of about 100 km above sea level and then subsequently land safely and after some time take-off again. This type of travel lets passengers experience g-forces, the curvature of the Earth and weightlessness by taking them up to the Karman Line, which is at an altitude of about 100 km [6, 7]. Recently, a number of private companies all over the world are involved in the space tourism activity. The examples of suborbital reusable spaceplane projects could be RocketplaneXP (Rocketplane Kistler, USA, 2005), New Shepard (Blue Origin, USA, 2006), Space Plane (EADS Astrium, France, 2007), Lynx Mark (XCOR Aerospace, USA, 2008), SpaceShipTwo (Virgin Galactic, USA, 2010), Oduvanchik (Bauman MSTU, Russia, 2007), Reusable Suborbital Space Complex (Cosmocourse, Russia, 2014) and others [8].

In Bauman University [9-11] during some years the suborbital spaceplane Oduvanchik is being developed. The general scheme of spaceplane is presented in figure 1.

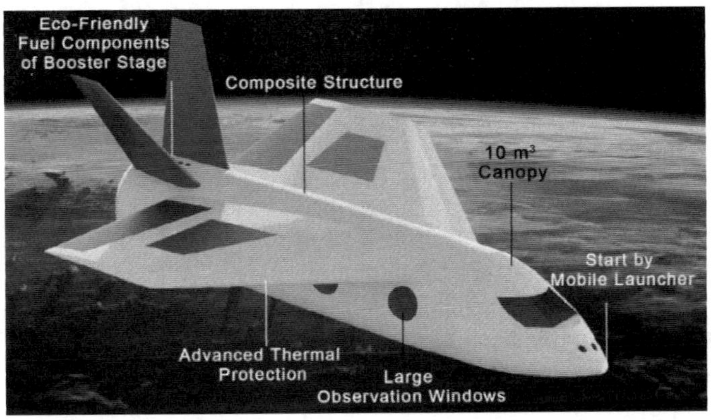

Figure 1. Three-dimensional geometric model of suborbital reusable space vehicle of tourist class.

The spaceplane flight profile includes the following phases: vertical lift-off by means of a booster stage, the spaceplane separation from a booster at an altitude of approximately 46 km, and the next flight under inertia forces up to an altitude of 102 km. At the apogee, tourists enjoy the view of Earth and experience the microgravity environment for 4 minutes as the spaceplane slowly decelerated before its descent into the atmosphere. At the end of the mission, the spaceplane lands horizontally.

The spaceplane structure contains 9 m fuselage length and V-shaped empennage that consists of two slanted control surfaces that operate as both vertical and horizontal stabilizers and perform the functions of rudder and elevator.

The essential structural element of the suborbital spaceplane that significantly determines its design appearance and weight efficiency is the wing. According to design data, the spaceplane consists of the trapezoidal wing that is an aerodynamic surface with leading edge sweep and straight trailing edge. High wing configuration offers better visibility below the vehicle and its stability. The span of 7.5 m and the area of 20 m^2 characterize the wing. The root and the tip chords are 4.3 m and 1.6 m respectively.

The wing skin is designed to minimize its weight by using composite materials and represents a sandwich panel made of lightweight aramid honeycomb core and hybrid polymer matrix composite (PMC) layers. For technological and economic reasons, hybrid composite layers consist of glass fibre fabric with harness-satin weave and unidirectional carbon fibre tape, as well as an epoxy resin are used. The hybrid PMC layers is mainly carrying tension and compression loads while the honeycomb bears compression and shear loads (figure 2) [12].

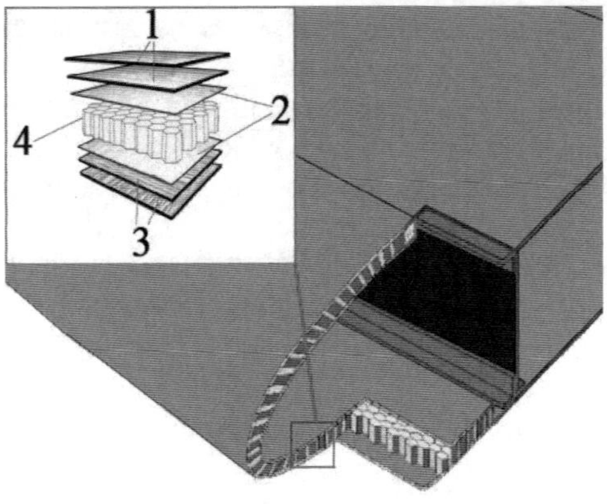

1 – outer hybrid PMC layers
2 – adhesive
3 – inner hybrid PMC layers
4 – honeycomb core

Figure 2. The representation of spaceplane wing skin structure.

2. Literature Review

The engineering design of the spaceplane wing is a complex scientific and technical issue. Because the wing of the suborbital reusable spaceplane is highly complex, interconnected, and operates in extreme conditions, its design represents a challenge of the highest order [13]. Spaceplane requires both very high cost and mass efficiency, sustainability, low g-forces, effective heat protection under the consideration of technological restraints. Low g-forces requirement is driven by medical reasons. High values of g-forces cause blood pooling in the extremities because the heart is unable to overcome the acceleration forces and complete the circulatory process [13]. Meeting the design challenges resulting from the combination of the preceding factors demands the inter-disciplinary approach and is presented relying on advanced methods of mathematical modelling [14-16].

The literature review shows that nowadays there are traditional [13] and recently developed [17-19] approaches to spaceplane design. The presented suborbital spaceplane occupies an intermediate position between hypersonic spaceplanes such as the Space Shuttle [20] and the Buran [21] and traditional supersonic aircraft. However, the traditional design methodologies that are used for aircraft or Space Shuttle are not applicable for the suborbital spaceplane due to significant differences that are described hereinafter. The velocity of the suborbital spaceplane is close to a jet fighter's velocity, but unfortunately, in open sources there is no information about detailed methodology characterized in that the spaceplane composite wing as a particular design object is considered.

As compared with aircraft the spaceplane gathers higher velocity. The significant thermal and force loads that occur during re-entry defines special requirements that were listed hereinbefore. Moreover, as the suborbital spaceplane is not provided with a powerplant, it glides into the atmosphere under gravity and aerodynamic forces. Hence, the lift should be sufficient for comfortable and safe landing [8]. To develop optimal wing configuration that provides sufficient aerodynamic quality whilst landing is a complex, iterative, and interconnected procedure. For instance, the change in wing configuration leads to changes in its aerodynamic characteristics that, in return, initiates changes in trajectory parameters. Wing configuration also defines the g-loads. The increase in the wing area leads to g-loads reduction, but it also initiates weight and cost rising. Moreover, the large-area wing is ineffective at high Mach. Note that spaceplane wing is made of PMC that allows to reduce the weight of the wing and to increase cost efficiency by obtaining the optimal layout of the multilayer wing skin and rational wing box of the irregular structure by means of topology optimization.

In contrast to Space Shuttle, the suborbital spaceplane has at least eight times lower re-entry velocity in the atmosphere and a significantly different flight profile. Furthermore, the structural frame of Space Shuttle mainly made of metal whilst load-carrying elements of suborbital vehicles can be manufactured from PMC that is connected with lower thermal fluxes occurring during re-entry [22]. Temperature of the Space Shuttle wing leading edges reached about 1650 °C during re-entry. It required the development of the combined thermal protection system that covered essentially the entire surface and consisted of several types of materials such as tiles, advanced flexible reusable surface insulation, reinforced carbon-carbon. In case of the suborbital spaceplane, such thermal protection measures are redundant as the thermal fluxes are much lower. In order to provide suborbital spaceplane thermal protection, the different approaches should be considered in terms of weight and cost. Furthermore, the methodologies developed in the 1980s can be amplified by advanced methods of mathematical modelling that are available nowadays.

The aim of the current study is to present an inter-disciplinary methodology relying on advanced methods of mathematical modelling for the suborbital spaceplane composite wing design in order to satisfy outlined requirements. The synthesis design methodology for spaceplane is a systematic or orderly way to the design of complex aerospace vehicle systems, where the interaction among several disciplines must be considered. The methodology couples key design disciplines such as trajectory, aerodynamics, weight, thermal, material science, economy, and performance. As a consequence, valuable feedback for design feasibility, sensitivities, and constraints can be obtained in a short time.

3. Design philosophy of the spaceplane wing

The design methodology of the spaceplane wing is a complex iterative procedure. The comprehensive illustration can be viewed in wheel-like fashion as shown in figure 3. The mainstream design process is broken down into nine stages: aerodynamic shape optimization, choice of rational trajectory parameters, aerodynamic flow analysis, obtaining of material properties by test/analysis approach, topology optimization of the wing box and thermal design.

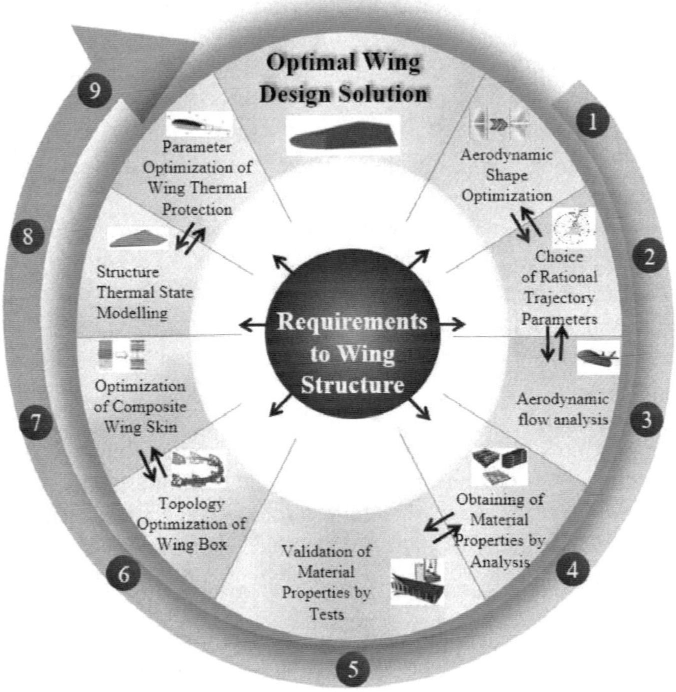

Figure 3. The representation of design procedure for suborbital spaceplane wing design.

The first step of the methodology is aerodynamic shape optimization that allows generating wing appearance based on the aerodynamic characteristics calculation [23, 24]. There are two design cases considered successively for shape optimization: subsonic and supersonic flight. Subsonic one is associated with the need to ensure the required lift level immediately before the landing and touchdown speed. In this case, the objective is to minimize wing square that drives mass reduction. The supersonic case involves maximization of aerodynamic quality of the wing at an altitude of approximately 30 km, which is characterized by maximum aerodynamic loads [25, 26]. For both stages as design variables the following geometry dimensions of the wing are considered: length, leading-edge sweep angle, taper ratio and a maximum thickness of root and tip chord. Note that constraint on the minimum wing sweep angle equal to 40 degrees is applied.

For obtained wing appearance the flight profile is developed. Firstly, the aerodynamic coefficients that define the aerodynamic performance of the unpowered spaceplane are acquired by CFD analysis in commercial software. Next, by the solution of the equations of three-dimensional gliding dynamics, the time-dependencies of basic trajectory parameters are obtained. Use both angle-of-attack and angle-of-bank changes to control the re-entry provides a set level of g-forces and heat flux during spaceplane re-entry into atmosphere. As the result, the time-dependencies of altitude, velocity, flight path angle, g-load factor variation, bank angle, dynamic pressure and heating rate variations are found. The developed flight profile is then used in the computation of thermal, strength loads for structure and thermal protection system designs.

According to flight profile, the suborbital spaceplane operates in a wide range of velocities (its maximum speed is 3.6 M at an altitude of approximately 40 km) and is exposed to intensive aerodynamic heating during re-entry into the atmosphere [27]. Therefore, the spaceplane design requires data on the thermal properties of hybrid composites applied in structure (thermal conductivity coefficients in different directions, heat capacity and density) in dependence on temperature. In addition to thermal properties, the reliable information on physical and mechanical characteristics is required in order to develop the composite wing box and to define the overall wing strength level. To acquire composite materials properties of the wing the analysis/test approach is applied. It comprises the multilevel modelling of material properties via generation of a representative volume element (RVE) of the composite. Obtaining of material properties includes the generation of a RVE, the application of load to its faces, and the subsequent calculation the values of both physical and mechanical properties such elastic and shear modulus, Poisson's ratio and thermal properties such thermal conductivity coefficients via finite element analysis.

To confirm the reliability of the numerical modelling results, its validation by comparison with the data obtained by experimental studies is conducted. As the result of the experimental research, thermal conductivity through-plane is obtained by laser flash method as well as thermal conductivity in-plane – by specialized heat equipment developed in Bauman University with a subsequent solution of the inverse heat conduction problem. The physical and mechanical properties are also validated by tensile strength tests.

To reduce the mass of the spaceplane wing, the topology optimization of the wing box is carried out in two stages. Firstly, the number and arrangement of the structural frame components are determined. In this case, an objective function is used to minimize the wing compliance with a constraint on the material volume fraction inside the wing and the total displacement. Secondly, the optimal layout of the composite structure elements is obtained [28]. To find the optimal layout of the obtained through previous optimization stage the topology optimization is conducted and formulated as follows: minimization of the weight with a constraint on the composite failure index of the wing skin, which value is chosen due to reusability of the vehicle, and the wing total displacement. To check optimization results and fulfilment of the applied constraints, the validation procedure is conducted in order to evaluate the composite failure index.

In addition, to enhancing mass efficiency of the wing structure, the optimization of its skin is conducted which uses both angles of orientations and thicknesses of composite monolayers as the variables with constraints on the strength and stiffness of wing skin. Described topology and structural optimizations allow not only design the lightweight structures but also realize the benefits of anisotropic composite microstructures.

Furthermore, the methodology includes thermal design due to high aerodynamic heating taking place during re-entry. Thermal loads are analysed by modelling the thermal state of the structure. The analysis showed some wing sections are exposed to the influence of the temperatures exceeding the allowable service temperature for polymer composites and need thermal protection (TP). Thus, the maximum service temperature for epoxy resin does not exceed 420 K. As the TP, coating based on glass microspheres is chosen. Its thickness through the wing is obtained in the result of parameter optimization in commercial software. Therefore, the leading edge as well as parts of the windward and leeward sides of the wing needs to be thermally protected. As the temperature distribution is uneven within the wing surface and thereby it is reasonable to apply different thicknesses of the TP layer on the different sections. Hence, it's necessary to find an optimum distribution of the TP within the RSV wing surface. However, parametric optimization of the TP thickness is necessary. Such optimization could be conducted by means of the introduction of the set of line-segment functions with apexes, which are coincident with the borders of the root and tip wing chords. Between these borders, the TP thickness varies linearly in both directions – along the chord length and along the wingspan.

To demonstrate the feasibility of the developed approach the solution of one of the methodology stages – aerodynamic coefficients determination for obtaining the mission profile is presented. In addition, pressure coefficients for subsonic and supersonic regimes are presented.

4. Solving of the formulated tasks

4.1. Aerodynamic coefficients determination

For trajectory design, the unpowered reusable spaceplane is assumed as a point mass. In order to obtain input aerodynamic coefficients versus Mach number, the modelling of aerodynamic flow field over the spaceplane surface for reference suborbital trajectory is conducted using the commercial software [29]. Baseline spaceplane configuration obtained in the result of shape optimization for subsonic and supersonic regimes is used [8, 30]. The complexity of the aerodynamic flow analysis is in the variety of flow conditions associated with wide ranges of the angle-of-attack and Mach. This required detailed tuning of each design case and careful control over the level of residuals, especially for supersonic gas flows.

The resulting dataset contains values for drag coefficient C_x and lift coefficient C_y over a range of angles-of-attack $10° \leq \alpha \leq 40°$and Mach numbers $0.1 \leq M \leq 4$. The datasets are depicted in figure 4.

It is shown that up to $M \approx 0.5$, the values of aerodynamic coefficients are practically constant since the compressibility of the medium (air) is practically not manifested. The transonic transition ("the sound barrier") is clearly visible. It is accompanied by an abrupt change in the flow phenomena around the spaceplane and a sharp increase in the drag coefficient. The values of the aerodynamic coefficients decrease with a further increase in the velocity M> 1. Obtained coefficients are used for further analysis of trajectory parameters and choice of the optimal wing configuration.

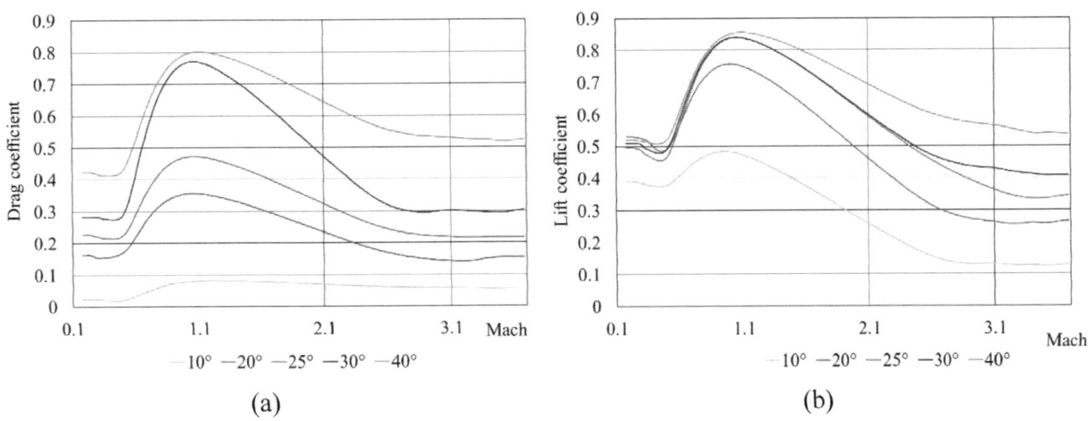

Figure 4. Aerodynamic coefficients: (a) C_x and (b) C_y *vs.* Mach number

4.2. Determination of pressure and temperature fields for the spaceplane wing

Analysis of spaceplane aerodynamic flow is performed by means of Fluid Flow module of FEA-package ANSYS Workbench. Modelling is conducted for landing and the most critically loaded trajectory interval – spaceplane re-entry with a high angle of attack at an altitude of approximately 30 km. To describe turbulence processes in the flow, two-equation k-ε turbulence model with standard wall functions is used. The resulting pressure coefficients distribution that demonstrates the difference in lift creation by over- and under-pressure between subsonic and supersonic flight is depicted in figure 5. Comparison of the velocity streamlines that demonstrates the flow phenomena for both velocity regimes is illustrated in figure 6.

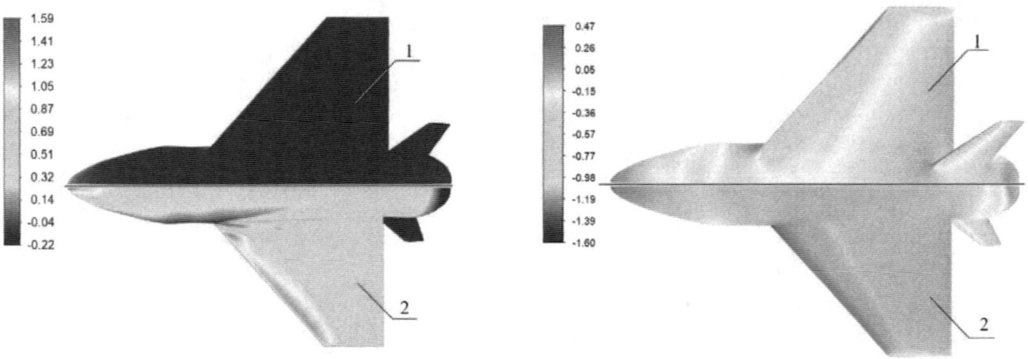

(a) (b)

Figure 5. Pressure coefficient contours for the spaceplane:
(a) at 30 km altitude at the angle-of-attack 40 °and
(b) landing at angle-of-attack 10 ° (1 – upper surface; 2 – lower surface).

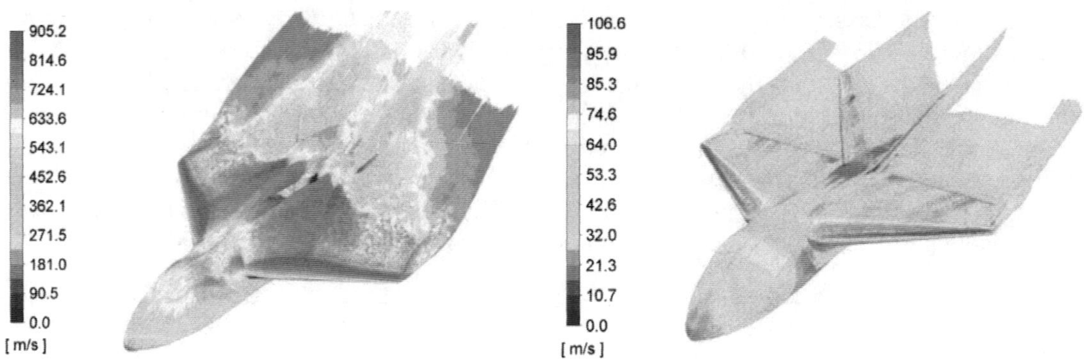

(a) (b)

Figure 6. Velocity streamlines for the spaceplane:
(a) at 30 km altitude at the angle-of-attack 40 ° and (b) landing at angle-of-attack 10 °.

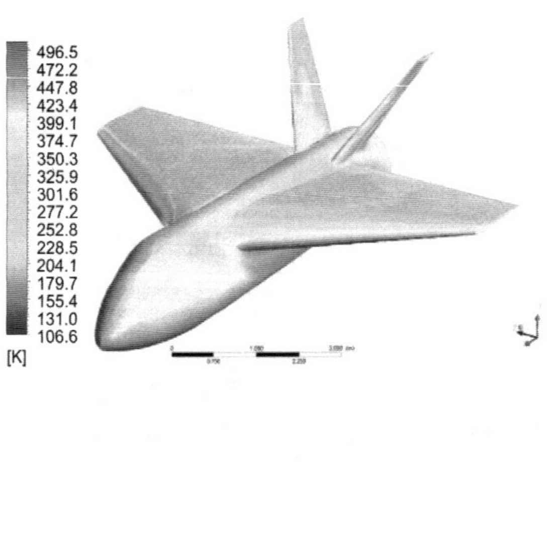

Figure 7. Temperature on the spaceplane surface at 30 km altitude at the angle-of-attack 40 ° [K].

Critical regions of the wing surface could be distinguished, *e.g.* the leading edge and neighbouring section in the windward side of the wing are subjected to the highest temperatures influence. Thus, the spaceplane surface temperature could reach 496.5 K (figure 7). As obtained in the result of CFD-analysis temperature exceeds the allowable service temperature of the matrix based on epoxy resin. As the TP the different materials are considered. In the result of weight analysis, the coating based on glass microspheres is chosen. Its thickness through the wing is obtained in the result of parameter optimization in commercial software [10].

5. Conclusion

In this study, we presented the interdisciplinary methodology of the spaceplane composite wing design. The methodology couples key design disciplines such as trajectory, aerodynamics, weight, thermal, economy, and performance. As a consequence, valuable feedback discussing design feasibility, sensitivities, boundaries, and constraints can be obtained in a short time from this top-level simulation process.

The feasibility of the developed methodology is illustrated with an example of a composite wing for spaceplane developed in Bauman University. By CFD analysis, the aerodynamic coefficients that define the aerodynamic performance of the unpowered glider are obtained. The results of aerodynamic flow analysis that are used for further analysis of trajectory parameters and choice of the optimal wing configuration, also in the strength and thermal spaceplane wing design are presented.

References

[1] Seedhouse E 2014 *Suborbital: Industry at the edge of space*: Springer Science & Business Media
[2] Denis G, Alary D, Pasco X, Pisot N, Texier D and Toulza S 2020 From new space to big space: How commercial space dream is becoming a reality *Acta Astronautica* **166** 431-43
[3] Crane L 2020 SpaceX's first crewed flight is a go *New Scientist* **246** 17
[4] Webber D 2013 Space tourism: Its history, future and importance *Acta Astronautica* **92** 138-43
[5] Chang Y-W 2015 The first decade of commercial space tourism *Acta Astronautica* **108** 79-91
[6] van Pelt M 2005 *Space Tourism*: (New York: Springer)
[7] Klemm G, Markkanen S 2011 *Space tourism in The Long Tail of Tourism* ed A Papathanassis (Wiesbaden: Gabler Verlag) ch 11 95-103
[8] Ashikhmina E and Prosuntsov P 2020 Coupled CFD-based Shape Optimization of a Wing of Reusable Space Vehicle of Tourist Class *IOP Conference Series: Materials Science and Engineering* **709** 022111
[9] Reznik S, Ageyeva T and Shafikova I 2012 Composite Wing for Reusable Space Vehicles *Acmtaa 2012: Proceedings of the Second International Conference on Advanced Composite Materials and Technologies for Aerospace Applications* 7-13
[10] Ashikhmina E R, Ageyeva T G and Prosuntsov P V 2019 Hybrid composites in reusable space vehicles wing structures *IOP Conf. Ser.: Mater. Sci. Eng.* **683** 012067
[11] Ageyeva T G, Ashikhmina E R and Prosuntsov P V 2018 Optimization of hybrid composite material structure for wing skin of tourist class reusable space vehicle *Herald of the Bauman Moscow State Technical University. Series Mechanical Engineering* **1** 4-19
[12] Reznik S V and Novikov A D 2016 Comparative analysis of the honeycomb and thin-shell space antenna reflectors *Thermophysical Basis of Energy Technologies (Tbet-2016)* **92**
[13] Hammond W E 2001 *Design Methodologies for Space Transportation Systems* (Reston, United States: American Institute of Aeronautics & Astronautics)
[14] Epikhin A S and Kalugin V T 2018 Features of numerical simulation of the unsteady vortex flow around aircraft considering airbrake *IOP Conf Ser: Mater Sci Eng* **468**
[15] Dimitrienko Y, Koryakov M and Zakharov A 2017 Numerical Modeling of Coupled Problems of External Aerothermodynamics and Internal Heat-and-Mass Transfer in High-Speed Vehicle Composite Constructions *Lecture Notes in Computer Science (including subseries Lecture Notes in Artificial Intelligence and Lecture Notes in Bioinformatics)* **10187** 294-301

[16] Reznik S V, Prosuntsov P V and Mikhailovskii K V 2019 Development of Elements of Reusable Heat Shields from a Carbon-Ceramic Composite Material 1 Theoretical Forecast *J. Eng. Phys. Thermophys.* **92** 89-94
[17] Frank C P 2016 *A Design Space Exploration Methodology to Support Decisions under Evolving Uncertainty in Requirments and Its Application to Advanced Vehicles* Georgia Institute of Technology
[18] Guerster M, Crawley E F 2018 Architectural options and optimization of suborbital space tourism vehicles 2018 *IEEE Aerospace Conference* (USA: Big Sky, MT) 1-18
[19] Huang X and Chudoba B 2005 A Trajectory Synthesis Simulation Program for the Conceptual Design of a Suborbital Tourism Vehicle. In: *AIAA Modeling and Simulation Technologies Conference and Exhibit*: American Institute of Aeronautics and Astronautics)
[20] Sivolella D 2017 *The Space Shuttle Program*: Springer International Publishing)
[21] Hendrickx B and Vis B 2007 *Energiya-Buran*: Praxis
[22] Reznik S, Prosuntsov P and Mikhaylovskiy K 2018 Development verification of coatings made from porous ceramic matrix composite materials *MATEC Web of Conferences* **224** 03019
[23] Ashikhmina E and Prosuntsov P 2020 Coupled CFD-based Shape Optimization of a Wing of Reusable Space Vehicle of Tourist Class *IOP Conf. Ser.: Mater. Sci. Eng.* **709** 022108
[24] Ashikhmina E and Prosuntsov P 2019 Wing airfoil selection and optimization for the tourist class reusable space vehicle *AIP Conference Proceedings* **2135** 020005
[25] Dirkx D and Mooij E 2017 *Aerothermodynamics in Conceptual Shape Optimization of Entry Vehicles* (Cham: Springer) ch 3 41-71
[26] Skinner S N and Zare-Behtash H 2018 State-of-the-art in aerodynamic shape optimisation methods *Applied Soft Computing* **62** 933-62
[27] Ageeva T G, Dudar E N and Reznik S V 2010 *Aviakosmicheskaya tekhnika i tekhnologiya.* [Aerospace Technic and Technology] **2** 3-8
[28] Ashikhmina E R and Prosuntsov P V 2019 Wing box topology optimization of tourist class reusable space vehicle *AIP Conference Proceedings* **2171** 030014
[29] Bui V T, Kalugin V T, Lapygin V I and Khlupnov A I 2017 Numerical study of the influence of flow blockage on the aerodynamic coefficients of models in low-speed wind tunnels *Thermophysics and Aeromechanics* **24** 857-66
[30] Ashikhmina E R and Prosuntsov P V 2019 Wing airfoil selection and optimization for the tourist class reusable space vehicle *AIP Conference Proceedings* **2135** 020005

IOP Conf. Series: Materials Science and Engineering 1060 (2021) 012022 doi:10.1088/1757-899X/1060/1/012022

Elastomeric composites for Arctic conditions: operation features and development experience

N N Petrova[1], V V Portniagina[1], V V Mukhin[1], A A Dyakonov[1,2], S F Ivanova[1], E N Timofeeva[1], A V Morozov[3], A A Okhlopkova[1]

[1] M. K. Ammosov North-Eastern Federal University, Yakutsk, Russian Federation

[2] V.P. Larionov Institute of the Physical-Technical Problems of the North of the Siberian Branch of the RAS, Yakutsk, Russian Federation

[3] Institute for Problems in Mechanics RAS, Moscow, Russian Federation

E-mail: pnn2002@mail.ru

Abstract. Operation of elastomeric materials in the extreme climatic conditions of the North is complex and expensive, since not always existing materials can provide the required level of low-temperature characteristics (down to - 60 °C). This leads to failure of machines and mechanisms, equipment downtime, additional costs for repair or replacement of elastomeric parts. This study investigates the features of elastomeric composites operation in cold climates and the results of full-scale tests with simultaneous exposure to naturally low temperatures and hydrocarbon environments in Yakutia (Siberia, Russia). It presents the main principles of creating elastomeric composites for sealing purposes with high frost resistance and examples of recent developments. To obtain elastomeric materials of high frost and oil resistance new rubbers of high frost resistance as Hydrin T6000 epichlorohydrin rubber (T_g = - 60 °C), and traditional rubbers (nitrile etc) were used. For elastomer modification we used carbon nanotubes, collagen hydrolysate and polymer mixtures. Different type of developed materials is a double-layer material based on ultra-high-molecular-weight polyethene (UHMWPE) and nitrile-butadiene rubber-based elastomeric composite. The effect of diphenylguanidine on the interaction of UHMWPE with elastomeric composite based on nitrile butadiene rubber and adhesion at the interface was investigated. The operating properties were studied by standard methods and the structure of the obtained materials was studied using electron microscopy.

1. Introduction

Elastomers are the class of polymers with a peculiar set of properties that differ from most of the known thermoplastics, thermosetting plastics, or polymer fibres. They characterized by high elasticity, *i.e.* the ability to undergo significant (up to a thousand per cent or more) reversible deformations at low values of stresses, low modulus combined with mechanical strength and performance over a wide temperature range [1]. As a result, elastomer materials (elastomeric composites) are unsurpassed materials for sealing purposes. They are applied widely to produce various types of seals, gaskets, cuffs, damping products. Machines and mechanisms are completed with these products. They also used as functional sealing parts, sealing joints and joints of parts of technological equipment, as part of friction units. The performance of the entire unit or the mechanism largely depends on the properties of elastomeric seals and their ability to seal joints.

Sealing purpose elastomeric materials must have several properties, including a certain level of physical and mechanical characteristics (products must withstand operating pressures and loads), resistance in working environments, most often of a hydrocarbon nature (oil, fuel), frost resistance required for operation in various climatic zones, the ability of products to restore their shape during operation, which is most often assessed by compression set, a certain level of wear resistance. All these properties are different and are the result of the various processes in an elastomeric composite. So, frost resistance determined by the flexibility and mobility of macromolecular chains of rubber, characterized by the degree of development of highly elastic deformation at the operating temperature. Its lower limit of manifestation is the glass transition temperature (T_g), which determines the transition of an elastomeric material from a glassy state to a highly elastic state of relaxation nature. Compression set is also a consequence of the manifestation of the relaxation properties of the material. The resistance of rubbers in hydrocarbon media is determined by the presence of polar fragments in the elastomer structure, which, however, reduce the flexibility and mobility of macromolecules, *i.e.* reduce the frost resistance of the material. Tensile strength and wear resistance are complex indicators that characterize both the structure and mechanical properties of the material, and the presence of various defects. The entire volume of the sample makes a greater contribution to the manifestation of physical and mechanical properties. The structure and composition of the surface layers of the material makes the greatest contribution to the wear resistance of rubbers. Thus, when developing elastomeric composites for sealing purposes, it is necessary to provide an optimal combination of all these characteristics, which, of course, is not an easy task. This combination of various properties in one material should ensure long-term operation of products made of elastomeric materials, minimal leakage of working fluids, and trouble-free operation in real conditions.

Many works devoted to the study of the operation of elastomeric seals at low temperatures [2-6]. There are woks studying an interfacial leakage of O-rings in sealing joints [2,3], and factors influencing low temperature flexibility of elastomers [4]. However, the properties of the seals are considered down to -25 ÷ -30 °C. In general, lower winter operating temperatures not realized in European countries. Still, further advancement into the Arctic (offshore oil and other minerals production) makes the development of elastomeric materials of unique frost resistance urgent. This article is devoted to this subject.

When an elastomeric composite operated in cold climates, for example, in the Arctic zone of the Russian Federation (Republic of Sakha (Yakutia)), the requirements for materials become more stringent, especially in terms of frost resistance. The climate of Yakutia is characterized by extremely low temperatures (up to - 60 °C) in winter, their frequent changes with a transition through 0 °C with an amplitude of up to 30 ÷ 40 °C in the autumn-spring period, high content of ultraviolet radiation due to the high transparency of the atmosphere and a large number of sunny days. All this leads to the fact that traditional industrial rubbers do not correspond to the operating conditions, due to rapid ageing and low temperatures, they become hard, not elastic, which leads to a decrease in the tightness of the connection, leaks of the working medium, and destruction of sealing parts. As the practice of operating machines and mechanisms in the Republic of Sakha (Yakutia) has shown, up to 30% of the total number of equipment failures in winter is caused by insufficient frost resistance of elastomeric products, which leads to equipment downtime, the need for repair work, and can initiate disasters and negative consequences for the environment [7].

We have carried out comprehensive tests of industrial elastomeric materials based on nitrile butadiene, propylene oxide, epichlorohydrin rubbers under the conditions of natural exposure of samples in the climatic conditions of Yakutia in a hydrocarbon environment (oil) [8-10]. Nitrile butadiene rubber with a low content of acrylonitrile (BNKS-18, T_g = - 50 °C) is of limited use under these conditions. As a rule, significant amounts of plasticizers - substances that improve the flexibility and mobility of rubber macromolecules - are introduced into industrial elastomeric composites based on BNKS-18. However, this method of increasing frost resistance turns out to be ineffective for elastomeric composites in contact with carbon media during operation because plasticizers are quickly washed out by the medium. Elastomeric composites based on rubbers of high frost resistance, for example, propylene oxide (Russia, T_g = - 74 °C) or epichlorohydrin rubber Hydrin T6000 (Japan, Zeon, T_g = - 60 °C) have excellent prospects for operation in extreme conditions of the Arctic [11].

Over the years, we have developed basic approaches to creating elastomeric composites that combine an increased level of frost and oil resistance [12]. They are systematized in the diagram shown in figure 1. There are two main ways to obtain such materials: using one rubber or a mixture of polymers to achieve a compromise combination of basic properties. In the first case, it is necessary to use mainly rubbers of high frost resistance as a basis; if BNKS-18 is used, then measures should be taken to slow down the washout of the plasticizers. In the second case, we are talking about the creation of composite materials based on mixtures of polymers with a certain phase morphology and increased interaction of rubbers at the interface, and in the extreme case, their co-vulcanization. If in either case the achieved level of oil resistance of materials while maintaining frost resistance is insufficient, then a surface modification of the elastomers should be carried out to protect the surface from the effects of environments. At the same time, for the reliable operation of products, it is important to securely attach the coating to the substrate, *i.e.* ensuring increased adhesion at the border of their contact.

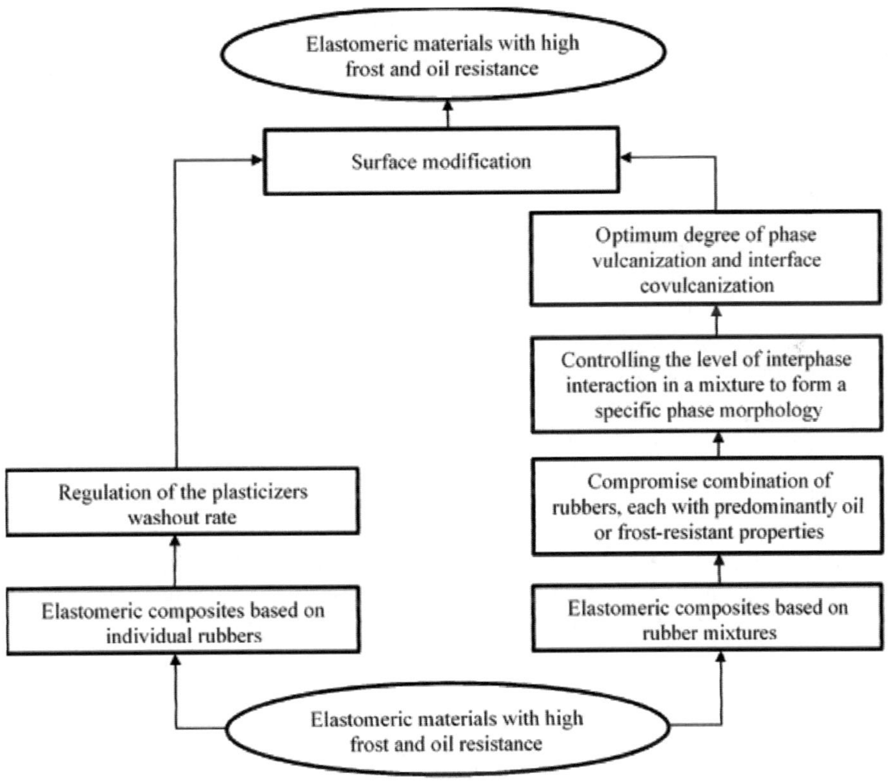

Figure 1. Ways to create materials with a high level of frost and oil resistance.

The proposed principles of creating frost-resistant materials have been tested and have shown their effectiveness. We have developed several elastomeric composites of various compositions, for which patents of the Russian Federation and the USA have been obtained. In this study, some recent developments are presented that confirm the correctness of the proposed scheme. In the first case, elastomeric composites based on epichlorohydrin rubber Hydrin T6000, modified with carbon nanotubes or a specially obtained environmentally friendly additive of a collagen nature, are considered.

Blends based on mixtures of diene and nitrile butadiene rubbers are considered as an example of the use of the compositional principle to produce elastomeric materials for Arctic use. In conclusion, a method for producing two-layer materials based on nitrile butadiene rubber and ultra-high molecular weight polyethylene is presented.

2. Materials and methods

The following rubbers were used in this work: Hydrin T6000 epichlorohydrin rubber, a terpolymer of epichlorohydrin, propylene oxide and allyl glycidyl ether (GPCO) [13]; nitrile-butadiene rubber (NBR), a copolymer of butadiene and nitrile of acrylic acid (BNKS-18, BNKS-26); and also butadiene (SKD) and isoprene (SKI-3) rubbers. Nitrile-butadiene rubbers are one of the most common rubbers to produce elastomeric composites for sealing purposes. They have a balanced set of properties, BNKS-18 combines frost resistance (T_g = - 50 °C) and oil resistance, but both properties need to be improved. Hydrin T6000 has high frost resistance (T_g = - 60 °C), oil resistance at the level of BNKS-18, but the property usually needs to be increased [14]. The vulcanizates of Hydrin T6000 has low wear resistance. GPCO epichlorohydrin and BNKS nitrile rubbers were used as the basis for the individual rubber elastomeric composite. Subsequently, they were modified with carbon nanotubes and collagen hydrolysate.

The properties of elastomeric composites based on rubber mixtures can be not only additive but also exceed the properties predicted by the ratio of rubbers in the mixture. This is also true in relation to frost resistance. When ingredients are introduced into a mixture of rubbers, the structure and properties of the obtained elastomeric composites can change significantly, which makes it possible to create materials with the required level of properties. Nitrile butadiene rubbers BNKS-18 and BNKS-26 were chosen as components of the mixture responsible for resistance to aggressive working environments, and diene rubbers were chosen as components responsible for frost resistance and wear resistance: butadiene (SKD) and isoprene (SKI-3) rubbers. SKD has a low glass transition temperature (T_g = < - 100 °C) but it is a highly crystallizing rubber, which significantly reduces its long-term frost resistance. Due to thermodynamic incompatibility, elastomeric composites based on mixtures of the two rubbers have two glass transition temperatures, respectively. For crystallizable rubbers, the addition of the second component can slow down the crystallization rate. A typical example of such system is a mixture of SKI-3 isoprene rubber with SKD butadiene rubber, where SKI-3 suppresses the crystallization rate of SKD [15]. Low-temperature performance of the elastomers based on this mixture is better than elastomers based on individual SKD and SKI-3 rubbers due to lower crystallization rate. When combined with nitrile rubbers, the resulting structure can impede the crystallization of diene rubbers at low temperatures.

Multi-walled carbon nanotubes (MWCNTs) manufactured by the Research Center for Radiation Application (China) were used as a carbon nanofiller for Hydrin T6000. These nanofillers were chosen after preliminary investigation of the effect of nanotubes of various compositions: functionalized (MWCNT) and not functionalized with oxygen groups (single-walled CNTs manufactured by OSiAl, Russia).

To determine the optimal content of the nanofiller, we studied rubber blends containing from 0.05 to 10 phr (parts per hundred parts of rubber) MWCNTs. Carbon nanotubes are one of the promising nanofillers that used in many types of elastomeric composites. However, MWCNTs was not previously used to modify GPCO epichlorohydrin rubber. The elastomeric composites were prepared using traditional technology. The developed elastomeric material contained all the necessary ingredients: dispersants, vulcanization activators and accelerators, antioxidant, 50 phr (parts per hundred parts of rubber) of P803 carbon black, sulphur as vulcanization agent. The blends additionally contained from 0.1 up to phr of MWCNTs. The mixtures were prepared in a standard way in a closed "Brabender" plasticorder at 50 °C with 40 rpm shaft rotation speed in 15 minutes, then vulcanized in a vulcanization press.

Hydrin T6000 epichlorohydrin rubber, as well as BNKS-18, BNKS-26 containing 18, 26 % by weight of acrylonitrile respectively, were also modified with collagen hydrolysate obtained from the swim bladder of northern fish, which considered as a multifunctional environmentally friendly modifier. Collagen hydrolysate was obtained from the swim bladder of northern species of fish with alkali-salt hydrolysis followed by freeze-drying. Collagen hydrolysate was added in a quantity of 1, 2, 3, 4, 5 phr with sulphur vulcanization system.

The rubber blends were prepared in a Brabender plasticorder, they contained all the necessary ingredients (accelerators and activators of vulcanization, sulphur, an active filler - carbon black, etc.), then they were vulcanized in an electric press.

To obtain double-layer materials a substrate based on BNKS-18 was used, on which a coating of ultra-high-molecular-weight polyethene (UHMWPE) was applied. UHMWPE has high physical and mechanical properties, practically does not wear out. A high degree of polymer crystallinity determines its low degree of swelling in hydrocarbon media. We also studied the effect of diphenylguanidine (DPG) on the adhesive interaction between GUR-4113 grade UHMWPE with an average molecular weight of 3.9 million g / mol and elastomeric composite based on nitrile butadiene rubber. The technological process of manufacturing products from UHMWPE and elastomeric composite is carried out by hot pressing. It is known [16,17] that the processing temperatures of elastomers and UHMWPE are in the same temperature range from 145 °C to 180 °C, which made it possible to develop a technology for applying it to the surface of an elastomeric material.

The kinetics of vulcanization was studied on a Monsanto rheometer (GOST 12535-84) to identify the effect of modifying additives (MWCNT, collagen hydrolysate) on the properties of Hydrin T6000 and BNKS based elastomeric materials. The kinetic parameters of the crosslinking process of rubber compounds were determined, such as the optimum vulcanization and the rate of crosslinking of vulcanizates.

The main operational characteristics of elastomeric composites were studied using standard methods: physical and mechanical properties by GOST 270-84, swelling degree by GOST 9.030-74, compression set by GOST 9.029-74, wear resistance by GOST 426-77. The abrasion resistance elastomeric composites were studied using MI-2 friction machine and an AP-40 device. The structure of materials was investigated using a JSM-7800FX LV electron microscope (JEOL, Japan) equipped with an Oxford X-ray spectroscope.

The study of the strength of the adhesive bond between elastomers and UHMWPE was carried out by GOST 6768-75 on a Shimadzu Autograph tensile testing machine (Shimadzu, Japan). Investigation of the structure of the interphase boundary layer of the elastomer and UHMWPE was carried out on a JEOL JSM-7800F scanning electron microscope (JEOL, Japan).

3. Results and discussion

3.1. Epichlorohydrin rubber based elastomeric composites modified with carbon nanotubes

Problems associated with improving the properties of elastomeric composites can be solved in various ways. Most often by changing the composition of rubber blend. To achieve the desired level of the properties of elastomeric composites based on Hydrin T6000 epichlorohydrin rubber we used carbon MWCNTs.

Electron microscopic studies of MWCNTs (figure 2) showed that MWCNTs were agglomerated with a fine structure in the form of nanofibers with a diameter of 20-47 nm and a length of 8-30 μm. The surface of carbon nanotubes was investigated using an X-max20 energy-dispersive attachment from OXFORD. It was found that in addition to carbon (88% by weight), it also contains up to 10% oxygen (figure 3). Functionalization of multi-walled carbon nanotubes with oxygen gives a definite advantage over analogues because oxygen atoms are also present in the main chain of GPCO epichlorohydrin rubbers. It should contribute to the uniform distribution of MWCNTs in the bulk of the rubber blend at the mixing stage. The presence of Co and Al elements is explained by the residues of the catalytic system used in the synthesis of nanotubes.

The analysis of the rheometric curves showed that the introduction of MWCNTs does not significantly affect the technological parameters: the viscosity of modified rubbers, the crosslinking rate and the optimum vulcanization do not undergo significant changes.

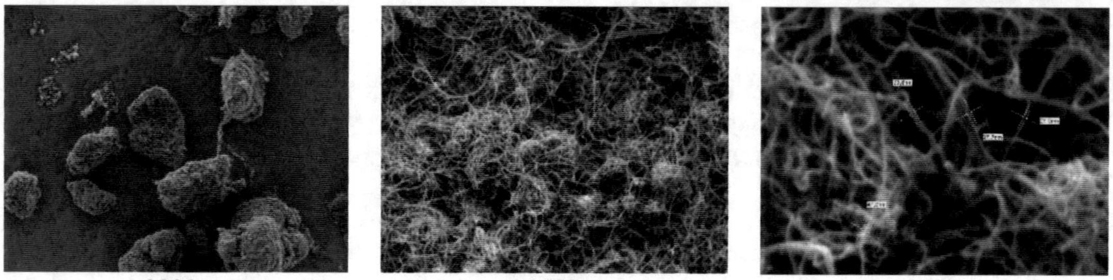

x1000 x10000 x50000

Figure 2. Electron micrographs of MWCNTs.

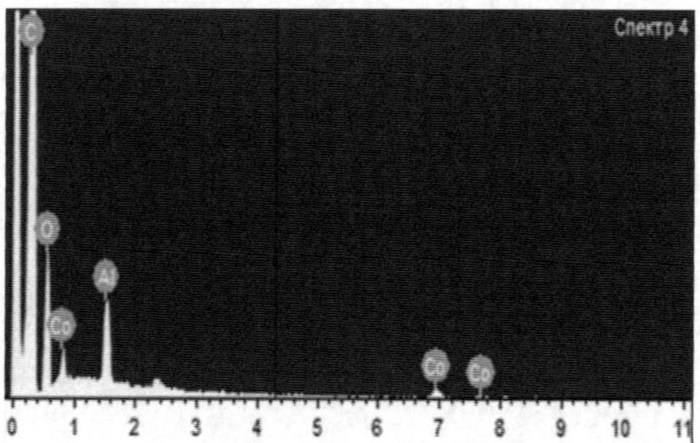

Element	Mass %	Atomic %
C	88.34	91.73
O	9.89	7.71
Al	0.76	0.35
Co	1.01	0.21

Figure 3. Elemental composition of MWCNTs.

Table 1 shows the properties of Hydrin T6000 epichlorohydrin rubber based elastomeric composites modified with nanotubes.

Table 1. Properties of elastomeric composites based on epichlorohydrin rubber modified with MWCNTs

Hydrin T6000+ MWCNTs phr	ε_p [a] (%)	f_P [b] (MPa)	f_{100} [c] (MPa)	C [d] (%)	ΔQ [e] (%)	ΔV [f] (cm^3)
GPCO	364	11.0	3.6	61.2	35.5	1.30
GPCO / 0,05	346	10.0	3.2	63.2	34.8	0.90
GPCO / 0,1	368	10.8	3.2	62.2	35.2	0.87
GPCO / 0,2	339	10.8	3.3	61.2	36.6	0.92
GPCO / 0,3	339	11.0	3.2	63.0	35.0	0.84
GPCO / 0,5	339	11.0	3.9	61.2	33.3	0.76
GPCO / 1	328	11.0	3.3	59.5	32.7	0.78
GPCO / 2	312	11.0	4.6	64.0	20.1	0.75
GPCO / 5	229	11.0	5.7	67.1	21.2	0.72
GPCO / 10	183	12.0	8.0	69.3	17.3	0.69

[a] elongation at break.
[b] tensile strength.
[c] 100% modulus.
[d] compression set at 100 °C and 72 h.
[e] swelling degree at 70 °C and 72 h in oil.
[f] volumetric wear.

Our study of physical and mechanical properties showed that as the content of MWCNTs increased, the strength characteristics of elastomeric composites based on Hydrin T6000 GPCO tend to increase. However, at high concentrations of MWCNTs, the elongation decreased significantly (by 49%), and the modulus doubled. It was found that the introduction of MWCNTs had the strongest effect on the wear resistance and the degree of swelling. Thus, the volumetric wear of the obtained elastomeric composites with an increase in the concentration of nanotubes decreased to 47%, which was apparently associated with a decrease in the values of the statistical and dynamic coefficients of friction of elastomeric composites based on Hydrin T600 GPCO modified with MWCNTs [18]. The degree of swelling at high contents of MWCNTs was also reduced two times.

Among all considered contents of nanotubes, the optimum properties were observed with the introduction of 1 phr of MWCNTs. These elastomeric composites, as was shown using electron microscopy, were characterized by the most uniform distribution of MWCNTs in the elastomeric matrix, high wear resistance (by 40%), lower compression set and the degree of swelling compared to the initial unfilled elastomeric composite, while maintaining high elasticity and strength characteristics.

3.2. Epichlorohydrin rubber based elastomeric composites modified with collagen hydrolysate

Recently there has been a development in the manufacturing of new biodegradable polymeric materials. The introduction of biologically active compounds such as keratin, lignin, starch, amino acids, and phospholipids into the polymer matrix allows obtaining elastomeric composites with improved properties. In this work, we obtained a collagen hydrolysate from the swimming bladder of northern species of fish. It was used as a modifier of elastomeric compounds such as Hydrin T6000 GPCO epichlorohydrin rubber, nitrile-butadiene–rubbers (BNKS-18, BNKS-26). Collagen hydrolysate was obtained from the swim bladder of northern species of fish with alkali-salt hydrolysis followed by freeze-drying. The substance retained its collagen nature and a set of functional groups inherent in proteins during hydrolysis. Collagen hydrolysate was added in a quantity of 1, 2, 3, 4, 5 phr with sulphur vulcanization system. Its effect was investigated on model non-carbon black filled blends based on BNKS-18, BNKS-26.

It was found that collagen hydrolysate affects the structuring process of the studied rubbers. The data of vulcanization kinetics showed that the introduction of the collagen hydrolysate into the rubber compound had a significant effect on the duration of their cure time and the cure rate index. Comparing results of the cure rate index for the control samples and the modified samples, showed that the cure rate index increases with the introduction of a collagen hydrolysate by 216%, 359%, 83% based on BNKS-18, BNKS-26 and Hydrin T6000 GPCO, respectively. Moreover, a collagen hydrolysate reduced the cure time vulcanization. It can be assumed that the collagen hydrolysate was involved in the crosslinking of the elastomeric material, contributing to the formation of additional cross-links between the macromolecules.

The qualitative analysis of the IR spectroscopy data indicated new absorption bands appear in the spectra of NBR compounds containing hydrolysate. The presence of a new absorption band at 1562 cm−1 which corresponds vibrations of NH–, C = N– bonds of collagen, may indicate a chemical interaction of the studied rubbers with hydrolysate. Table 2 presents the results of the effect collagen hydrolysate on mechanical properties of blends.

The study of physical and mechanical properties showed that the addition of the collagen hydrolysate promoted an increase in tensile strength of the NBR rubber. It was explained by the cross-link density increase of rubber with the introduction of collagen [19]. It should be noted that all blends of NBR rubber filled with collagen hydrolysate were characterized by a higher cross-linking density compared with control samples which did not contain hydrolysate.

At the same time, the modifier did not cause a significant effect on the physical and mechanical properties of Hydrin T6000 GPCO blends. The tensile strength of the Hydrin T6000 GPCO filled with collagen hydrolysate remained almost at the level of the control samples. It was found that the use of collagen hydrolysate as a filler of blends caused a reduction of the level of compression set accumulation of elastomeric composites based on GPCO and NBR. The compression set of NBR and GPCO were decreased with increasing collagen hydrolysate content in blends. The compression set decreased by 27% and 20% compared with the control samples based on NBR, GPCO, respectively.

It can be assumed that the collagen hydrolysate may be a secondary curing agent. To verify that the collagen was involved in cross-linking processes, the structures of blends were investigated using electron microscopy. Scanning electron microscopy was used to observe the morphology of the control samples of NBR and blends filled only collagen hydrolysate. These blends were prepared in Brabender plasticorder and subjected to heating in an electric press at a vulcanization temperature at the same conditions as described above but did not contain the sulphur vulcanization systems. The images are shown in figure 4.

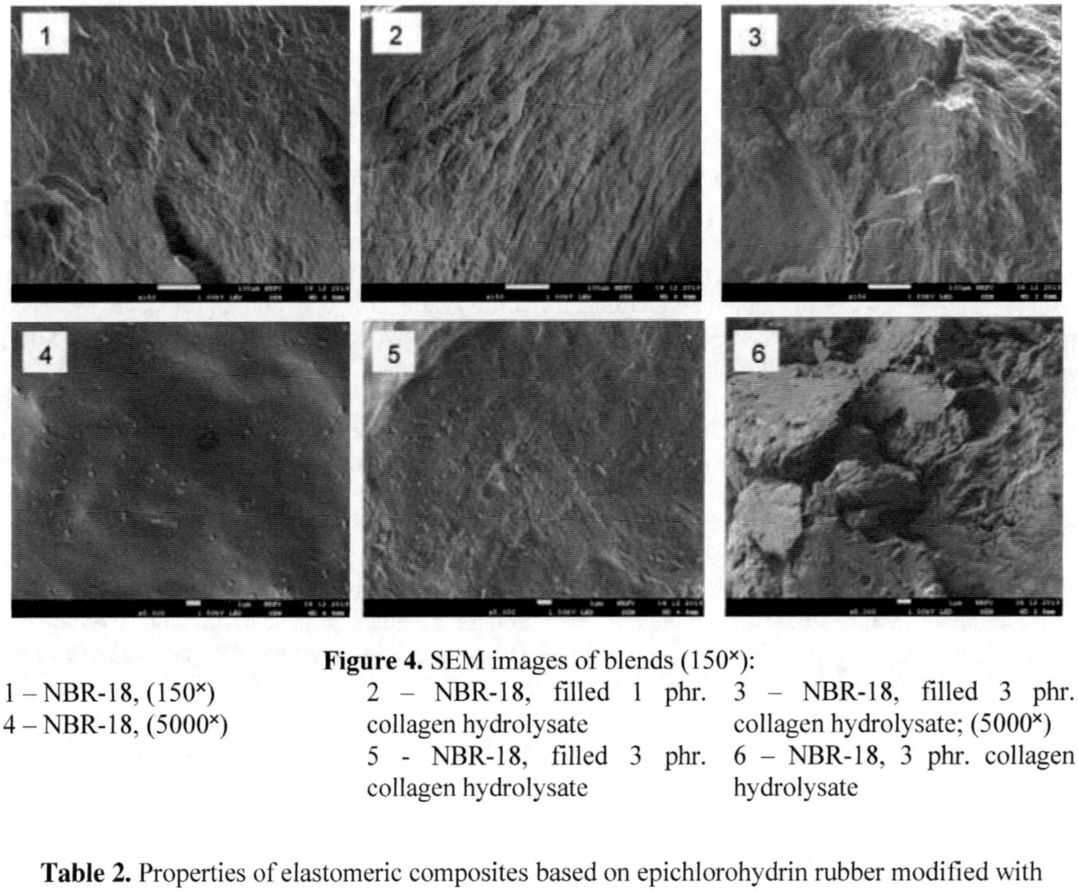

Figure 4. SEM images of blends (150ˣ):
1 – NBR-18, (150ˣ)
2 – NBR-18, filled 1 phr. collagen hydrolysate
3 – NBR-18, filled 3 phr. collagen hydrolysate; (5000ˣ)
4 – NBR-18, (5000ˣ)
5 - NBR-18, filled 3 phr. collagen hydrolysate
6 – NBR-18, 3 phr. collagen hydrolysate

Table 2. Properties of elastomeric composites based on epichlorohydrin rubber modified with MWCNTs

Collagen hydrolysate, phr	BNKS-18			BNKS-26			Hydrin T6000 GPCO		
	f_P [a] (MPa)	C [b] (%)	ν [c] (mol/cm^3)	f_P [a] (MPa)	C [b] (%)	ν [c] (mol/cm^3)	f_P [a] (MPa)	C [b] (%)	ν [c] (mol/cm^3)
0	3.5	80	0.96	5.3	80	1.26	7.6	65	3.5
1	3.2	70	1.39	4.2	66	1.37	8.2	64	3.2
2	3.8	66	1.53	4.1	66	1.72	7.6	60	3.8
3	4.5	68	1.57	4.6	63	1.50	7.9	57	4.5
4	4.7	68	1.55	3.5	76	1.28	6.9	53	4.7
5	5.2	66	1.24	-	-	-	7.0	61	5.2

[a] tensile strength.
[b] compression set at 100 °C and 72 h.
[c] crosslink density

The surface of elastomeric composite containing collagen was embossed, which can indicate that the collagen hydrolysate could be involved in the formation of new cross-links with the rubber macromolecule. This explained with the introduction of hydrolysate in the rubber formulation, the density of the network increases (table 2), which may be a consequence of additional crosslinking of rubber macromolecules. Crosslink density of compounds was determined by the solvent swelling method using the Flory–Rehner equation. For this purpose, the insoluble blends gel remaining after extraction was dried to constant weight, and then it was subjected to swelling in chloroform at room temperature for 18 h.

The introduction of small amounts of collagen hydrolysate into rubber blends led to a significant increase in the cure rate index, a reduction of the level of compression set accumulation while increasing or maintaining the mechanical properties of the modified rubbers, which is associated with the chemical modification of elastomers by collagen hydrolysate and an increase in the density of the network of vulcanization bonds. Thus, the collagen hydrolysate obtained from swim bladder is a new promising rubber compound modifier.

3.3. Elastomeric composites based on mixtures of diene and nitrile butadiene rubbers

The ratio between nitrile butadiene (BNKS-18, BNKS-26) and diene rubbers (SKD, SKI-3) was 70:30. The nitrile-butadiene phase prevailed to provide acceptable aggressive media resistance in non-polar hydrocarbon mineral environments. Stability in synthetic polar media was provided by diene rubbers.

The composition of elastomeric composites also included: vulcanization activators, vulcanization accelerators, antioxidant, carbon black filler, plasticizer, vulcanizing agent. The rubber blend was obtained based on nitrile butadiene rubbers BNKS-18 and BNKS-26, into which 40% of the total carbon black and the plasticizer were added, which is then combined with the diene rubbers and the remaining ingredients (figure 5).

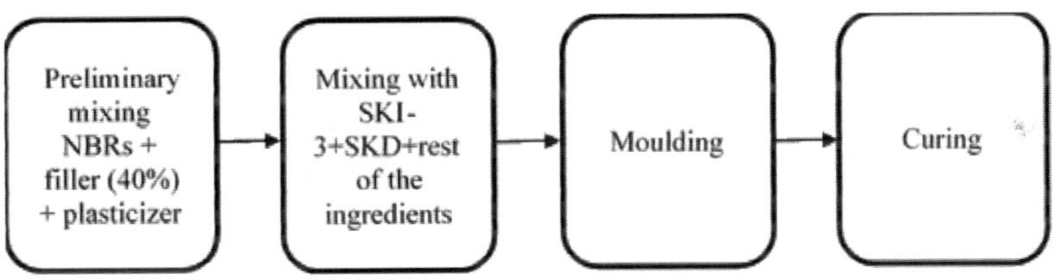

Figure 5. Elastomeric composites mixing technology based on mixture of rubbers

Table 3. Properties of elastomeric composites based on mixture of rubbers

Hydrin T600+ MWCNTs phr	ε_p [a] (%)	f_P [b] (MPa)	f_{100} [c] (MPa)	C [d] (%)	ΔQ [e] (%)	T_g [f] (°C)	ΔV [g] (cm^3)
V-14	223	9.0	5.1	74.0	2.19	-45.3	1.297
NBR+diene	428.1	10.0	2.2	66.2	18.6	-60.2	0.697

[a] elongation at break.
[b] tensile strength.
[c] 100% modulus.
[d] compression set at 100 °C and 72 h.
[e] swelling degree at 70 °C and 72 h in hydraulic oil.
[f] glass transition temperature.
[g] volumetric wear.

This method of combining provides better mixing due to a different ratio of the viscosities of the diene and nitrile rubber phases and a more uniform distribution of the components in the mixture. After mixing the rubber blend in the Brabender plasticorder it was cured in an electric hydraulic press. The obtained properties are shown in table 3.

For comparison, the characteristics of industrial rubber V-14, based on rubber BNKS-18, are given. Tensile strength of the investigated elastomeric composite based of rubber mixture was comparable to the V-14 rubber. The elasticity of the investigated elastomeric composite was 1.9 times higher than V-14. The compression set values for the investigated elastomeric composite were 8% lower than that of V-14. Lower values of the compression set indicate better relaxation properties of the material (better recoverability after unloading after accelerated ageing), which are an important indicator for elastomeric composite for sealing purposes. The elastomeric composite based of rubber mixture had increased wear resistance: volumetric wear was 1.9 times lower than for V-14 rubber. The glass transition temperature was -60.2 °C. The value is 14.9 °C lower than that of V-14 rubber. It should also be noted that the high frost resistance of the elastomeric composite based on mixtures of butadiene-nitrile and diene rubbers was due not to the presence of plasticizers, but the polymer base having high low-temperature characteristics. This elastomeric composite based on mixture of rubbers was patented (RU 2 705 069 C1).

3.4. Effect of diphenylguanidine on the adhesive interaction of ultra-high molecular weight polyethylene and elastomeric composite based on nitrile butadiene rubber

Currently, polymer composite materials are increasingly finding widespread use in various industries due to their unique qualities and relatively simple processing process. Polymer products are used as structural and functional materials in critical units and joints in mechanical engineering, automotive and aircraft construction [20,21]. Elastomeric materials remain unsurpassed sealing materials, but their seals perform at lower loads and sliding speeds than other polymer classes. In this regard, there is a constant search and development of new polymeric materials, which are subject to even higher reliability requirements during operation. One of the most effective methods for manufacturing materials with improved performance is the application of thermoplastic coatings on elastomeric materials or the creation of multilayer polymeric materials. The advantage of the method is the ability to combine the properties of different materials in one product - rigidity, elasticity, hardness, and damping properties. One of the possible ways of creating a double-layer material is a combination of thermosetting plastic with a thermoplastic. It can be elastomer and polyethene (figure 6). These materials are thermodynamically incompatible, *i.e.* adhesion at the contact boundary is low. It is necessary to take several measures to increase the reliability of the operation of such materials.

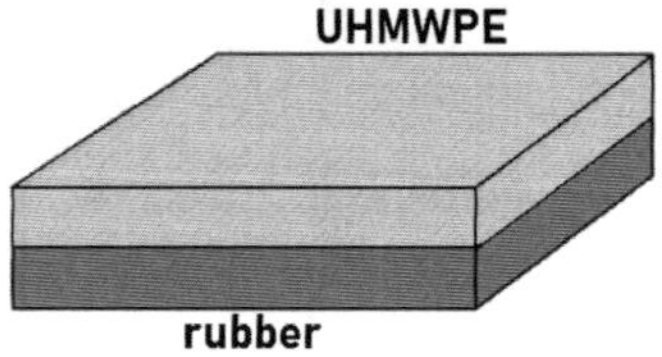

Figure 6. Double-layer material based on UHMWPE and rubber.

We investigated the effect of DPG, an accelerator of sulphur vulcanization of medium activity, on the adhesive interaction between GUR-4113 UHMWPE and elastomeric composite based on nitrile butadiene rubber (BNKS-18). DPG used as secondary accelerators in combination with higher activity primary accelerators. As it was found in [22], the addition of DPG with sulphur to polyethene leads to the formation of sulphide bonds between polyethene macromolecules. Since UHMWPE is also polyethene by chemical structure, the same chemical processes of interaction with sulphur apply to it.

We have developed a technology for producing double-layer materials, based on joint hot pressing at a temperature of 155 °C, which creates conditions for good interaction of the components. The study of the strength of the adhesive bond between elastomers and UHMWPE was carried out by analogy with GOST 6768-75 on a Shimadzu Autograph tensile machine. Investigation of the structure of the interphase boundary layer of the elastomer and UHMWPE was carried out on the scanning electron microscope. Table 4 shows the results of a study of the effect of adding 0.3 phr of DPG in the process of mixing the rubber blend ingredients to adhesion between the elastomer and UHMWPE.

Table 4. Adhesion of UHMWPE with an elastomer based on NBR

Sample	Adhesion, N/cm
UHMWPE / NBR	56
UHMWPE / NBR + 0.3 DPG	148

In the process of delamination of double-layer samples based on BNKS-18 rubber with DPG, destruction occurred on the elastomer. This indicates that the adhesive interaction between materials exceeded the cohesive strength of rubber. When a double-layer material without DPG was delaminated, destruction occurs along the contact boundary.

The introduction of DPG into the rubber blend led to an increase in the intensity of crosslinking of rubber macromolecules with UHMWPE (covulcanization) and a change in the supramolecular structure, which can be seen in micrographs of the interaction interface (Figure 7, b). The process of formation of an adhesive bond between UHMWPE and elastomer occurs according to different mechanisms, depending on the ingredients introduced into the rubber mixture.

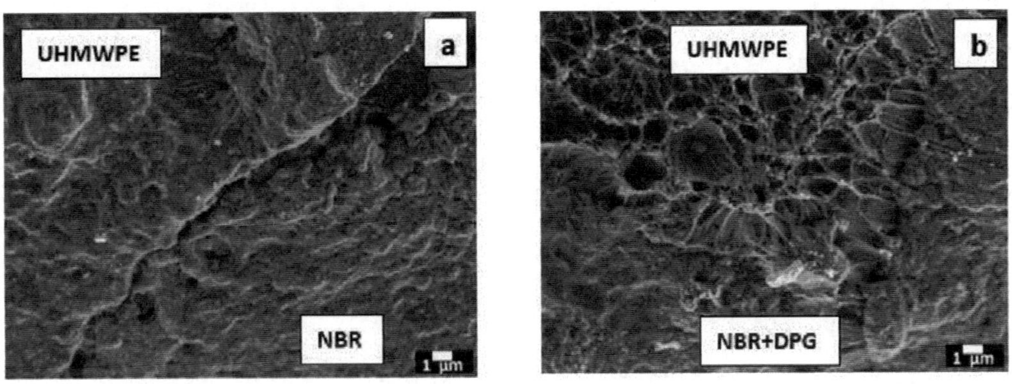

Figure 7. Micrographs of UHMWPE transitional layer with elastomers: a) NBR, b) NBR + 0.3 DPG phr.

Explaining the adhesive interaction between polymers, several theories were considered: mechanical, electronic, diffusion, adsorption, and chemical. In this case, the most preferable is the formation of chemical bonds at the interaction interface was confirmed by IR spectroscopy. Thus, the strength of the adhesive interaction between elastomers and UHMWPE largely depends on the composition of the vulcanization group. With the correct combination of vulcanization accelerators, sulphide bonds are formed between the rubber macromolecules and UHMWPE. It leads to an interaction between the elastomer and UHMWPE at the chemical level. Such joints have high strength, and such materials recommended for widespread use in industry.

4. Conclusions

The development of frost-resistant elastomeric materials for sealing purposes for Arctic applications is a complex process that requires many factors determined by the operating conditions (climate, product type, working environment, load, etc.). Based on years of research, we developed approaches that have made it possible to create materials with high frost resistance in combination with oil and wear resistance (patented materials: US 8,841,370 B1; US 8,822,579 B1; RU 2 705 069 C1; RU 2 615 416 C2; RU 2 641 816 C2; RU 2 664 405 C1). The materials were successfully tested in the climatic conditions of the Yakutia. In this study, there are examples of similar recently developed elastomeric composites. The materials were created based on both traditional BNKS NBR materials and the new frost resistant Hydrin T6000 GPCO rubber. By comprehensively affecting the structure and vulcanization parameters of rubbers by introducing oxygen-functionalized carbon nanotubes or specially prepared additives of natural origin, it was possible to improve the operational properties, wear resistance, and some other properties of the obtained elastomeric composites maintaining high frost resistance. The materials with extended functionality were developed by combining several different polymers in one elastomer. We obtained material based on NBR elastomer and wear-resistant UHMWPE plastic with combined properties. The thickness of the UHMWPE layer can be varied within wide limits – from applying thin coatings on a finished rubber product to obtaining duplicated materials, where the thickness of the elastomeric material and thermoplastics comparable. This expands the range of this method. In this case, high adhesion between layers achieved by the introduction of DPG, which ensures the crosslinking of materials at the interface. All investigated materials were recommended for implementation on the territory of the Russian Federation.

References

[1] 2005 *Science and Technology of Rubber* ed Mark E J, Erman B et al. (Burlington: Elsevier Academic Press)
[2] Akulichev A G, Echtermeyer A T and Persson B N J 2018 *Int. J. of Pressure Vessels and Piping* **160** 14
[3] Grelle T and Wolff D, Jaunich M 2017 *Pol. Testing* **58** 219
[4] Warren P 2008 *Sealing Technol.* **9** 7
[5] Zhang J, Wang L and Zhao Y 2013 *Mat. and Des.* **50** 322
[6] Jaunich M, Wolff D and Stark W 2013 *Kautsch. Gummi Kunstst.* **66** 26
[7] Chersky I N and Morov V A 1990 *Kautsch. Gummi Kunstst.* **43** 128–9
[8] Petrova N N, Popova A F and Fedotova E S 2003 *Int. Pol. Sci. and Technol.* **30(2)** 7-10
[9] Petrova N N, Popova A F and Startsev O V 2005 *Progr. in Chemometrics Res.* ed. Pomerantsev A (NY: Novascience Publ.) pp 265-271
[10] Petrova N N and Fyodorova A F 2005 *Trends in Mol. and High Mol. Sci.* ed. Zaikov G E (NY: Novascience Publ.) pp 301-6
[11] Petrova N N, Portnyagina V V, Mukhin V V, Timofeeva E N, Matveeva N V and Ivanova S F 2018 *Mat. Sci. For.* **945** 412-6
[12] Petrova N N 2006 *Int. Pol. Sci. and Technol* **33** pp 33-8
[13] 2008 *Handbook of Specialty Elastomers* ed. Klingender R C (Boca Ranton: CRC Press)
[14] 2012 *Handbook of Rubber Technologist* vol 1, ed. Reznichenko S V and Morozov Y L (Moscow: Izdatelskii Tsentr "Tekhinform" MAI)
[15] Bukhina M F and Kurlyand SK 2007 *Low-Temperature Behaviour of Elastomers* (CRC Press)
[16] Andreyeva I N, Veselovskaya Ye V and Nalivayko Ye I 1982 Ultrahigh molecular weight polyethylene of high density. (L: Chemistry, V. 5)
[17] Chaykun A M, Yeliseyev O A, Naumov I S and Venediktova M A 2013 Proceedings of VIAM
[18] Bukovskij P O, Morozov A V, Petrova N N *et al.* 2019 *Izvestiya RAN, Mekhanika Tverdogo Tela* **6** 148
[19] Sruthi P R and Anas S. 2020 *J. Polym. Sci.* **58** 1039
[20] Carraher C E 2018 Carraher's Polymer Chemistry (Boca Raton: CRC Press, Taylor & Francis Group)
[21] Kickelbick G 2007 Hybrid Materials: Synthesis, Characterization, and Applications (Weinheim: Wiley - VCH)
[22] Karaseva Y S, Bashkatova T V, Cherezova E N and Husainov A D 2006 *Kazan Technol. Univ. Bull.* **5** 57-62

Numerical analysis of an electrostatically formed membrane mirror

O S Volchek

Faculty of Fundamental Sciences, Bauman Moscow State Technical University, 2nd Baumanskaya Street, 5/1, Moscow, 105005, Russia

E-mail: volcheko@bmstu.ru

Abstract. The paper presents a numerical algorithm developed to determine the deformed shape of a membrane mirror controlled by electrostatic forces. Deformable mirrors are key components that are used in combination with wavefront sensors and real-time control systems in adaptive optics. The electrostatic membrane mirror concept implies using a thin conductive reflective membrane stretched over a solid flat frame and deformed electrostatically by applying control voltages to electrostatic actuators positioned behind the membrane. The proposed algorithm implies solving a coupled structural-electrostatic problem by using finite element and boundary element methods. Small deflections of a membrane are described by Poisson's equation. The electric charge distribution over the membrane surface having a prescribed potential is governed by a Fredholm integral equation of the first kind. The coupled problem is solved iteratively, and a criterion for terminating iterations when searching for a steady-state solution is presented. The distinctive feature of this approach is that it allows us to take into account electrical edge effects typical for conducting thin-walled structures of very small thickness. Illustrative examples are provided to show the applicability and validity of the proposed method as well as its advantage over some existing techniques.

1. Introduction

Electrostatic formation and control of membrane mirrors has received much attention since the late 1970s as a promising adaptive optics technology that benefits both Earth and space applications [1, 2]. The key feature of any adaptive optical system is that it uses optical elements that can change their shape when outside control signal is applied [3]. Figure 1 shows two typical designs of an electrostatically formed membrane mirror that have been under intensive theoretical and experimental study in recent decades. In the first design, the mirror is considered to be a two-membrane capacitor. When the electrical voltage is applied, electrostatic forces deform the membranes so that their shape is close to the shape of a parabolic reflector. The focus length can be changed by changing the voltage. In the second design, a rigid segmented electrode is used to provide the surface shape control by means of voltage variation on the segments. The typical membrane mirror is made of a thin lightweight flexible polymer film coated with a very thin layer of metal, usually aluminum, but gold and silver can also be used. To provide a satisfactory performance of the mirror, membrane materials must have high elasticity modulus, low density, small thickness, high thermal stability, low thermal expansion coefficients, and, for applications in space, strong space radiation resistance. In [4], two candidate materials for future space applications are considered. It is shown that polyimide films have many advantages over polyester films as they have a larger modulus of elasticity, higher tensile strength, lower thermal expansion coefficient, smaller elongation at break, and stronger anti-ultraviolet radiation ability.

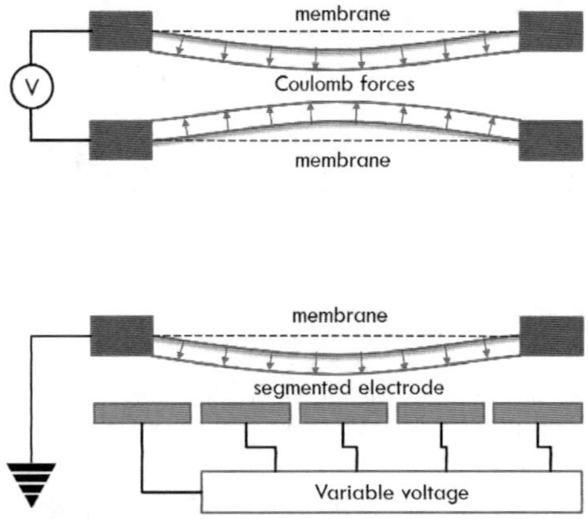

Figure 1. Basic designes of an electrostatically formed membrane mirror.

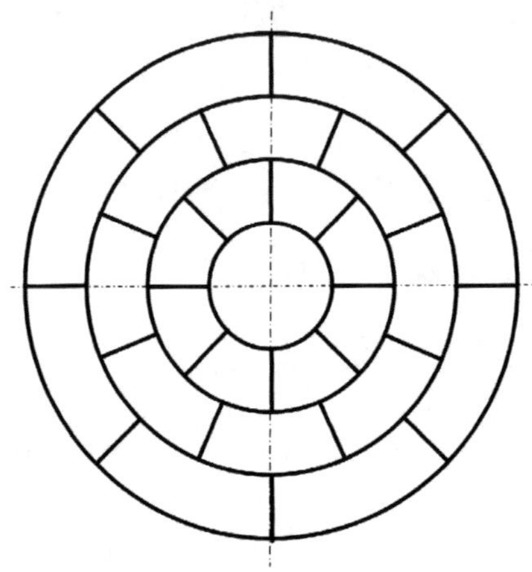

Figure 2. A segmented electrode.

To determine the shape of an electrostatically formed membrane mirror, the coupled structural-electrostatic problem is to be solved. The main challenge is to compute electrostatic pressure acting on a conducting membrane. In [5], the first configuration of an electrostatically formed membrane mirror was considered. Two elastic membranes with preliminary tension were fixed along their contours. To determine the membranes' shape under electrostatic pressure, the theory of large deflections of plates [6] was applied, and the nonlinear coupled problem was solved numerically using the shooting method [7]. The electric field between the membranes was evaluated by a simple formula

$$E = \frac{\Delta u}{H - 2w}, \tag{1}$$

where Δu is the applied electrical voltage, H is the distance between the membranes, and w is the deflection of the membrane from its normal flat shape. The main limitation of formula (1) is that it does not take into account the so-called "edge effect", which occurs near the capacitor edges and plays an important part in the electromechanical interaction [8].

In [9], an attempt was made to obtain a more elaborate solution to the electrostatic problem by applying the finite element method to solve the Laplace equation in an open domain. In [10], a strongly coupled electro-mechanical finite element was introduced to model an electrostatic actuated deformable mirror. However, since the coupled problem is nonlinear, an iterative procedure should be applied, which makes the computational cost extremely high. To reduce the model complexity and decrease the computational cost, we suggest using integral equations to determine electrical charge distribution on a conducting surface having a prescribed potential. The suggested approach has been stimulated by recent advances in integral equations [11], boundary elment methods [12], and new applications of the optic technology [13, 14, 15, 16, 17]. Our objective is to analyze the influence of the geometry of the segmented electrode and electric edge effects on the deformed membrane profile by means of computer simulations. The main focus is on the numerical aspects of modelling the electrostatic formation of metallized polymer films of arbitrary shapes. The simulation tool was developed and tested using the Matlab environment. A parallel-plate capacitor with a metal-coated membrane and a rigid metal segmented electrode is considered as an example.

IOP Conf. Series: Materials Science and Engineering 1060 (2021) 012023 doi:10.1088/1757-899X/1060/1/012023

2. Mathematical model

The mathematical model of an electrostatically formed membrane mirror includes equations of soft-shell structures and equations of electrostatics of thin-walled conductors that must be solved simultaneously. The problem is coupled because the value of electrostatic pressure acting on a flexible conductor is determined by the electric charge distribution over its surface. When the conductor is deformed, the charge distribution changes depending on the surface curvature.

The fundamentals of the theory of large deformations of soft shells used in the stress analysis of structures made of composite materials were presented in [18, 19]. In [20], strict nonlinear equations of soft shells in an electrostatic field were deduced. These equations form the basis of numerical algorithms for computing the formation of flexible conductors of complex shapes. They can also be used to construct solutions in simpler cases.

In this study, the mathematical model includes Poisson's equation to describe small deflections of a membrane and a Fredholm integral equation of the first kind to describe a surface charge distribution over the membrane surface [21].

2.1. Properties of a conductor in an electrostatic field

Owing to the mobility of charges, the electric field within the volume of a conductor is always zero. In all points on the surface the tangential components of the field are also equal to zero. Therefore, outside a conductor, the electric field lines are perpendicular to its surface, ending or beginning on charges on the surface. The entire volume of a conductor is equipotential because the electric field strength $\mathbf{E}$ is the negative gradient of the electrostatic potential ϕ

$$\mathbf{E} = -\nabla\phi. \tag{2}$$

For a conductor in a homogeneous and isotropic dielectric, we have

$$E = \frac{\sigma}{\varepsilon\varepsilon_0}, \tag{3}$$

where σ is the local surface charge density of a conductor, $\varepsilon_0 = 8.854 \cdot 10^{-12} F/m$ is the vacuum permittivity, and ε is the relative permittivity, which is equal to unity in this study. Distribution of electric charges on the conductor surface depends on the surface curvature: the charge density is greater where the radius of curvature is smaller.

The potential of the electrostatic field created by charged surfaces (surface field potential), given that $\phi(\infty) = 0$, is equal to

$$\phi = \frac{1}{4\pi\varepsilon\varepsilon_0} \int_S \frac{\sigma dS}{r}, \tag{4}$$

where r is the distance from the surface charge σdS to the field point in question. Integration is performed over the surfaces where the electric charges are distributed.

2.2. Pressure on the surface of a conductor due to a surface charge distribution

The electric force dF acting on a surface area dS is directed along the normal to the surface and is determined by the formula

$$dF = \frac{\sigma^2}{2\varepsilon\varepsilon_0} dS = \frac{\varepsilon\varepsilon_0 E^2}{2} dS. \tag{5}$$

Hence, the force per unit area or electrostatic pressure is

$$p = \frac{dF}{dS} = \frac{\sigma^2}{2\varepsilon\varepsilon_0} = \frac{\varepsilon\varepsilon_0 E^2}{2}. \tag{6}$$

This pressure tends to draw the conductor into the field, regardless of the sign of the surface charge.

2.3. Deformation of a flexible membrane

To perform the stress-strain analysis of thin-walled structures with low flexural rigidity, the theory of soft shells is used [18]. In accordance with this theory, the soft shell is momentless and unsuited for absorbing compressive forces. On the surface, where compressive forces may occur, folds appear. We consider a flexible metallized polymer membrane, which is prestressed to maintain a tensile state and to avoid folds under applied electrostatic pressure. It is assumed that the membrane is made of a homogeneous linearly elastic material. Small normal deflections $w(x, y)$ of a prestressed metallized polymer membrane are described by Poisson's equation

$$T_x \frac{\partial^2 w}{\partial x^2} + T_y \frac{\partial^2 w}{\partial y^2} = p(x, y), \tag{7}$$

where T_x and T_y are prestressing forces per unit length acting in the x and y directions, and $p(x, y)$ is electrostatic pressure depending on the deflection of the membrane and defined by relationship (6).

3. Numerical algorithm

To determine the form of a metallized membrane loaded by electrostatic forces, we need to know the charge distribution over the conducting surface. The surface charge density uniquely defines electrostatic pressure acting on the membrane and thus determines its shape. The problem is coupled and nonlinear because the deformed shape of the membrane is not known in advance but depends on the surface charge distribution which in turn depends on the shape of the membrane.

All solutions shown here are based on the partitioned scheme where individual problems are solved independently and the interaction information is transferred between them at every stage of the solution process. The models and the software developed for computing the charge distribution are based on the boundary element technique considered in [22], and those for the membrane deformation are based on the finite element method described in detail in [23, 24].

3.1. Synthesis of an electric field due to a surface potential

To explain the main points of the approach, let us consider a rectangular conducting plate of negligible thickness having a prescribed potential ϕ_0 as shown in figure 3. According to [21], [25], the electric potential synthesis problem in this case is reduced to the solution of a Fredholm integral equation of the first kind

$$\frac{1}{4\pi\varepsilon\varepsilon_0} \int_{-a}^{a} \int_{-b}^{b} \frac{\sigma(x', y')dx'dy'}{R} = \phi_0, \tag{8}$$

where $R = \sqrt{(x - x')^2 + (y - y')^2}$ is the distance between a source point (x', y') and the field point (x, y). To solve the problem, the method of subsections with collocation is implemented. According to this method, the rectangular plate is divided into N rectangular elements

$$x_{i+1} - x_i = \Delta x, \quad y_{i+1} - y_i = \Delta y. \tag{9}$$

Collocating the charge densities σ_j at the centers of the elements (x_{cj}, y_{cj}) gives the system of N linear equations

$$\sum_{j=1}^{N} a_{ij}\sigma_j = \phi_i, \tag{10}$$

where $\phi_i = \phi_0$, and

$$a_{ij} = \frac{1}{4\pi\varepsilon\varepsilon_0} \int_{x_i}^{x_{i+1}} \int_{y_i}^{y_{i+1}} \frac{dx'dy'}{\sqrt{(x_{cj} - x')^2 + (y_{cj} - y')^2}}. \tag{11}$$

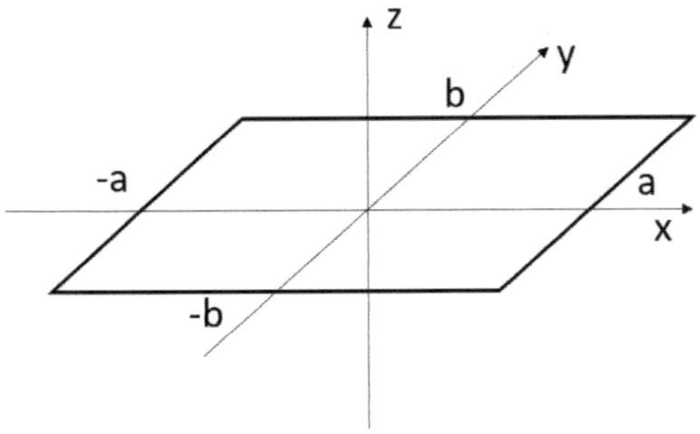

Figure 3. A rectangular conducting plate of negligible thickness.

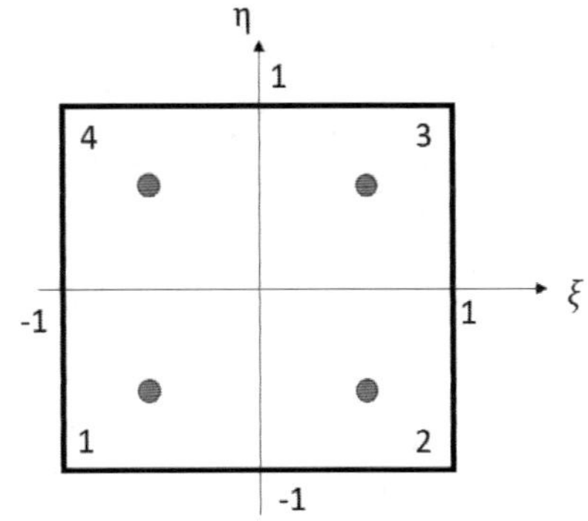

Figure 4. An isoparametric finite element with integration points.

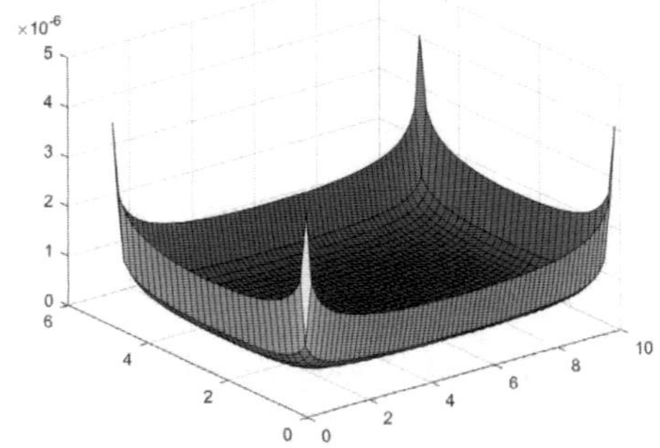

Figure 5. Surface charge density distribution on a rectangular metal plate.

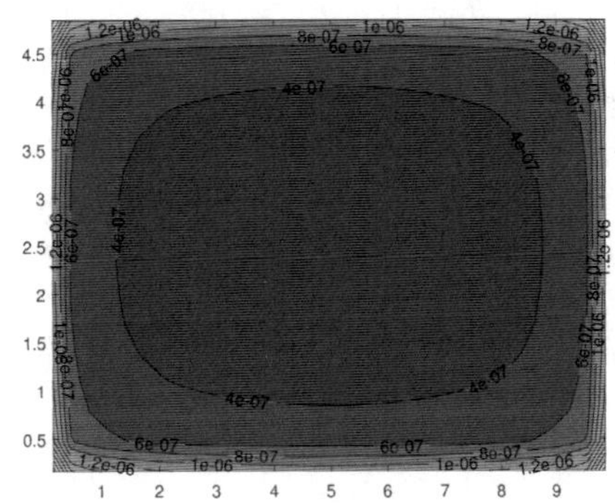

Figure 6. Isolines of surface charge density distribution on a rectangular metal plate.

The diagonal elements a_{ii} of the system matrix are calculated analytically, while the off-diagonal elements are computed numerically [26]. It is essential that this electrostatic problem is solvable and has only one solution. Figures 5 and 6 show the surface charge density distribution on a rectangular metal plate divided into 64×32 square elements. The plate is assumed to have the length $2a = 10m$, the width $2b = 5m$, and the surface potential $\phi_0 = 10^5 V$. The results are in good agreement with those presented in [21].

This technique can be applied with relevant modifications to the calculation of the surface charge density distribution on a metal circular disk. The charge density of an infinitely thin circular disk is given analytically by

$$\sigma(r) = \frac{Q}{2\pi a} \frac{1}{\sqrt{a^2 - r^2}}, \tag{12}$$

where Q is a total charge, a is the radius of the disk, and r is the radial distance from the center of the disk [27]. Note the special values $\sigma(0) = \frac{Q}{2\pi a^2}$ and $\sigma(r \to a) = \infty$.

Figure 7 shows 2700 collocation points on the conducting circular disk of radius $a = 1m$, and figure 8 displays the surface charge density distribution on it, given the disk potential $\phi_0 = 20kV$.

The convergence of the results of computing the surface charge density on the circular disk is shown in table 1. A double-ended estimate is used to verify the solution.

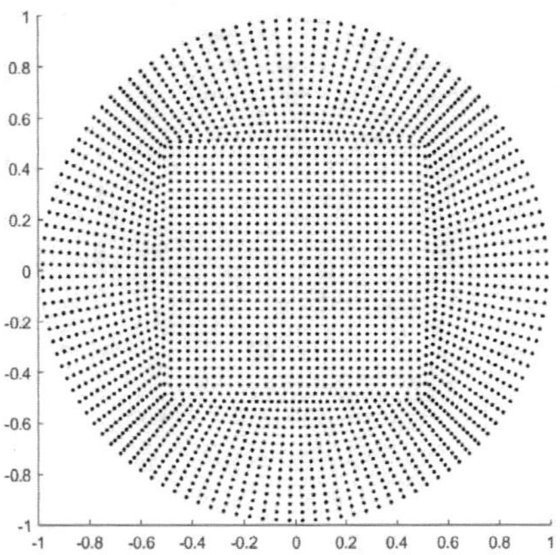

Figure 7. Collocation points on a circular disk.

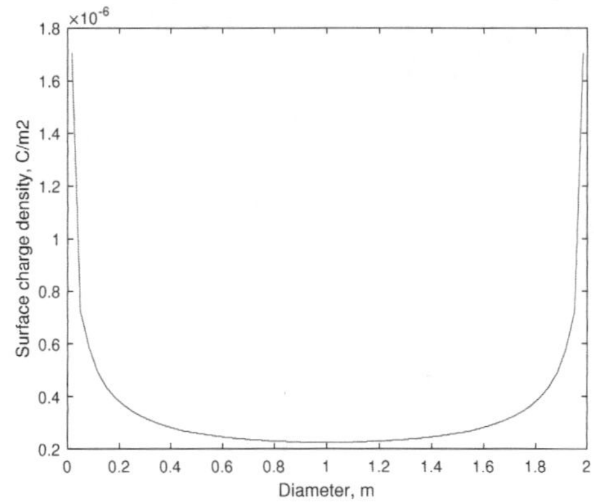

Figure 8. Surface charge density distribution on a circular disk.

Table 1. Convergence of the results of computing the surface charge density on a circular disk.

Num. of FE	Num. of nodes	Total charge Q, C	$\sigma_0, C/m^2$	$\frac{Q}{2\pi a^2}, C/m^2$	$\delta, \%$
12	17	$1.25819 \cdot 10^{-6}$	$0.24606 \cdot 10^{-6}$	$0.20025 \cdot 10^{-6}$	18.617
48	57	$1.35669 \cdot 10^{-6}$	$0.24110 \cdot 10^{-6}$	$0.21592 \cdot 10^{-6}$	10.444
108	121	$1.38253 \cdot 10^{-6}$	$0.23391 \cdot 10^{-6}$	$0.22004 \cdot 10^{-6}$	5.930
300	321	$1.39960 \cdot 10^{-6}$	$0.22963 \cdot 10^{-6}$	$0.22275 \cdot 10^{-6}$	2.996
588	617	$1.40574 \cdot 10^{-6}$	$0.22813 \cdot 10^{-6}$	$0.22373 \cdot 10^{-6}$	1.929
1200	1241	$1.40980 \cdot 10^{-6}$	$0.22715 \cdot 10^{-6}$	$0.22437 \cdot 10^{-6}$	1.223
2700	2761	$1.41258 \cdot 10^{-6}$	$0.22648 \cdot 10^{-6}$	$0.22482 \cdot 10^{-6}$	0.733
4800	4881	$1.41382 \cdot 10^{-6}$	$0.22618 \cdot 10^{-6}$	$0.22501 \cdot 10^{-6}$	0.517

We can see that the computed values of the surface charge density at the center of the disk σ_0 converge as the number of finite elements (subdomains) increases. This corresponds to the upper curve in figure 9. The lower curve displays the values of $\sigma(0)$ obtained on the basis of analytical formula (12). The total charge of the disk is computed by summing up the charges of subdomains $Q = \sum \sigma_i S_i$. It is then used to determine the charge density at the center of the disk $\sigma(0)$ by formula (12). The last column of the table contains the values of the relative error $\delta = \left|\frac{\sigma_0 - \sigma(0)}{\sigma_0}\right|$.

It should be noted that in both examples adequate simulation is also provided in boundary areas where the edge effect can be easily observed. It will be shown in what follows that edge effects become especially important when computing multiple electrode structures.

3.2. Finite element analysis of the membrane deformation

The finite element method is implemented to find the deformed shape of a flexible membrane. The variational formulation for this problem has the form

$$\int_{-a}^{a}\int_{-b}^{b} T_x \frac{\partial w}{\partial x}\frac{\partial \delta w}{\partial x} + T_y \frac{\partial w}{\partial y}\frac{\partial \delta w}{\partial y} dxdy = \int_{-a}^{a}\int_{-b}^{b} p(x,y)\delta w dxdy. \quad (13)$$

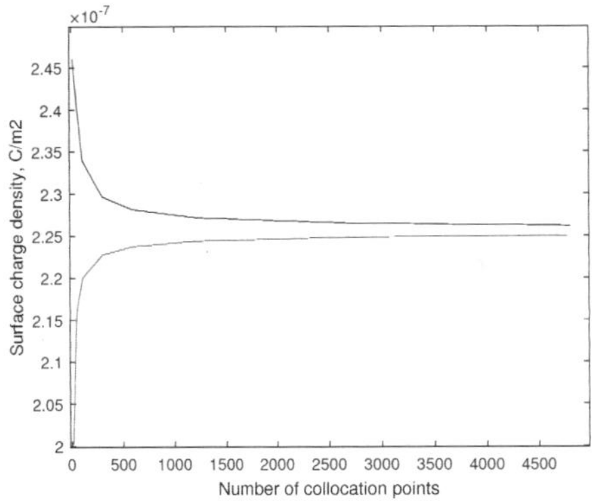

Figure 9. Convergence of the surface charge density at the center of the disk.

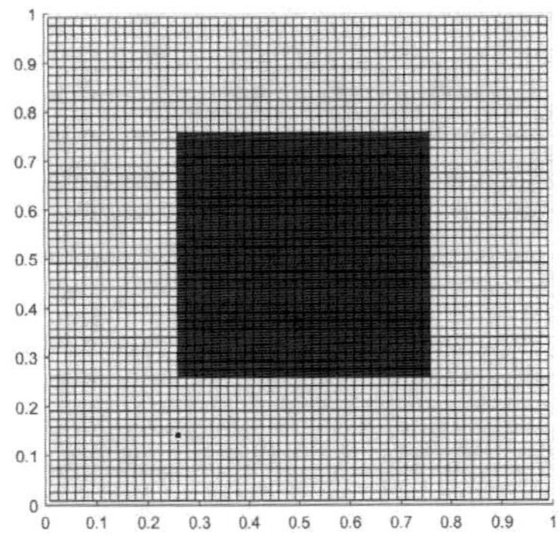

Figure 10. A segmented square electrode.

A linear 4-node quadrilateral isoparametric membrane element is used to perform the finite element analysis. According to the isoparametric element concept [28], similar expressions are used to approximate the unknown function w and to determine the coordinates x and y within the finite element

$$w = N_i w_i, \quad x = N_i x_i, \quad y = N_i y_i,$$

where $w_i (i = 1, ..., 4)$ are nodal unknowns, $(x_i,\ y_i)$ are nodal coordinates, and N_i are shape functions defined in terms of non-dimensional coordinates $-1 \leq \xi, \eta \leq 1$ as follows:

$$N_1 = \frac{1}{4}(1-\xi)(1-\eta); N_2 = \frac{1}{4}(1+\xi)(1-\eta); N_3 = \frac{1}{4}(1+\xi)(1+\eta); N_4 = \frac{1}{4}(1-\xi)(1+\eta).$$

The numerical integration pattern shown in figure 4 is applied to form the stiffness matrix of a finite element.

3.3. Iterative procedure for solving a coupled problem

The solution of this coupled set of equations requires an iterative approach. Iteration continues until a convergence criterion of the form

$$\frac{||W^k||}{||W^{k-1}||} - 1 \leq \epsilon, \quad ||W|| = (W^T W)^{1/2} \tag{14}$$

is satisfied for some small tolerance ϵ. According to [23], the best choice is to take it equal to half machine precision $\epsilon = 10^{-8}$. Here W^k and W^{k-1} are deflection vectors of the membrane at two successive iteration steps.

4. Results and discussion

In this section, numerical results are presented for two types of membrane mirror configurations to prove the validity and efficiency of the proposed approach.

4.1. A parallel-plate capacitor with two flexible square membrane electrodes

In this example, the capacitor is composed of two parallel metallized flexible square membranes to which a certain voltage is applied. The parameters of the capacitor are presented in table 2.

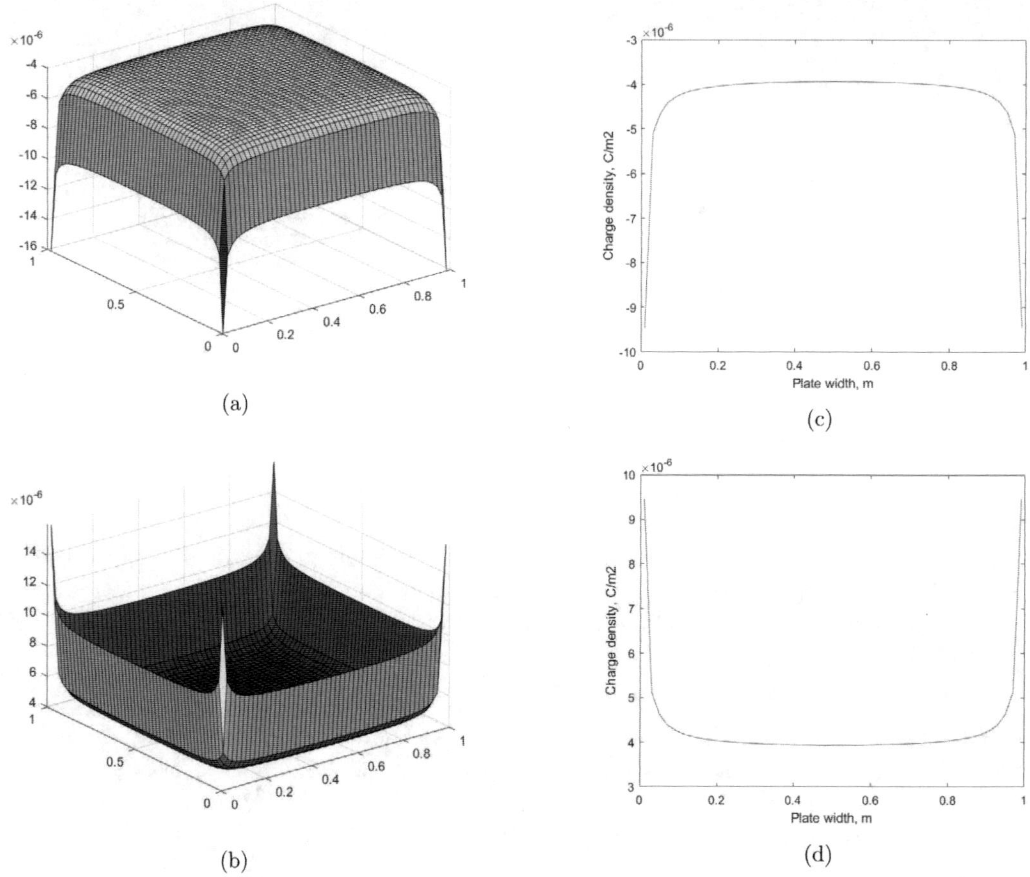

Figure 11. Surface charge density distribution over the upper (a) and bottom (b) electrodes of the capacitor and along the line $y = 0.5$m on the upper (c) and bottom (d) electrodes of the capacitor.

Table 2. Parameters of a parallel-plate capacitor with membrane electrodes.

Parameter	Notation	Value	Unit
Upper electrode potential	ϕ_u	$2 \cdot 10^4$	V
Bottom electrode potential	ϕ_b	$-2 \cdot 10^4$	V
Size of electrodes	$a \times a$	1×1	m
Distance between electrodes	H	0.01	m
Prestressing forces	T_x, T_y	100	N/m
Number of computational subsections	N	50×50	

The surface charge density distribution over the upper and bottom undeformed metallized square membranes and the corresponding surface charge density distribution along the line $y = 0.5$m are displayed in figure 11. The deformed shape of the square membrane electrode under electrostatic pressure and its profile w at $y = 0.5m$ are presented in figure 12 for two types of boundary conditions: all edges fixed (a, c) and two opposite edges fixed (b, d).

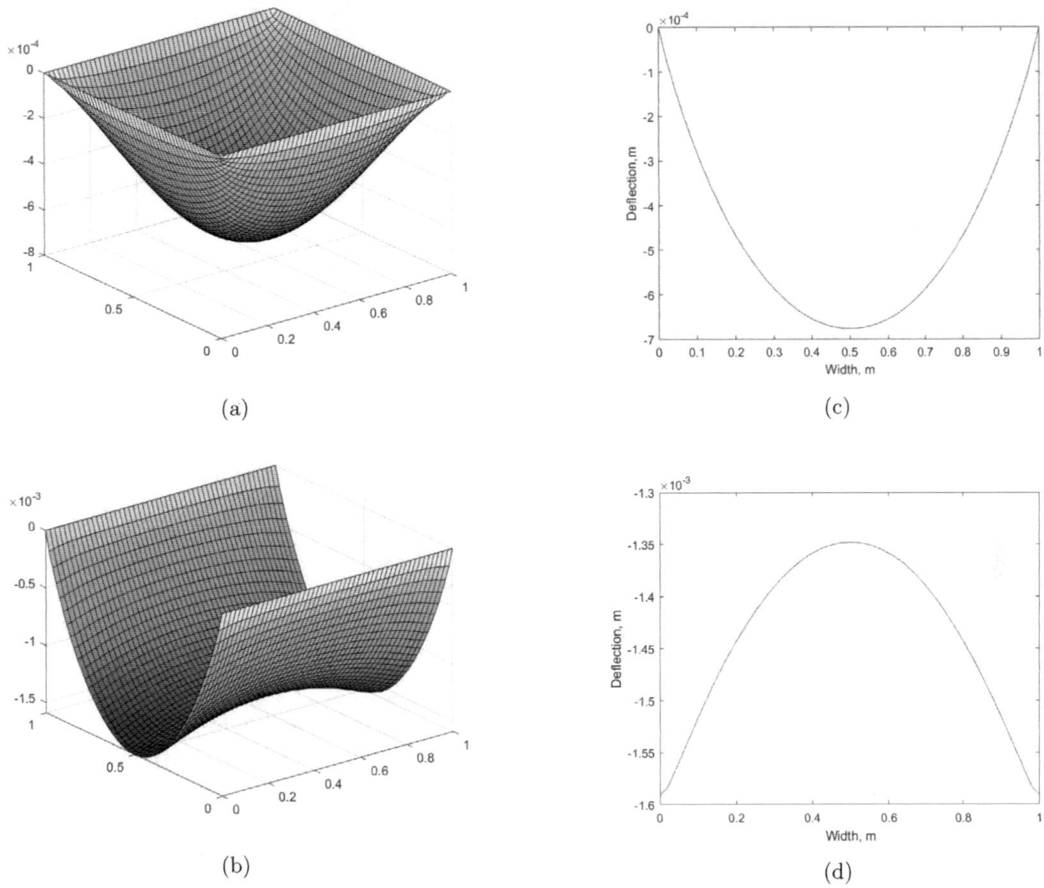

Figure 12. The deformed shape of a square membrane electrode and its profile w ($y = 0.5$m) for different types of boundary conditions: all edges fixed (a, c) and two opposite edges fixed (b, d).

To make sure that the proposed approach provides adequate simulation results and to estimate the influence of the capacitor edge effect, the membranes' shapes in the capacitor were also computed on the basis of simple formula (1). In figure 13, the corresponding deflection of a square membrane w ($y = 0.5$m) is shown for two types of boundary conditions: all edges fixed (a) and two opposite edges fixed (b), respectively.

Comparing the results presented in figures 12 and 13, we notice that, for the membrane with all its edges fixed, the maximum discrepancy of the results occurs in the center of the membrane and reaches almost 22%. As for the membrane with two opposite edges fixed, the maximum discrepancy of the results is 44%, which takes place in the middle of free edges. Furthermore, in the latter case, the deformed shape of the membrane is a double curved surface if the edge effect is counted, otherwise it is a parabolic cylinder. The presented results and their analysis prove that the electric field edge effect makes a substantial contribution to the formation of a flexible membrane in an electrostatic field.

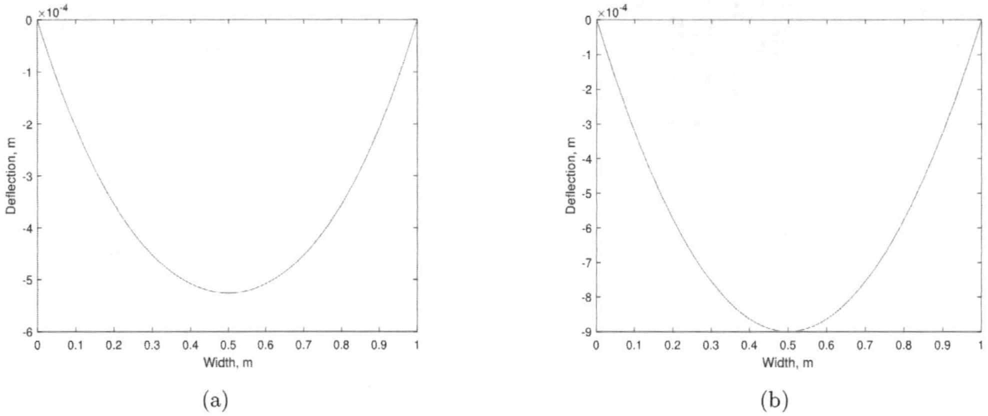

Figure 13. Deflection of a square membrane in the capacitor computed by using formula (1) for different boundary conditions: all edges fixed (a) and two opposite edges fixed (b), respectively.

4.2. A parallel-plate capacitor with a flexible membrane electrode and a rigid segmented electrode

In this example, the parameters of the capacitor are taken from table 2. The potential of the upper flexible membrane electrode is $-2 \cdot 10^4 V$. The bottom rigid square electrode is divided into two segments according to the pattern in figure 10. The potential of the central blue segment of size 0.5 m × 0.5 m is $10^4 V$, while the potential of the surrounding yellow segment is $2 \cdot 10^4 V$. Figure 14 illustrates the surface charge density distribution over the upper (a) and bottom (b) electrodes of the capacitor and along the line $y = 0.5$m (c) and (d), respectively. As one can see in the pictures, the step form of the potential on the segmented electrode produces the step-like charge density distribution on the membrane surface. The corresponding deformed shape of the square membrane electrode (a) and its profile w ($y = 0.5$m) (b) are shown in figure 15.

To verify the results and estimate the influence of the edge effect on the deformed shape of the membrane in the capacitor, we also performed calculations of the membrane deformation using formular (1) for computing the electrostatic field strength. Figure 16 illustrates the profiles of the surface charge density (a) and the square membrane deflection (b) at $y = 0.5$m for this case.

A number of numerical experiments has been carried out on various meshes. Table 3 contains the values of the deflection of the central point of the square membrane in the capacitor computed on 20-fold meshes. The values w_0 reflect the presence of the edge effect in the model under consideration, and the values $w_0^{(1)}$ reflect its absence according to formula (1). The last column of the table contains percentages of the relative difference $\delta = \left|\frac{w_0 - w_0^{(1)}}{w_0}\right|$.

Table 3. Square membrane deflection convergence.

Mesh size	Deflection w_0, m	Deflection $w_0^{(1)}, m$	$\delta, \%$
20 × 20	$-0.32104 \cdot 10^{-3}$	$-0.22222 \cdot 10^{-3}$	22.221
40 × 40	$-0.30604 \cdot 10^{-3}$	$-0.22196 \cdot 10^{-3}$	22.195
60 × 60	$-0.30103 \cdot 10^{-3}$	$-0.22192 \cdot 10^{-3}$	22.191
80 × 80	$-0.29849 \cdot 10^{-3}$	$-0.22190 \cdot 10^{-3}$	22.189

The convergence of the results of computing the square membrane deflection in the capacitor

IOP Conf. Series: Materials Science and Engineering 1060 (2021) 012023 doi:10.1088/1757-899X/1060/1/012023

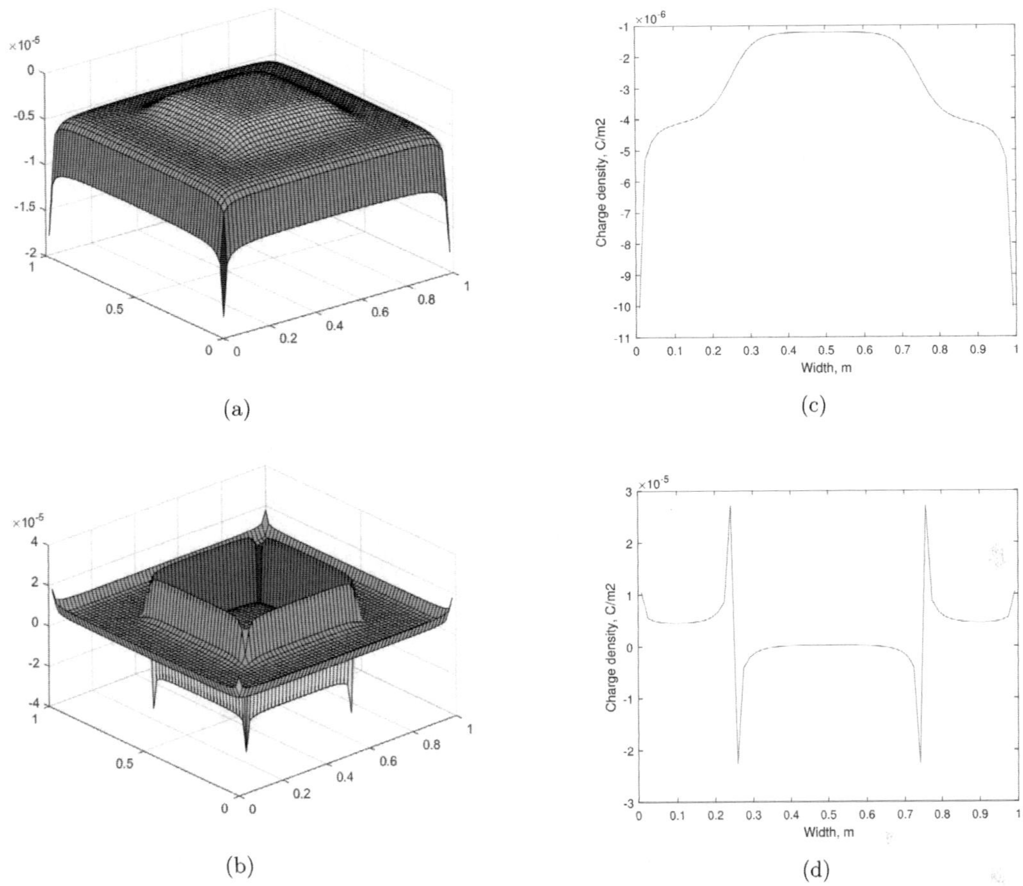

Figure 14. Surface charge density distribution over the upper (a) and bottom (b) electrodes of the capacitor and along the line $y = 0.5$m (c) and (d), respectively.

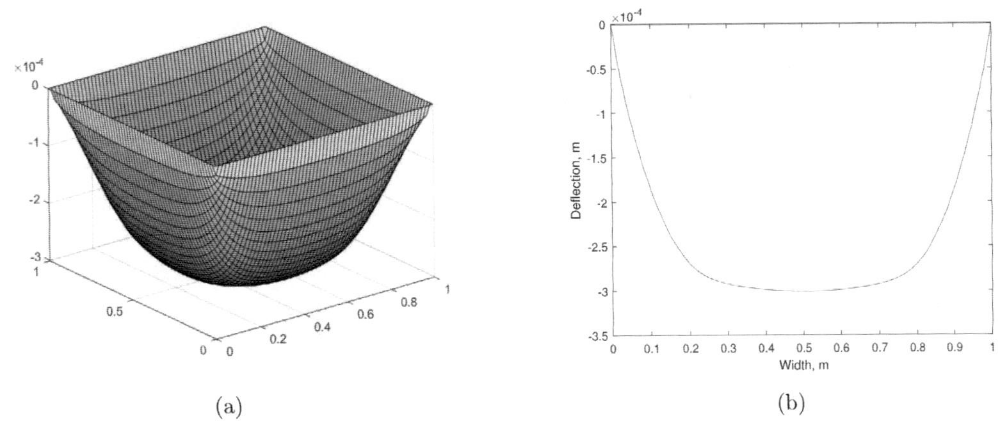

Figure 15. The deformed shape of a square membrane electrode (a) and its profile w ($y = 0.5$m) (b) in the capacitor with a segmented electrode.

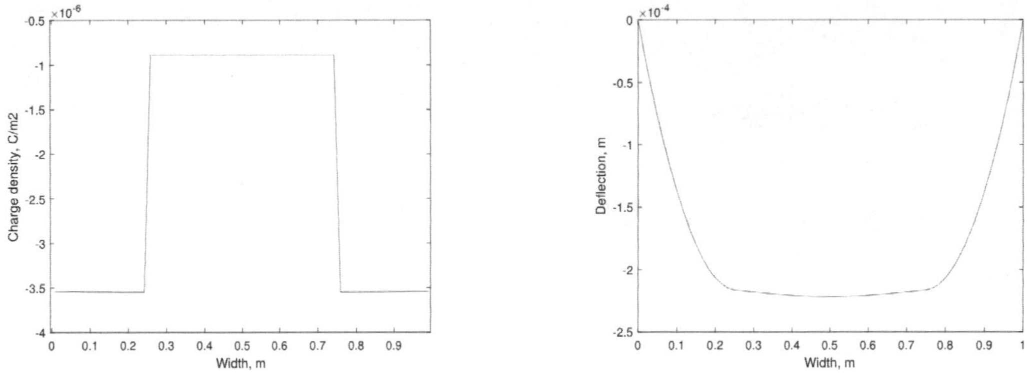

(a) (b)

Figure 16. The surface charge density (a) and the square membrane deflection (b) profiles at $y = 0.5$m computed by using formula (1).

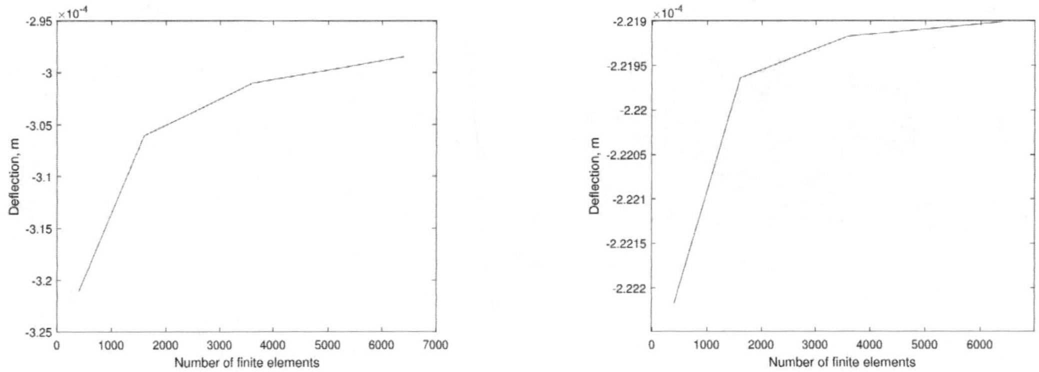

(a) (b)

Figure 17. Convergence of the results of computing the square membrane deflection w_0 (a) and $w_0^{(1)}$ (b) in the capacitor with a segmented electrode.

with a segmented electrode is also demonstrated by figure 17, where the plots visualize the data collected in the table. We can see that the computed values of the square membrane deflection at the center of the membrane w_0 (a) and $w_0^{(1)}$ (b) converge as the mesh refines. The relative differences between these values are very close to 22% for all mesh patterns.

The analysis of the presented results allows us to make a conclusion similar to that in the previous subsection about a significant influence of the electric field edge effect on the shape of a flexible metallized membrane loaded by electrostatic forces.

5. Conclusions

The developed mathematical model and numerical algorithm have shown their efficiency in determining the deformed shape of a metal-coated membrane in the capacitor with a segmented electrode. The mathematical formulation and the solution techniques for more general cases are straightforward. It is important that the proposed approach makes it possible to perform synthesis of an electrostatic field due to a surface potential for segmented electrodes of various configurations providing desirable shape control of deformable mirrors. A number of numerical experiments have been performed to shown that the edge electric field significantly affects the resulting deflection of the membrane. At the same time, there is still lack of sufficiently detailed information on the interaction between an electrostatic field and flexible membranes of complex configurations. The extension of the model implies including nonlinear deformation effects and structural dynamics [29]. The model will be very helpful for designing membrane elements in adaptive optics systems involving electrostatic forces. Having such a tool will make it possible to solve many practical problems of shaping metallized membranes by electrostatic forces.

References

[1] Grosso R P and Yellin M 1977 The membrane mirror as an adaptive optical element *J. Opt. Soc. Am.* **67** 399
[2] Mihora D J and Redmond P J 1980 Electrostatically formed antennas *J. Spacecraft and Rockets* **17(5)** 465
[3] Madec P-Y 2012 Overview of deformable mirror technologies for adaptive optics and astronomy *Proc. SPIE 8447 Adaptive Optics Systems* III doi.org/10.1117/12.924892
[4] Liu Z-Q, Qiu H, Li X and Yang S-L 2017 Review of large spacecraft deployable membrane antenna structures *Chinese J Mech Eng* **30** 1447 doi:10.1007/s10033-017-0198-x
[5] Volchek O S 1986 Calculation of an electrostatic film concentrator *Appl. Sol. Energy* **22** 29
[6] Vol'mir A S 1967 *Flexible Plates and Shells* (Florida: University of Florida)
[7] Na T Y 1979 *Computational Methods in Engineering Boundary Value Problems* (New York: Academic Press)
[8] Aleshin Y K and Sivkova M A 2016 The determination of edge electric field effect on measurements carried out by piezoquartz resonators *Lomonosov MSU: Memoirs of the Faculty of Physics* **5**
[9] Volchek O S 1987 Static and dynamic characteristics of a membrane in an electrostatic field *Statics and Dynamics of Flexible Systems* ed V V Vasil'ev *et al* (Moscow: Nauka) 139
[10] Rochus V, Golinval J-C, Louis C, Mendez C and Klapka I 2007 Model of electrostatic actuated deformable mirror using strongly coupled electro-mechanical finite element *Analog Integrated Circuits and Signal Processing* **53(2)** 129 doi:10.1007/s10470-007-9066-4
[11] Bulnes F 2019 *Recent Advances in Integral Equations* (London: IntechOpen)
[12] Manolis G D and Polyzos D 2009 *Recent Advances in Boundary Element Methods* (Patras: Springer)
[13] Bonora S, Bortolozzo U, Naletto G and Residori S 2011 Innovative membrane deformable mirrors *Topics in Adaptive Optics* ed R K Tyson (Rijeka:InTech) 231
[14] Banerjee K, Rajaeipour P, Ataman Ç and Zappe H 2018 A low-cost 25-actuator electrostatic deformable mirror with polyimide membrane for adaptive optics microscopy *Proc. of SPIE* **10545**
[15] Kronast W, Mescheder U, Müller B and Huster R 2014 Development of a focusing micromirror device with an in-plane stress relief structure in silicon-on-insulator technology *J. Micro/Nanolith. MEMS MOEMS* **13(1)**
[16] Lieber M, Kendrick S, Lipscy S, Ebbets D, Acton S and Knight S 2013 New paradigm for rapid production of large precision optics: frozen membrane mirror technology *Optical Engineering* **52(9)**
[17] Jallouli A, Kacem N and Lardies J 2018 Investigations of the effects of geometric imperfections on the nonlinear static and dynamic behavior of capacitive micromachined ultrasonic transducers *Micromachines* **9(575)** doi:10.3390/mi9110575
[18] Usyukin V I 1988 *Structural Mechanics of Space Engineering Structures* (Moscow: Mashinostroenie)
[19] Usyukin V I 1990 Fundamentals of the theory and calculation of soft thin-walled structures of composite materials *Mech Compos Mater* **26** 357 doi.org/10.1007/BF00613114
[20] Volchek O S and Usyukin V I 1994 Equations of a flexible shell in an electrostatic field *Mech. Solids* **1** 103
[21] Neittaanmäki P, Rudnicki M and Savini A 1996 *Inverse Problems and Optimal Design in Electricity and Magnetism* (Oxford: Clarendon Press)
[22] Aliabadi M H 2002 *The Boundary Element Method Vol 2: Applications in Solids and Structures* (Chichester: John Wiley & Sons)
[23] Zienkiewicz O C and Taylor R L 2000 *The Finite Element Method Vol 2: Solid Mechanics* (Bristol: Butterworth-Heinemann)
[24] Smith I M and Griffiths D V 2004 *Programming the Finite Element Method* (Chichester: John Wiley & Sons)
[25] Harrington R F 1993 *Field Computation by Moment Methods* (New York: IEEE Press)
[26] Krylov A N 1950 *Lectures on Approximate Calculations* (Moscow: GITTL)
[27] Given J A, Hwang C, and Mascagni M 2002 First- and last-passage Monte Carlo algorithms for the charge density distribution on a conducting surface *Phys. Rev.* E **66**
[28] Petyt M 1990 *Introduction to Finite Element Vibration Analysis* (Cambridge: Cambridge University Press)
[29] Volchek O S and Usyukin V I 1997 Analysis of the dynamic stability of an electrically controlled membrane reflector *Mech. Solids* **2** 176

IOP Conf. Series: Materials Science and Engineering 1060 (2021) 012024 doi:10.1088/1757-899X/1060/1/012024

Evolution, features, and prospects of materials science innovations in the development of space communication systems: A short survey

S V Reznik* and A D Novikov

Bauman Moscow State Technical University, 2nd Baumanskaya str., b. 5, Moscow, 105005, Russia

*corresponding author: sreznik@bmstu.ru

Abstract. Since the launch of the first artificial satellites of the Earth, significant changes in the space communication engineering have taken place over the years. The paper considers a brief history of shapes of onboard space antennas and antenna materials. In the literature, there are only scattered data on the materials of space antennas that do not pretend to generalize and do not convey the reasons and results of the evolution of space antenna designs in relation to the materials used. A number of examples are based on Soviet and Russian experience, which is less well known to foreign readers and little covered in foreign literature. The paper focuses on the use of composite materials in reflectors of onboard high-gain antennas with a solid non-deployable surface. The paper presents the results of selecting a thin-walled CFRP shell reflector structures with a ribbed convex surface, which is rational in terms of mass and shape stability.

1. Introduction.

Space communications play a crucial role in the economy, science, culture, defence, and security. The Soviet artificial earth satellite "Sputnik", launched on October 4, 1957, transmitted the first radio signals from space. The logic of space exploration led to the need to use increasingly more advanced onboard space antennas. Instead of omnidirectional metal rod antennas, high-gain antennas with a focused, narrow radio wave beam width became increasingly common.

In recent years, the Ka band (26.5-40.0 GHz) has been actively developed for inter-satellite communication, and the V band (40.0-65.0 GHz) has begun to be used. In these ranges, the permissible deviations of the reflector shape under operating loads should not exceed fractions of a millimetre. According to ESA data, the linear density of rigid reflectors with a three-layer structure has decreased to 3.5 kg/m^2. However, traditional three-layer honeycomb-filled configurations appear to be close to their weight efficiency limit. Approaches to creating a reflector with a linear density of less than 2 kg/m^2 are considered. The main feature of this configuration is a thin smooth carbon fibre shell with ribs on the convex surface. The results of modelling the temperature and stress-strain state are presented, justifying the choice of design solutions.

2. A brief history of onboard space antennas

First onboard space antennas were omnidirectional and had the simplest configuration. They were made out of straight or curved hollow rods. These antennas were installed on the first Soviet and American satellites, as well as first-generation moon stations (figure 1) [1, 2]. Their antennas were not designed for large data transmission rate; the rate ranged from 1 to 10^4 bps.

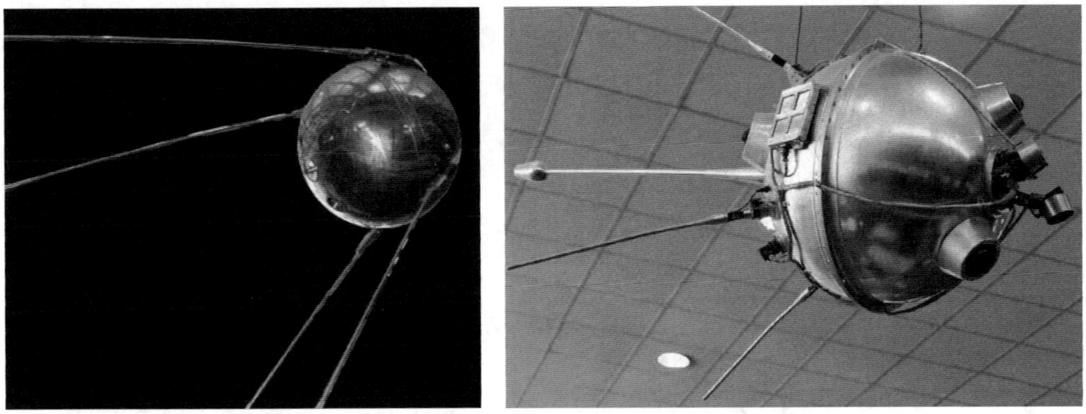

(a) (b)

Figure 1. First spacecraft generation has rod omnidirectional antennas
a) First artificial Earth satellite "Sputnik", 2 dual-rod antennas; ν = 20.005 and 40.002 MHz
b) Automatic moon station "Luna-2" (E-1A) First hard landing on the Moon, 4 rod antennas; ν = 19.993 (uplink); 39.986 (telemetry); 183.3 MHz (downlink).

In 1960 and 1964, the USA attempted to make passive repeater satellites [3]. For this, the USA launched the satellites "Echo-1" and "Echo-2" in the form of inflatable large-diameter aluminium-coated polymer film spheres (figure 2). It was assumed that other receivers would receive the radio waves directed at the satellites after the waves are reflected from its mirror surface. It was expected that multipath radio waves reflected from satellites positioned at the angle of 120° would enable global radio communication. From the standpoint of mastering new technologies, these investigations were interesting; however, they were cancelled due to weak reflected signals.

Figure 2. NASA "Echo" project for transmitting radio waves by balloon satellites with thin wall from polymer material and aluminium skin: "Echo-1".

D=30.5 m
m=76 kg
wall h=12.7 μm
ν = 960-2390 MHz.

The first television satellite "Telstar" launched in 1962 had a helical antenna and horn antennas (figure 3). Relatively weak onboard transmitters and signal density resulted in the need for huge ground transmitter/receiver antennas with parabolic reflectors for the concentration of radio waves. For example, the receiver antenna constructed at the south of Great Britain in Goonhilly Downs near Helston on the Lizard peninsula in Cornwall had a diameter of 25.9 m and the mass around 1.118 t.

The first two-way communication system first implemented in 1961 on the automatic interplanetary space station "Venus-1". Besides non-directional antennas, "Venus-1" had a two-meter parabolic antenna with a transformable metal mesh reflector (figure 4).

(a)

(b)

Figure 3. Bell Telephone Laboratories, USA "Telstar"
a) An array of antennas around "equator" of satellite and helical antenna for control commands
ν = 6 GHz (uplink) and 4 GHz (downlink – retransmitting)
b) Goonhilly Satellite Earth Station with ground Antenna One (dubbed "Arthur").

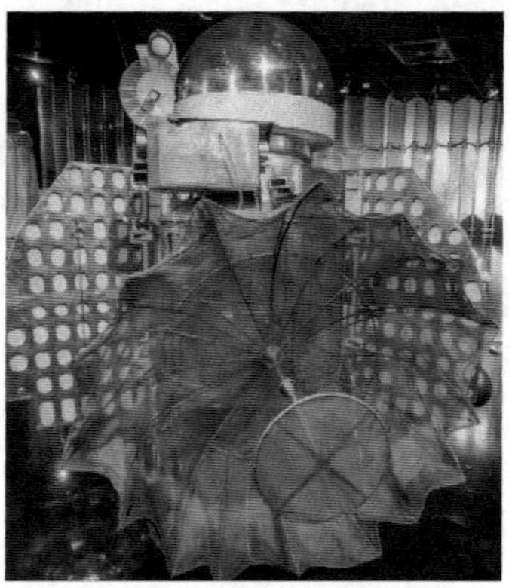

(a)

(b)

Figure 4. First USSR automatic interplanetary station "Venus-1"
a) The antenna parabolic reflector from metal mesh
d = 2 m, ν = 700 (uplink); 900 MHz (downlink);
b) The ground antenna ADU-1000 8x8 m diameter. Evpatoria, Crimea.

Reflector antennas had become an indispensable part of automatic interplanetary stations and satellites operating at the near-Earth orbits [1, 2] (table 1).

In 1965, the USSR put into operation a communication system with "Molniya" satellites, launched into high elliptical orbits with an apogee of about 40,000 km above the Northern hemisphere. The transmitter-receiver antennas of the "Molniya" satellites had the umbrella structure with the reflector made out of metal mesh sheet (figure 5). As the applications became more complex, the operating frequencies and traffic volumes increased.

In the middle of the 1970s, the active development of space systems with the global transmission of digital data, television programs, and voice signals began. Satellites with several reflector antennas of different shape, diameter, and structures had started launching into orbit. Space communication ranges had started expanding towards higher frequencies, first taking the L band, then the Ku band, and, finally, Ka and V bands. As the frequency of radio signals grew, the requirements to the form and shape stability of antenna reflectors had become more rigorous. First, the shape deviation tolerance of Λ/8 (where Λ

IOP Conf. Series: Materials Science and Engineering 1060 (2021) 012024 doi:10.1088/1757-899X/1060/1/012024

is the radio signal wavelength) was permissible. Later, the tolerance of Λ/16 became common, and now the tolerance can be as low as Λ/50. With the increased number of gain antennas onboard of spacecraft and larger reflector sizes, the issue of reducing linear density (mass divided by aperture area) of reflector became relevant. Conventional metals and alloys did not meet the ever-growing requirements, so they were gradually replaced with composite materials, mainly carbon fibre reinforced plastic (CFRP) having the unique combination of specific strength, stiffness, low linear thermal expansion coefficients, and controlled thermal conductivity.

Table 1. Characteristics of reflector antennas of Soviet/Russian automatic interplanetary stations [1].

Parameters	Generation of spacecrafts (operation period)			
	1st generation (1960-1969)	2nd generation (1969-1988)	3rd generation (1988-1998)	4th generation (1998-2008)
Frequency range, (GHz)				
Earth-SC	0.7	0.7	0.7; 5.0	5.0; 7.2
SC-Earth	0.9	0.9; 1.6; 6.0	0.9, 6.0; 8.4	6.0; 8.4
Transmitter power in the range, (W)				
0.9 GHz	20	100	10	–
0.7 GHz	–	24	50	50
Diameter of the antenna reflector, (m)	1.6; 2.0	2.8	0.9; 1.6	1.5; 2.8

(a) (b)

Figure 5. The telecommunication satellite "Molniya-1"
a) 2 umbrella-type reflector antennas D=1.4 m ν = 800 MHz (uplink) and 1000 MHz (downlink), 3.4/4.1 GHz for TV broadcasting;
b) A Soviet/Russian system "Orbita" ground antenna for broadcasting and delivering TV signals via "Molniya" satellites.

The development of current telecommunication systems is developing in two directions: low-orbit and high-orbit systems. The low-orbit systems do not need reflector antennas, and they are fitted with horn, helical, and rod antennas. Helical antennas utilize copper and copper alloys. Shape stability is ensured by cylindrical and conical polymer mandrels. Horn antennas are made out of copper alloys and composite materials; their interior surface is coated with silver for reducing electrical losses. Satellites for high orbits (geostationary, geosynchronous, and high-apogee orbits) are fitted with two types of

IOP Conf. Series: Materials Science and Engineering 1060 (2021) 012024 doi:10.1088/1757-899X/1060/1/012024

antenna reflectors: the reflectors with the solid non-deployable shape and the transformable reflectors taking the required shape at the operation orbit (table 2).

Table 2. Materials of onboard space antennas.

The shape of the antenna	Materials	Radio band
	Linear (omnidirectional antennas)	
Rod	Aluminium alloys, materials with shape memory from sheet bronze, CFRP	**HF** (3-30 MHz/10-100 m)
Spiral Spiral with reflector	Copper, Copper+polymer mandrel Copper+aluminium (reflector)	**P** (< 300 MHz > 1 m), **UHF** (300-1000 MHz/ 0.3-1 m), **S** (2-4 GHz/7.5-15 sm)
	Aperture (gain antennas)	
Horn	Copper with silver/gold coating Fiberglass with metallization, CFRP	**S** (2-4 GHz/7.5-15 sm) **X** (8-12 GHz/2.5-3.75 sm)
Solid non-deployable form	Aluminium, CFRP, Kevlar-49 Epoxy+ Aluminium or Nomex honeycomb, Carbon-Carbon, Silicon carbide	**L** (1-2 GHz/15-30 sm), **X** (8-12 GHz/2.5-3.75 sm) **Ku** (12-18 GHz/1.67-2.5 sm) **Ka** (27-40 GHz/7.5-11.1 mm), **V** (40-75 GHz/4.0-7.5 mm) – only for the non-deployable form
Transformable: • Umbrella type • Cable-Rimmed	CFRP (rods, spokes)+Steel or Tungsten or Molybdenum mesh with Gold coating, polymer ropes	
• Truss	Same Aluminium, CFRP (rods)+ Steel mesh with Nickel coating, polymer ropes	
• Petal	CFRP plates+ Aluminium honeycombs	
• Inflatable (sphere, lens, torus)	Polyethylene terephthalate, Polyimide (films) with Aluminium coating	

3. Current requirements to onboard reflector antennas with a solid non-deployable shape

The improvement of the structure of onboard space antenna reflectors aims at reducing their linear density and shape deviation due to temperature gradients caused by heat flows from the Sun and the Earth. The main approaches for producing efficient engineering solutions were formulated as far as 40 years ago [4-9], but their implementation and the development of ever-improving constructions are still active [10-29].

For solid non-deployable reflectors, there are several structural layouts enabling reflector stiffness: the three-layer layout (the honeycomb-filled sandwich type), the double rim layout, the layout with frame reinforcements, and the ribbed layout. According to ESA, the current linear density of best specimens of such sandwich structures is at the level of 3.5 kg/m^2, and the shape deviation is less than 0.2 mm.

Does it mean that there are no more ways of achieving better characteristics? To answer this question, one has to take into account that the reflector is an open-profile parabolic shell whose mass and shape stability depend on the thickness. However, these dependencies can have different ratios. The thinner the shell is, the lighter it is. However, it is less stiff and has a lower shape stability. The shell must be thick enough to withstand potential loads, and it cannot be thinner than its original components, such as fibres, fabrics, and the polymer adhesive layers. Evidently, a smooth shell would be the lightest, but it must be thick enough to have adequate stiffness for ensuring shape stability. Among options of increasing shape stiffness, the most promising are the options with rib reinforcement on the convex side of the shell.

As for values of optimal reflector characteristics, this paper aims to achieve a linear density of 2.0 kg/m^2 and a maximum shape deviation of 0.1 mm. It was assumed that this goal is feasible if CFRPs from high-modulus carbon fibres and epoxy adhesives are used. These materials have high stiffness (over 140 GPa), high strength (from 1.3 GPa to 2.1 GPa), low values of linear thermal expansion coefficient ($\pm 0.5 \cdot 10^{-6} K^{-1}$) and density (1500-1650 kg/m^3). Compared to metals (in particular, aluminium alloys, which are common in rocket and space technology), CFRP has a significantly lower density at a higher Young's modulus. The disadvantages of CFRP include sensitivity to ultraviolet light and limited reflectivity at high frequencies, but these effects can be mitigated by using polymer and metallized coatings. The main advantage of CFRP can be called near-zero linear thermal expansion coefficient, which makes it possible to avoid the effect of warping of structures with uneven heating.

4. Determining rational configuration of the antenna reflectors with a ribbed convex surface

To determine the rational structural reflector layout, the following design method was used:

- Selecting multiple design options for design
- Defining the thermal loads on the reflector in GEO;
- Creating the geometric and finite element model of the reflector
- Determining thermal and mechanical characteristics of the material on samples having physical similarity with designed structure
- FE modelling of the thermal state using "Siemens Space System Analysis"
- Selecting the method for fixing the reflector to the spacecraft
- FE modelling of the stress-strain state using "Siemens NX Nastran"
- Determining the maximum thermal deformation and displacement of the reflecting surface
- Analysing the results and comparing them with other design layout variants.

The temperature state was modelled for the geostationary orbits in the periods of autumn and spring equinox and solstice (direct solar radiation flux 1377 W/m^2, space temperature 4 K, Stefan-Boltzmann constant 5.67 10^{-8} W/(m$^2 \cdot$K^4)). Special attention was given to a number of orbit segments: before entering the shadow zone, the shadow zone, and the segment when the angle between the direction to the Sun and the axis of rotation of the parabolic reflector shell is 270°. In the FEA, 3-node elements with the side length of 60 mm were used; the number of elements varied from 6000 to 10000 depending on the structural layout. The element size was based on considerations of modelling accuracy and time. The analysis of the models with element sizes of 30 mm and 15 mm revealed that that the results vary within calculation error of 5%, whereas modelling time increased from 8 hours to 20 and 50 hours, accordingly.

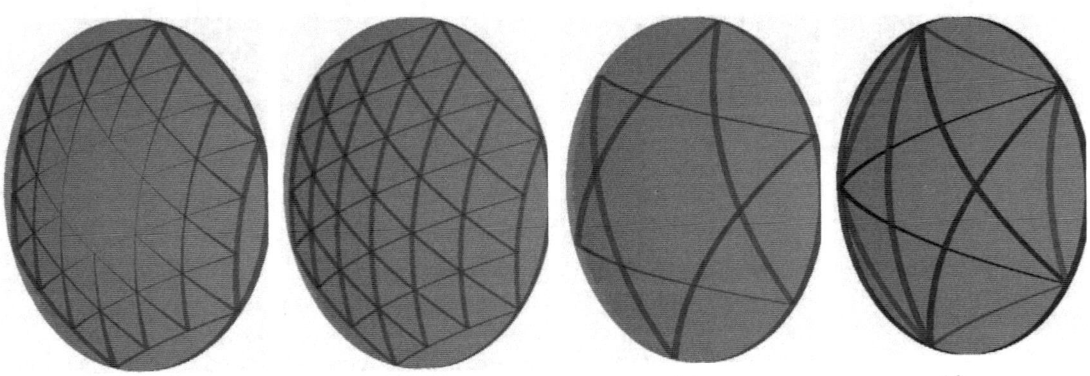

(a) (b) (c) (d)

Figure 6. Variants of ribbed reflectors
(a) Izogrid with constant rib height; (b) Izogrid with variable rib height;
(c) Six-pointed star; (d) Five-pointed star.

Four structural layouts of the 1200 mm diameter and 500 mm focal distance reflector were considered (figure 6). The thickness of the structural elements was varied from 0.4 to 1.2 mm. Layers of carbon fabrics were positioned at the angles of 0° and 90°. The rib reinforcement height was varied from 20 mm to 90 mm. Figures 7 and 8 show results of thermal modelling of layouts with five- and six-pointed star reinforcements.

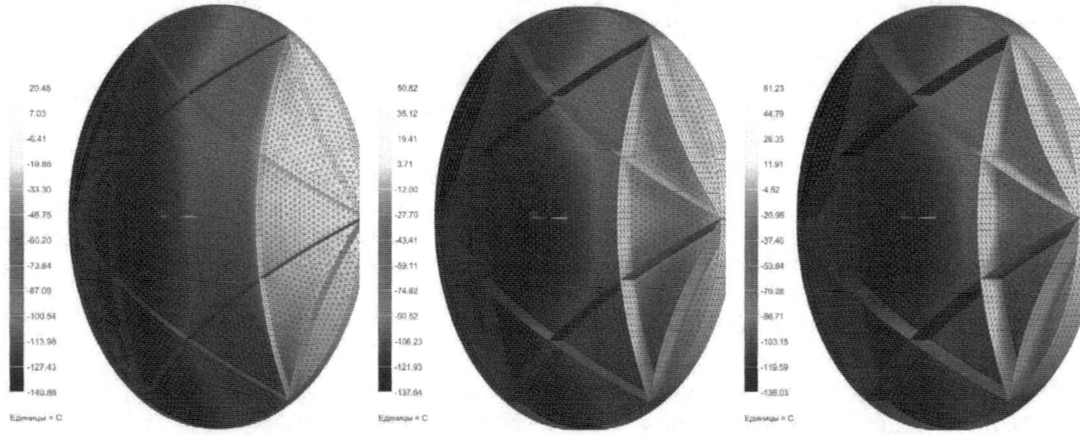

(a) (b) (c)

Figure 7. Temperature state of the reflector (degrees Celsius) for:
(a) Star6_25_12_06 (ΔT=~120 °C); (b) Star6_60_06_06 (ΔT=~87 °C);
(c) Star6_90_06_04 (ΔT=~75 °C); with 270° angle between reflector rotation axis and the Sun – Earth line.

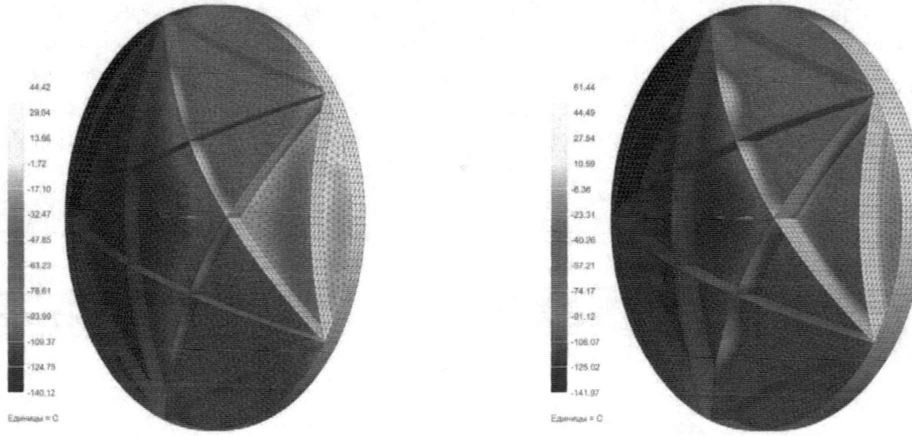

(a) (b)

Figure 8. Temperature state of the reflector (degrees Celsius) for:
(a) Star5_53_06_06 (ΔT=~96 °C); (b) Star5_90_06_04 (ΔT=~81 °C);
with 270° angle between the reflector rotation axis and the Sun-Earth line.

After determining the temperature state of the structures on the geostationary orbit, the displacements of the reflecting surface due to non-uniform loading were calculated. Table 3 shows that the layouts with the "five-pointed star" and "six-pointed star" reinforcement have the biggest stiffness. The following design calculations were limited to these two structural layouts.

The analysis of the ways of fixing the reflector to the spacecraft was done for six fixture variants. (figure 9). It was established that the minimum displacement of 0.041 mm and the linear density of 1.6 kg/m^2 correspond to the reflector with the rib height of 90 mm and the rib and reflector shell thickness of 0.6 mm fixed at the central polygon of the "six-pointed star".

Table 3. Comparison of ribbed design layouts

Scheme of ribbed surfaces	Shell and ribs thickness, (mm)	Mass of the reflector, (kg)	Maximum displacement of reflecting surface, (mm)	Linear density, (kg/m^2)
Six-pointed star	0.6	1.523	0.14	1.354
Five-pointed star	0.6	1.502	0.12	1.328
Izogrid with constant rib height	0.6	1.549	0.185	1.369
Izogrid with variable rib height	0.6	1.479	0.18	1.308

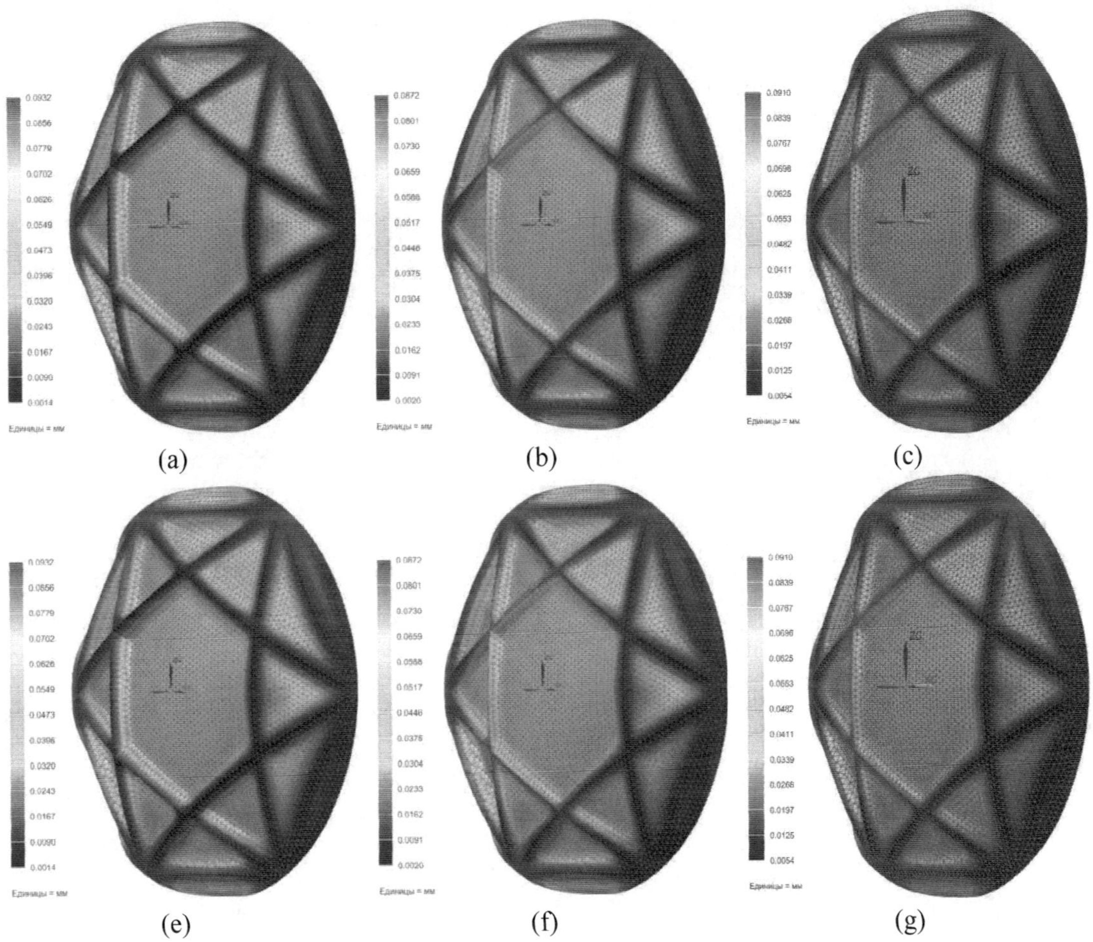

Figure 9. Displacement of reflecting surface (mm) under the action of thermal loads for Star6_90_06_04:
(a) circular; (b) prop; (c) centre; (d) beams; (e) frame; (f) nodes.

Table 4. Comparison of reflector design layout

Design layout	Displacement of reflecting surface under the thermal loads, (mm)	Linear density, (kg/m^2)
Sandwich	0.238	2.3
Double shell	0.047	3.5
Ribbed shell "six-pointed star"	0.041	1.6

It is important to note that the ribs are of constant height, which is due to both the manufacturability of the structure and the rigidity. Table 4 shows the comparison of this structural layout against conventional layouts.

The reflector manufactured from the flattened fabric Aspro-A80 and epoxy adhesive ED-20 by vacuum infusion had a record-breaking linear density of 1.92 kg/m^2 (figure 10).

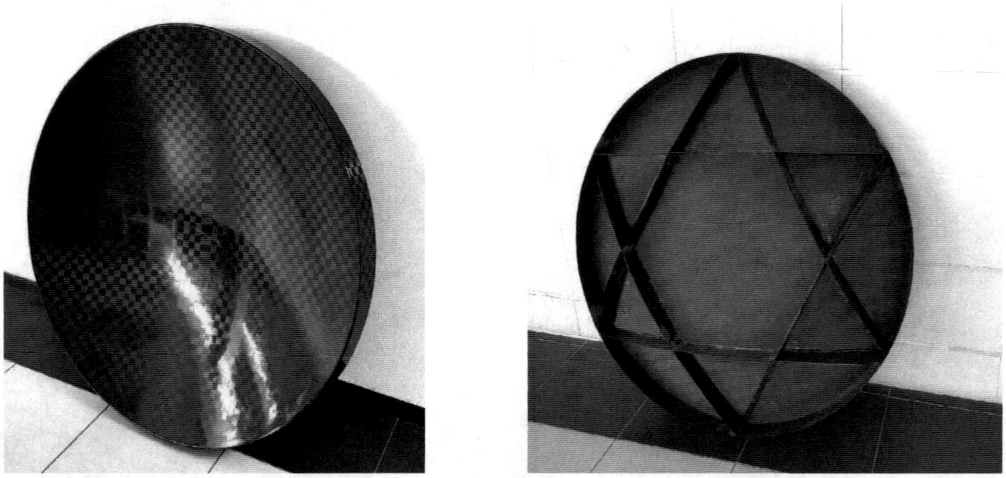

Figure 10. Bauman MSTU space antenna reflector.

5. Conclusion

Several reflector design layouts with ribbed convex surface were analysed, showing the superiority of the six-pointed star pattern, thin-walled shell configuration. Simulation of temperature and stress-strain state showed that for the GEO, the maximum temperature gradient could be up to 220ºC, and the maximum displacement would not exceed 0.041 mm.

Comparison with the conventional configuration (honeycomb and double shell) showed a significant (up to 2 times) reduction in the linear density of the structure combined with the best shape and size stability.

References

[1] Bakitko R V, Vasilev M B, Vinitsky A S et al 1993 *Radio System of Interplanetary Spacecrafts*, ed. A S Vinitsky (Moscow: Radio and Communications) p 328
[2] *Spaceborne Antennas for Planetary Exploration* 2008 ed W A Imbriale (NJ.: John Wiley and Sons, p 592
[3] Wilson A 1981 A history of balloon satellites *J. Br. Interplanet. Soc.* **34** 10–22
[4] Ingerson P G and Wong W C 1972 The analysis of deployable umbrella parabolic reflectors *IEEE Trans* **AP–2**0 (4) 409–14
[5] Powell R V and Hibbs A R 1977 An entre for large space antennas *Astronautics and Aeronautics* **15** (11) 58–64
[6] Tankersley B C 1979 Maypole (Hoop/Column) deployable reflector consept development for 30 to 100 meter antenna *Proc. AIAA/NASA Conf. Adv. Technol Future Space Syst.* (Hampton, Virginia) 446-458
[7] Archer J S 1980 High-perfomance parabolic antenna reflectors *J. Spacecr. Rockets* **17**(1) 20–6
[8] Fager J A 1980 Large space erectable antenna stiffness requirements *J. Spacecr. Rockets* **17**(2) 86–92
[9] Hedgepeth J M 1982 Accuracy potentials for large space antenna reflectors with passive structure *J. Spacecr. Rockets* **19**(3) 211–7

[10] Thornton E A, Dechaumphai P and Pandey A K 1986 Finite-element thermal-structural analyses of a cable-stiffened orbiting antenna *J. Spacecr. Rockets* **23**(6) 620–4

[11] Takamatsu K A and Onoda J 1991 New deployable truss concepts for large antenna structures or solar concentrates *J. Spacecr. Rockets* **28**(3) 330–8

[12] Guest S D and Pellegrino S A 1996 A new concept for solid surface deplyable antennas *Acta Astronaut.* **2** 103–13

[13] Tsunoda H, Harin K, Kawakami Y, *et al.* 2000 Design evaluation of a large deployable mesh reflector *J. Spacecr. Rockets* **37**(1) 108–13

[14] Thomson M W 2002 AstroMesh deployable reflectors for Ku-and Ka-band commercial satellites *AIAA Papers* 2002-2032 9 p

[15] Shintate K, Terada K, Usui M, Tsujihata A and Miyasaka A 2003 A large deployable reflector (LDR) *J. Nat. Inst. Informat. and Communicat. Technol.* **50** (3/4) 33–9

[16] Meguro A, Harada S and Watanabe M 2003 Key technologies for high-accuracy large mesh antenna reflector *Acta Astronaut.* **53** 899–908

[17] Bei N A and Zimin V N 2005 Transformable larger antennas for geostationary satellites *Antennas* 10 (101) 24–7

[18] Datashvili L, Baier H, Schimitschek J, Lang M and Huber M 2007 High precision large deployable space reflector based on pillow-effect-free technology *AIAA Paper* 2007-2186

[19] Datashvili L, Endler S, Wei B, Baier H, Tsignadze N and Santiago-Prowald J 2013 Study of mechanical architectures of large deployable space antenna apertures: from design to tests *CEAS Space J.* 5 169–184

[20] Ponomarev S V 2011 Transformable reflectors of spacecraft antennas *Vestn. Tomsk. Gos. Univ. Mat. Mekh.* [Bulletin Tomsk State University. Mathematics and Mechanics] 4 (16) 110–9.

[21] Reznik S V, Prosuntsov P V, Mikhailovsky K V and Shafikova I R 2016 Material science problems of building space antennas with a transformable reflector 100 m in diameter *IOP Conf. Ser.: Mater. Sci. Eng.* **153** 012001 10 p DOI: 10.1088/1757-899X/153/1/012001

[22] Sayapin S N and Shkapov P M 2016 Kinematics of deployment of petal-type large space antenna reflectors with axisymmetric petal packaging *J. Mach. Manuf. Reliab.* **45** (5) 387–97

[23] Kotik A, Usyukin V, Vinogradov I and Arkhipov M 2006 Simulation of reflecting surface deviations of centimeter-band parabolic space radiotelescope (SRT) with the large-size mirror *Proc. SPIE-Int. Soc. Opt. Eng.* A **10567** 6

[24] Reznik S, Prosuntsov P, Denisov O, Denisova L, Zavaryev V and Mikhailovskiy K 2019 Development of transformable reflectors of mirror space antennas *ITM Web of Conferences* **30**, 05026 DOI: 10.1051/itmconf/20193005026

[25] Bitkina E V, Denisov A V and Bitkin V E 2012 Design engineering methods of creating of dimensionally stable space structures of integrated type made of composite materials *Izvestiya Samarskogo Nauchnogo Centra Rossiiskoi' Akademii Nauk* **14** 4(2) 555–60

[26] Reznik S V, Prosuntsov P V and Azarov A V 2015 Substantiation of the structural-layout scheme of the mirror-space-antenna reflector with a high shape stability and a low density per unit length *J. Eng. Phys. Thermophys.* **88**(3) 699–705 DOI:10.1007/s10891-015-1239-x

[27] Reznik S V, Prosuntsov P V and Azarov A V 2015 Modeling of the temperature and stressed-strained states of the reflector of a mirror space antenna *J. Eng. Phys. Thermophys.* **88**(4) 978–83 DOI:10.1007/s10891-015-1273-8

[28] Prosuntsov P V, Reznik S V, Mikhailovsky K V, Novikov A D and Zaw Ye Aung 2016 Study variants of hard CFRP reflector for intersatellite communication *IOP Conf. Ser.: Mater. Sci. Eng.* **153** (1) 012012 7 p DOI: 10.1088/1757-899X/153/1/012012

[29] Reznik S V, Novikov A D 2017 Comparative analysis of the honeycomb and thin-shell space antenna reflectors *MATEC Web Conf. 01012* DOI: 10.1051/matecconf/20179201012

Increasing the Wear Resistance of the Friction Surfaces of Rotating Parts Made of Bearing Steels Through Hardening Processing Methods

A Tkachuk, O Zabolotnyi, T Chetverzhuk, K Svirzhevskyi and A Kononenko

Lutsk National Technical University, 75, Lvivska street, 43018 Lutsk, Ukraine

E-mail: a.tkachuk@lntu.edu.ua

Abstract. During the study, it was found that the basic technological processes of forming cylindrical working surfaces of friction pairs do not fully provide a rational combination of micro geometric characteristics and physical and mechanical properties of the surface layer. Therefore, we proposed to evaluate the wear resistance of the working surfaces of parts to use a complex indicator that takes into account the influence of technological factors on the properties of the material and the performance characteristics of the surface layer for the running-in and continuous operation modes. Plots of distribution of equivalent stresses by indentors of different geometric configurations are constructed and the optimal shape and material of the deforming element are established. A model of the dynamics of the deformation of the workpiece with a single-axis elastic tool has been developed, which can be used to calculate the radial force in the contact zone. The calculation scheme corresponds to the scheme of the axial cam mechanism. Based on the analysis of the influence of the geometric characteristics of the deforming element on the parameters of the deformation process, a barrel-type indenter was developed for fixing which a slider-type holder was developed with the ability to adjust the stiffness of the dynamic system.

1. Introduction

During the period when the bearing is applied, there is intense local wear of the working surfaces of its parts. Touching surfaces tend to acquire a geometric shape and roughness that will best meet the specific operating conditions [1]. During this process, there are increased contact stresses, which causes thermal processes in the surface layer of the metal, and a decrease in its physical and mechanical properties. In places where the metal surface layer is most deformed, contact stresses increase to critical values [2]. However, if the optimal operational microgeometry of the profile and physical and mechanical properties of the surface layer is provided during the rolling surface shaping, the run-in period will be reduced and the moment of their acquisition of a stable operational state will be accelerated [3]. Basic technologies for forming the working surfaces of bearing rings do not fully provide a rational combination of micro geometric characteristics and physical and mechanical properties of the surface layer. Analysis of manufacturing defects of functional surfaces showed that finishing technologies do not provide the required quality of their manufacture at a sufficient level, which reduces the wear resistance of the roller bearing as a whole [4]. Significantly greater opportunities in technological quality management of surface layers appear during the application of strengthening by smoothing processing methods based on surface plastic deformation (SPD), due to the appearance of favourable technological factors [4,5]. Qualitative and quantitative evaluation of interaction methods of shaping surfaces of rotation details and operational requirements to the surfaces of rotation of the rings, served as a basis to model the relationships of structural and technological factors, technological operations-enhancing processing performance [6].

To assess the wear resistance of the working surfaces of the rings, we used a complex indicator, J, the wear intensity, which takes into account the influence of technological factors on the material properties and operational characteristics of the rings, and derived dependencies for the run-in mode:

$$J_p = \frac{1.2Ra^{\frac{2}{3}}}{n\ \lambda\ Sm\ t_p^{\frac{3}{2}}}\left(\frac{p}{H\mu_0}\right)^{\frac{7}{6}}\sqrt{15\pi(2\pi\ Wz\ H_{max})^{\frac{1}{3}}\left[1+\frac{2\pi\ H\mu_0\ (1-\mu^2)}{E}\right]} \quad (1)$$

and set operating mode:

$$J_p = \frac{1.2\pi\ p^{\frac{7}{6}}}{n\ \lambda\ t_p^{\frac{3}{2}}\ H\mu_0^{\frac{2}{3}}}\sqrt{\frac{30(1-\mu^2)\cdot(2\pi\ Ra\ Wz\ H\mu_{max})^{\frac{1}{3}}}{E\cdot Sm}} \quad (2)$$

where Ra – the arithmetical mean deviation of roughness profile of the surface; n – is the number of cycles which leads to manifestations destruction of the material; λ – coefficient taking into account influence of surface residual stresses on the process of wear; Sm is the average pitch of the irregularities of roughness profile; t_p – relative length of the reference line of the roughness profile; p – the working pressure at the contact surface; $H\mu_0$ – surface microhardness; Wz – wave height; H_{max} – maximum height of macro deviations; μ and E are Poisson's ratio and the modulus of elasticity of the material of the part. The analysis of theoretical and empirical dependences of the operational properties of roller bearing parts showed that they depend on the system of parameters of the quality of functional surfaces: macro deviations – H_{max}, Hp; undulations – Wz, Wp, Smw; roughness – Ra, Rz, R_{max}, Rp, Sm, t_p; roughness – Ra', Sm'; physical and mechanical properties $-\sigma_0$ (surface residual stresses); $h\sigma_0$ (depth of surface residual stresses); $h\mu_0$, h_l (depth of the hardened layer); I_s (grain size); ρ_D (dislocation density).

Elaboration of models of processing by blade, abrasive and smoothing tools allowed to establish sequence of formation of operational parameters in a technological cycle of manufacturing of rings of roller bearings [7–9]. It is found that during processing of the blade formation of roughness depends on the initial roughness Rz and fluctuations in the initial microhardness of the surface layer $H\mu_{max}$, $H\mu_{min}$. The average pitch of the Sm irregularities depends on its initial value, on the Rz and changes in the surface microhardness $H\mu$. Initial macro-deflections are partially inherited during the next machining and depend on the initial physical and mechanical parameters of the surface layer, namely microhardness and depth of residual stress ($H\mu_0$, $h\mu$, σ_0, $h\sigma_0$). The formation of physical and mechanical properties of the surface layer during machining largely depends on their initial state. Based on the analysis of the influence of the quality parameters of the surface layer with the processing conditions, the hereditary and consequential nature of the relationship between the quality indicators of the surface layer and the performance characteristics is established [10,11].

2. Materials and Methods

The Abbott-Firestone curve of the surface roughness profile is selected for the analysis of performance characteristics. Analysis of different profiles showed that the bearing capacity of the surface layer at a constant height $R(Rz)_{max}$ and the value of Ra is greater, the lower the smoothing height Rp (the distance from the line of protrusions to the middle line). If the values Rp and $R(Rz)_{max}$ are the same, the higher the load-bearing capacity of the surface, the greater the parameter Ra. With a decrease in the height of the R_{max} micrometer, the Ra and Rp parameters decrease, and the load-bearing capacity increases. It was found that the load-bearing capacity of surface microcavities depends on the value of the parameter $R(Rz)_{max}$, the smoothing height Rp and the arithmetic mean deviation of the profile

Ra. The optimal shape of the deforming element is determined based on the constructed plots of the equivalent stress distribution (figure 1) by indentors of various geometric configurations and on the basis of the calculations performed, figure 1 (a).

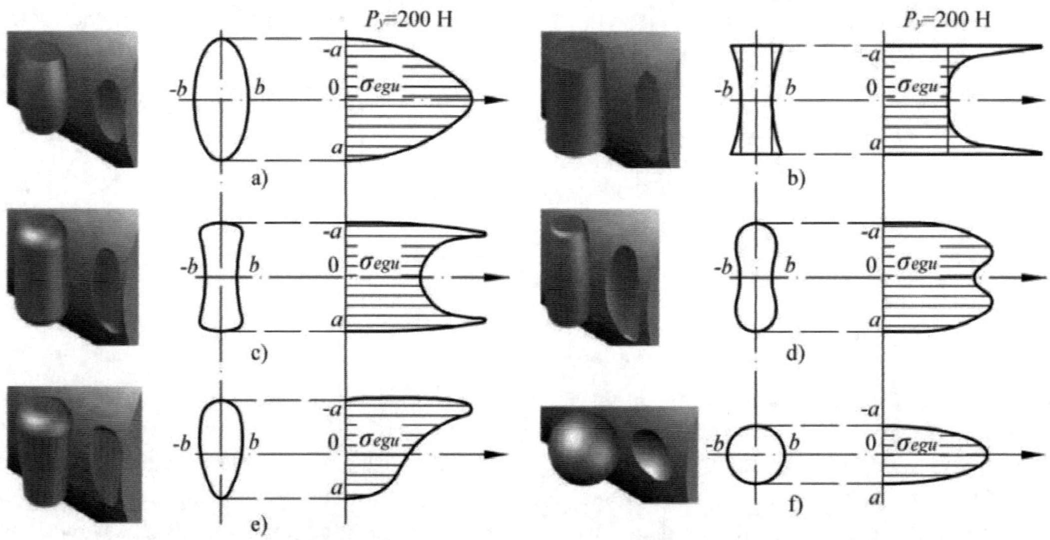

Figure 1. Changing the equivalent voltage distribution plots depending on the shape of the indenter.

The stress state in the contact zone of the part with the indenter is determined not by the original, but by the plastic deformed shape of its surface as a result of indentation. Considering this, when calculating the stresses arising in the part, not only the initial but also the curvature formed by the plastically deformed surface must be taken into account. In addition, during plastic deformation of the part material, the equivalent stress plot is curved within the deformed layer, reflecting the load history. As the deformation proceeds, the maximum yield point increases and deepens until it reaches the position corresponding to the maximum load. In the process of metal hardening, this result is usually fixed in the cross-section of the surface layer, and the microhardness equalization point determines the limit of the plastically deformed layer, *i.e.* the depth of strengthening $h\sigma$. For the general case of contact of bodies of arbitrary curvature, the main stresses σ_x, σ_y, σ_z, planes perpendicular to the coordinate axes, of which z is normal to the contact surface, and x coincides with the main axis of the elliptical contact zone, are defined as:

$$\left.\begin{aligned}
\sigma_x &= -P_y \frac{\frac{b}{a}}{1-\frac{b^2}{a^2}}\left\{-\sqrt{\frac{\frac{b^2}{a^2}+\frac{z^2}{a^2}}{1+\frac{z^2}{a^2}}} + 2\frac{z}{a}(L-K) - 2\mu\left[1-\frac{a^2}{b^2}\sqrt{\frac{\frac{b^2}{a^2}+\frac{z^2}{a^2}}{1+\frac{z^2}{a^2}}} + \frac{z}{a}\left(\frac{a^2}{b^2}L-K\right)\right]\right\} \\
\sigma_y &= -P_y \frac{\frac{b}{a}}{1-\frac{b^2}{a^2}}\left\{\frac{1+\frac{z^2}{a^2}\left(2\frac{a^2}{b^2}-1\right)}{\sqrt{1+\frac{z^2}{a^2}}\sqrt{\frac{b^2}{a^2}+\frac{z^2}{a^2}}} - 2\frac{z}{a}\left(\frac{a^2}{b^2}L-K\right) - 1 + 2\mu\left[1-\sqrt{\frac{\frac{b^2}{a^2}-\frac{z^2}{a^2}}{1+\frac{z^2}{a^2}}} + \frac{z}{a}(L-K)\right]\right\} \\
\sigma_z &= -P_y \frac{1}{\sqrt{1+\frac{z^2}{a^2}}\sqrt{1+\frac{z^2}{b^2}}}
\end{aligned}\right\}, \quad (3)$$

where P_y – radial load; a, b are the semi-axes of the elliptical contact in the axial and cross sections, respectively.

The main stresses are expressed as functions of the ratio of the half-axes b and a of the print, the ratio of the z coordinate of the occurrence of the point under study to the half-axes of the ellipse, and the greatest pressure P_y in the centre of the contact zone. The formula (3) for σ_x, σ_y includes Poisson coefficients μ and elliptic integrals of the first kind, $K(e,\theta) = \int_0^\theta \frac{d\theta}{\sqrt{1-e^2 \sin^2\theta}}$, and the second kind, $L(e,\theta) = \int_0^\theta \sqrt{1-e^2 \sin\theta}\ d\theta$, depending on the parameter and eccentricity of the ellipse, $e = \sqrt{1-(b/a)^2}$. For 100Cr6 steel coefficient $\mu = 0.3$, hence it is possible to determine for a number of values $\frac{b}{a}$ and $\frac{z}{b}$ the main (3) and equivalent voltage (4):

$$\sigma_{equ} = \sqrt{\frac{1}{2}\left[\left(\sigma_x - \sigma_y\right)^2 + \left(\sigma_y - \sigma_z\right)^2 + (\sigma_z - \sigma_x)^2\right]}. \quad (4)$$

After equating, according to the Huber-Mises condition, the average tangent stresses σ_{equ} to the yield point of the part material in the initial state, the coordinate $z = h \cdot \sigma$ of the boundary of the plastically deformed layer is found.

$$\sigma_{equ} = \frac{3}{2} \cdot \frac{P_y}{\pi ab} \sqrt{\frac{1}{2{P_y}^3}\left[\left(\sigma_x - \sigma_y\right)^2 + \left(\sigma_y - \sigma_z\right)^2 + (\sigma_z - \sigma_x)^2\right]}. \quad (5)$$

3. Results

A model of the dynamics of deformation of the workpiece with a single-indenter elastic tool allows determining the radial force P_y in the contact zone "surface-indenter". Design scheme (figure 2) corresponds to the axial Cam mechanism, where 1 – Body, 2 – Movable slider with the Indenter 3 is pressed against the Workpiece 5 by a Spring 4 with a stiffness, with the force $P_{y_0} = Q_0 = cY_0$ thanks to the Nut 6. Nominal cross-section of the workpiece has the shape of a circle 7 with radius ρ_0. The actual section of the billet is represented as a wavy curve that has a maximum height of W_{max}, the angular pitch of the wave ϕ_S, and the Nominal Section Profile 7 is the mean line m_w of the Billet 8, which determines the actual radius of the billet $\rho(\phi)$.

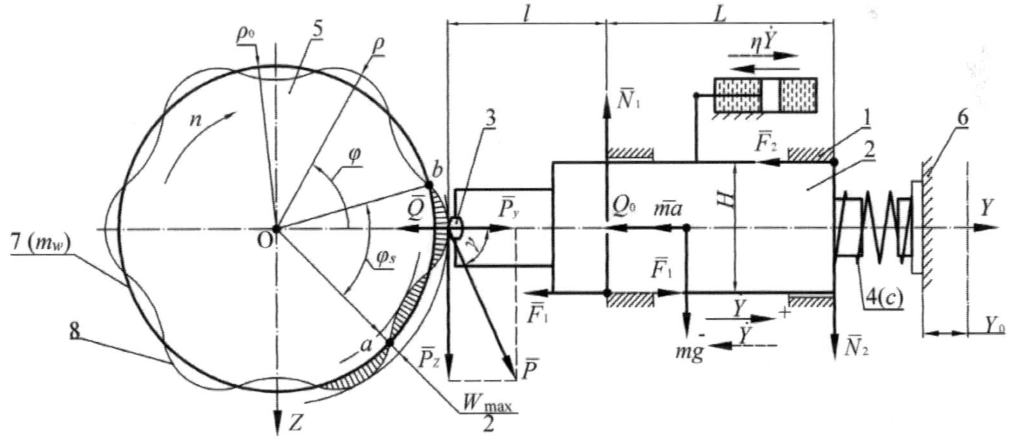

Figure 2. Calculation scheme of the process of combined smoothing of rotation surfaces that have deviations from roundness.
1 – Body, 2 – Movable slider, 3 – Indenter, 4 – Spring, 5 – Workpiece,
6 – Nut, 7 – Nominal Cross-section Profile, 8 – Billet.

In the design scheme provided by the following forces: $P_{y_0} = Q_0 = cY_0$ – the nominal value of the smoothing force; $P_y = Q$ – the resulting impact force of the indenter on the surface in the contact zone; $\overline{m\ddot{y}}$ – the inertial force of moving parts of the tool; $\bar{P}_y$ – the reaction force of the workpiece on the indenter; $\bar{P}_z\bar{P}_z$ – strength of resistance to smoothing; $\bar{P}$ – resultant force smoothing; $\overline{mg}$ – the weight of the moving parts of the tool; $\bar{N}_1$, $\bar{N}_2$ the force reactions in the bearings of the slider; $\bar{F}_1$, $\bar{F}_2$ – the friction force-sliding that occurs during the movement of the slider; $\overline{\eta\dot{Y}}$ – the moment of resistance proportional to the speed of movement of the slider caused by the presence of lubricating and cooling fluid (coolant).

A number of issues have been solved using mathematical modelling:

- The necessary conditions for ensuring stable contact of the indenter and the workpiece are determined, that is, the possibility of separation of the indenter and the occurrence of impact of the indenter on the surface of the workpiece is excluded:

$$f(\omega t) = f(\phi) = \rho \tag{6}$$

- The range of smoothing speeds $(v_{\min}, v_{\max})$ is defined, within which the smoothing force Q will be stored in the allowed range $(Qmax_{min})$, which must be maintained in order to form the quality parameters of the surface layer with regulated reliability:

$$Q_0 - \Delta Q \leq Q_0 + cf(\omega t) + M\omega^2 f''(\omega t) \leq Q_0 + \Delta Q \tag{7}$$

- A number of additional operating conditions of the tool are determined in order to select the optimal technological modes of processing, namely: slider acceleration,

$$\ddot{y}(t) = \omega^2 f''(\phi) = \omega^2 f''(\omega t) \tag{8}$$

- Slider jamming condition

$$tg\gamma_{jc} = \frac{1}{f\left(1 + \frac{2h}{L}\right)} \tag{9}$$

- Equation of forced vibrations

$$\ddot{Y} + 2n\dot{Y} + \omega_0^2 Y = \frac{cf(t)}{m} \tag{10}$$

- The coefficient of dynamic

$$\mu = \frac{a}{Y_{st}} = \frac{1}{\sqrt{\left(1 - \frac{\omega^2}{\omega_0^2}\right)^2 + \frac{4\omega^2 n^2}{\omega_0^4}}} \tag{11}$$

- Oscillation amplitude

$$a = \frac{Y_{st}}{\sqrt{\left(1 - \frac{\omega^2}{\omega_0^2}\right)^2 + \left(\frac{ka^{n-1}}{c}\right)^2}} \tag{12}$$

Based on the analysis of the geometric characteristics of the deforming element on the parameters of the deformation process for the developed tool (figure 3), a barrel-type indenter made of a hard alloy MC241 was made. To fix the deforming body, a slider-type assembly was developed consisting of: the Housing 1, the Adjusting Washer 2, the Cover 3, three Screws 4, and the Indenter 5. In this node, a groove is cut in order to ensure that it is accurately based relative to the workpiece, with the ability to move relative to the tool axis. The housing 1 has an elliptical recess in which the indenter 5 can perform rotational movements around its axis. The washer 2 is designed to regulate the accuracy of fixing the indenter since after a while there are wear in the housing 1, which can be eliminated by replacing it. Cover 3 secures the indenter in the node. Housing, adjustment washer, and the lid made of heat-resistant material CuZn37Al1-C.

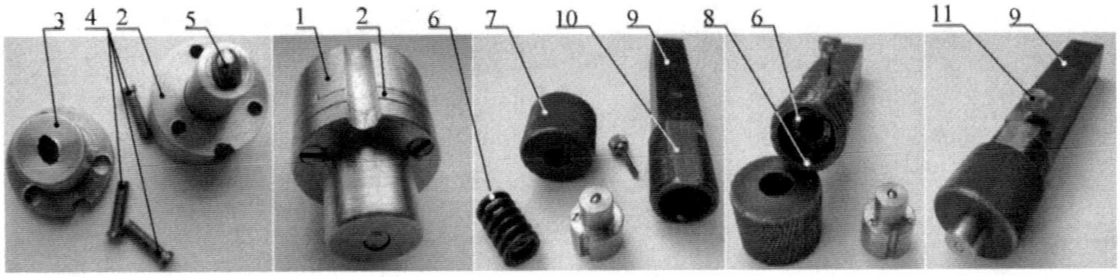

Figure 3. Consolidation of the indenter and the main structural elements of the holder.
1 – Housing, 2 – Adjusting washer, 3 – Cover 3, 4 – Screws, 5 – Indenter 5,
6 – Spring 7 – Nut, 8 – Guide, 9 – Holder, 10 – Scale, 11 – Screw.

The preload is created by compressing the Spring 6 with the Nut 7. To determine the preload force Py_0, a Scale 10 is applied to the Holder 9. The slider in the holder 9 is placed along with the guide 8. The holder 9 has a hole where the inductive Converter model 221 is mounted and secured with a Screw 11. Measurements of the smoothing force active control of the process of smoothing the rolling surfaces of the rings were carried out using a sensor, the signal from which was transmitted to the primary converter, and later the signal was transmitted in a unified form to the analogue-to-digital converter "L-Card E-154" and a PC.

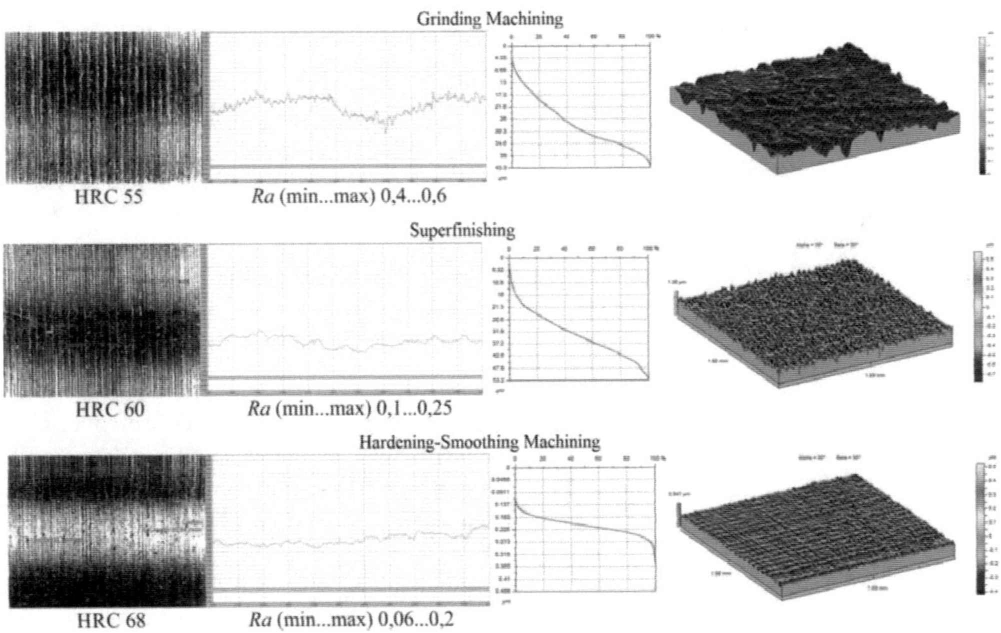

Figure 4. Analysis of microgeometric parameters of functional surfaces using Mountains Map Universal software.

4. Discussion

Experimental studies of the influence of surface layer quality parameters on the wear resistance of the working surfaces of roller bearing rings during strengthening and smoothing operations have been conducted. The influence of the smoothing force P_y on the depth, width, height of inflows, and the area of irregularities in the cross-section was determined by the method of test passes. Fragments of the studied surfaces and their indicators are shown in figure 4. The study of the surface microgeometry was carried out using the Rank Taylor Hobson Talyskan 150.

Study of parameters of quality of surfaces of rolling operations carried out by SPD regressing multi-factor model reflecting a quantitative relationship between radial force smoothing P_y, feed S, spindle speed n, source parameters of surface layer quality Ra_{in} (arithmetical mean deviation of the profile, source) S_{min} (the average step roughness profile, baseline) and quality parameters of the surface layer ($Ra,\ Sm,\ H\mu$), on the basis of the developed experimental setup.

The influence of technological factors on the geometric parameters of the microrelief and wear resistance was studied on samples made of steel 100Cr6 in accordance with the developed methodology. These samples were roughed, the hardness is 55-60 HRC, the initial surface roughness is equal to Ra_{in} = 0.5-0.7 microns.

The influence of technological factors (smoothing forces P_y, the radius of the deforming element R_d, tool feed S S , speed of rotation of the spindle n) on the geometric parameters of the microrelief (depth h_σ, width b_n, the height of swells h_{hs}, relative reference area t_p, the average pitch of profile irregularities Sm, arithmetic mean deviation of the profile Ra) and physical and mechanical properties of the surface layer (surface microhardness $H\mu$) was studied. Based on the results of experimental studies and mathematical modeling, empirical dependencies are obtained (13), (14), (15) parameters of the surface layer after combined smoothing from the modes P_y, S, n for strengthening and smoothing operation using a single-tool elastic action tool.

$$\begin{gathered} Ra = 2.761 \cdot 0{,}0225 \cdot P_y + 7.01 \cdot S - 0.0184 \cdot n - 1.255 \cdot Ra_{in} - 0.0415 \cdot Sm - \\ -0.000009 \cdot P_y \cdot n + 0.00676 \cdot P_y \cdot Ra_{in} + 0.00018 \cdot P_y \cdot Sm + 0.0277 \cdot S \cdot n - 0.178 \cdot S \cdot Sm \\ +0.00744 \cdot Ra_{in} \cdot n + 0.000143 \cdot n \cdot Sm + 0.01492 \cdot Ra_{in} \cdot Sm + 0.59 \cdot P_y \cdot S \end{gathered} \quad (13)$$

$$\begin{gathered} Sm = -116.7 + 0.032 \cdot P_y - 106 \cdot S \cdot n + (-0.275 \cdot n) + 35.5 \cdot Ra_{in} + \\ +5.787 \cdot Sm_{in} - 0.00069 \cdot P_y \cdot n + 429 \cdot S \cdot Ra_{in} - 23.3 \cdot S \cdot Sm_{in} + \\ +0.0896 \cdot Ra_{in} \cdot n - 2.118 \cdot Ra_{in} \cdot Sm_{in} + 3.17 \cdot n \cdot S - 0.00245 \cdot n \cdot Sm_{in} \end{gathered} \quad (14)$$

$$\begin{gathered} H\mu = 3601 + 77.4 \cdot P_y - 3910 \cdot S - 2.04 \cdot n - 11.93 \cdot Ra_{in} - 41 \cdot Sm - 0.019 \cdot P_y \cdot n - \\ -1.75 \cdot P_y \cdot Ra_{in} + 10\ 185 \cdot S \cdot Ra_{in} + 100 \cdot S \cdot Sm + 0.032 \cdot n \cdot Sm + \\ +16.4 \cdot Ra_{in} \cdot Sm - 626 \cdot P_y \cdot S \end{gathered} \quad (15)$$

On the basis of experimental and theoretical studies, graphical dependencies of microgeometry parameters of the treated surfaces on the parameters of the P_y, S, n, smoothing modes are constructed, which allow optimizing the technological factors of for strengthening and smoothing operation to obtain surfaces with predicted values of microgeometry parameters of the surface layer and improved performance properties.

In order to improve the performance of roller bearing rings and reduce the cost of manufacturing them, it is proposed to redistribute the allowance for grinding operations in the technological cycle of ring manufacturing and replace the superfinishing operation with combined smoothing.

Elaboration of empirical data (13), (14), (15) and graphical dependencies allowed us to establish that reducing Ra by 6 times reduces the intensity of surface wear also by 6 times, reducing Sm by 2.5 times reduces the intensity of surface wear by 4.4 times, increasing $H\mu$ by 5 times increases the wear resistance of the surface by 7.5 times (under all other identical conditions). Therefore, the wear resistance is most affected by the surface microhardness $H\mu$, the arithmetic means deviation of the profile Ra and the average pitch of the irregularities Sm.

Studies of the wear process and determination of the burn-in period were performed by the installation VNIPP-542 (figure 5). Tests were performed on new roller bearings. The experiment ended when the bearing began to collapse. This was determined by the period of stable operation of the roller bearing, depending on the method of finishing.

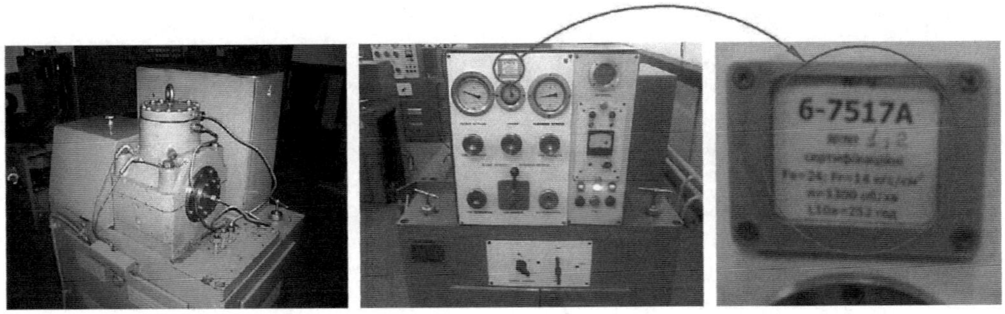

Figure 5. Installation of VNIPP-542 to test the wear resistance of bearings.

As the samples used were roller bearings of the 700 series (steel 100Cr6, hardness HRC 60-62) made according to various finishing technologies, with the microstructure of martensite with carbides. For testing 5 gradations of purity of treatment were taken. The first sample was made by the method of hardening-smoothing treatment (height of inequalities 0.2μ), the remaining 4 samples were made by basic technology – superfinishing with the height of micro-inequalities 0.8μ; 1.5μ; 3μ and 6μ. The inner and outer rings had the same roughness.

If the material densities of the reference and test samples are equal may be replaced by the ratio of absolute wear of the mass. It is allowed to list mass demolitions for linear ones, taking into account the density of the sample materials, when it is impossible to accurately measure the absolute linear wear, or when the density of the reference and test samples is different. The relative wear resistance calculated as the arithmetic mean of the results of two identical experiments is taken as the result of testing the material.

Studies of roller bearings treated with strengthening smoothing for wear resistance were carried out under the following conditions: F_a=24 kgf/cm^2, F_r=14 kgf/cm^2 (axial and radial loads, respectively); n =1300 rpm; T_{10}=252 hour. It was found that the wear resistance of the rolling surfaces of roller bearing rings increases by 15-20% with the use of hardening and smoothing technologies. The physical picture of the experimental fact of increasing the wear resistance of functional surfaces after hardening smoothing operations confirmed the validity of the theoretical positions.

5. Conclusions

The contact resistance of rolling surface is ensured not only by thermal and chemical-thermal effects, but also methods of a rational combination of grinding and firming-smoothing processing, such as directional influence on the smoothing of the peaks of the asperities of the surface, providing the effect of riveting the surface layer, and as a result of the reduction period, the running contact surfaces by 30 % and increase the resource of work at 10 %.

According to the research results, the technology of combined grinding and hardening-smoothing machining of the rolling surfaces of the rings has been developed, which provides an increased level of wear resistance of roller bearings by 15-20 % and reduces the cost of finishing operations by 12 %. A mathematical model of the process of smoothing with a single-indenter tool of elastic action on the basis of which the radial force in the contact zone "surface - indenter" P_y=200 N is determined.

The conditions for ensuring continuous contact of the indenter from the workpiece for the frequency range of the workpiece – 400-600 rpm, which is the basis for ensuring the stability of the smoothing process at the stage of technological design.

Based on modeling the dynamics of the smoothing process, taking into account the deformation gradient of the workpiece material, the geometric configuration of the deforming element (indenter) of an ellipsoid type with a profile radius R_d=20 mm and the ratio of the length of the vertical axis l_a to the horizontal l_b – 1÷0.63, respectively, and the design of the tool block with a mechanism for setting and stabilizing the radial smoothing force P_y.

References

[1] Tkachuk, A.; Zablotskyi, V.; Kononenko, A.; Moroz, S.; Prystupa, S.: Directed Formation of Quality, as a Way of Improving the Durability of Conjugated Parts of Friction Pairs. In: Ivanov V. et al. (eds) Advances in Design, Simulation and Manufacturing II. DSMIE 2019. Lecture Notes in Mechanical Engineering, pp. 370-377. Springer, Cham (2020).

[2] Zablotskyi, V.; Moroz, S.; Tkachuk, A.; Prystupa, S.; Zabolotnyi, O. Investigation of the Influence of Diamond Smoothering Treatment Power Parameters on Microgeometry of Working Surfaces of Conjugated Parts. In: Tonkonogyi V. et al. (eds) Grabchenko's International Conference on Advanced Manufacturing Processes InterPartner-2019, pp 372-381 (2020).

[3] Kindrachuk, M.; Radionenko, O.; Kryzhanovskyi, A.; Marchuk, V.: The friction mecha-nism between surfaces with regular micro grooves under boundary lubrication. Aviation 8(2), pp. 64–71 (2014).

[4] Gold, P.; Loos, J.: Wear resistance of PVD-coatings in roller bearings. Wear 253(3–4), pp. 465–472 (2002).

[5] Kukhar, V.; Vasylevskyi, O.: Experimental research of distribution of strains and stresses in work-piece at different modes of stretch forging with rotation in combined dies. Metall. & Min. Ind. 3, pp. 71–78 (2014).

[6] Chernets, M.: Prediction of the life of a sliding bearing based on a cumulative wear model taking into account the lobing of the shaft contour. J. Frict. Wear 36(2), pp. 163–169 (2015).

[7] Marchuk, V.; Kindrachuk, M.; Kryzhanovskyi, A.: System analysis of the properties of discrete and oriented structure surfaces. Aviation 18(4), pp. 161–165 (2014).

[8] Kaplun, P.; Dykha, O.; Gonchar, V.: Contact durability of 40 Kh steel in different media after ion nitriding and nitroquenching. Mater. Sci. 53(4), pp. 468–474 (2018).

[9] Etsion, I.: Improving tribological of mechanical components by laser surface texturing. Tribol. Lett. 17(4), pp. 733–737 (2004).

[10] Kun, H., Yang, W., Ye, X.: Adjustment of machining-induced residual stress based on parameter inversion. Int. J. Mech. Sci. 135, pp. 43–52 (2018).

[11] Kuznetsov,Yu., Dmitriev, D.: Realization of the conception of frame configuration machine-tools. Bulletin NTUU «KPI». vol. 3 (78), pp. 26-32 (2016).

The study of the applicability of polymer composite materials for the manufacture of the impeller of a centrifugal compressor

L A Martynyuk[1,2], D V Afanasiev[1], L V Bykov[2], A D Ezhov[2] and M A Mezintsev[1]

[1] Central Institute of Aviation Motors, Moscow, Russia
[2] Moscow Aviation Institute (National Research University), Moscow, Russia

E-mail: Larion-m@mail.ru

Abstract. The use of composite materials in modern designs is one of the promising areas. Despite a number of advantages, the use of composite materials imposes a number of restrictions on their use. One of them is a binder's degradation temperature. The revision of the existing methodology for determining the thermal state of the compressor wheel through the use of modern calculation tools and their experimental justification will allow further reliable thermal calculation taking into account the anisotropy of material properties. For carrying out computational research and testing manufacturing technology, a three-dimensional model of the impeller of a centrifugal compressor and the motor shaft was developed.

1. Introduction

The use of composite materials in modern designs is one of the promising areas [1-4]. Small specific gravity, high strength characteristics determining a composite material are an advantage when choosing small compressors (figure 1). Despite a number of advantages, the use of composite materials imposes a number of restrictions on their use, including binder degradation temperature. Based on this, a reliable calculation of the thermal state of the structure is the basis for the further use of composite material in the manufacture of the compressor impeller.

Figure 1. Centrifugal compressor impeller made of polymer composite materials.

The objectives of this work were: improving the methodology for determining the thermal state of the impeller of a centrifugal compressor through the use of modern calculation tools and their experimental justification; and studying the possibility of using polymer composite materials in highly loaded units of modern aircraft propulsion systems. The problems addressed:

- Three-dimensional models of the impeller of a centrifugal compressor and shaft, taking into account the upcoming technological processes, were created.
- The boundary conditions, temperature distribution T and convective heat transfer coefficients α along the profile of the blades and the back surface of the compressor depending on the gas flow rate for cooling Gy, were calculated.
- The thermophysical properties of carbon fibre based on carbon fibres UMT 49S and phthalonitrile binder PN-3M for unidirectional layer and laying used in the manufacture of compressor elements were experimentally determined.
- The thermal state of the impeller of a centrifugal compressor was calculated according to the previously defined boundary conditions.
- The technology of automated calculation of reinforcing material was developed and applied.

As the object of study, the impeller of a centrifugal compressor of a promising small-sized gas turbine engine was selected, having the following parameters: an engine life of 12 hours; a maximum duration of engine operation in high power mode of 3 hours; a maximum operation temperature in the centrifugal compressor of 560 K; and an outlet pressure of 7 bar. As a result of the analysis of the known methods for calculating the thermal state of the impellers of centrifugal compressors, it was determined that the existing methods are applicable only for the calculation of metal compressors and do not take into account the anisotropy of the thermophysical properties of composite materials [5]. A method for calculating the thermal state of the impeller of a centrifugal compressor was developed taking into account the anisotropy of the thermophysical properties of the composite material.

2. Results of three-dimensional modelling and numerical simulation

To conduct design studies and refine the manufacturing technology, a three-dimensional model of the impeller of a centrifugal compressor was developed in the Siemens NX software package, which consists of 20 segments, engine shaft and 5 power rings (figure 2).

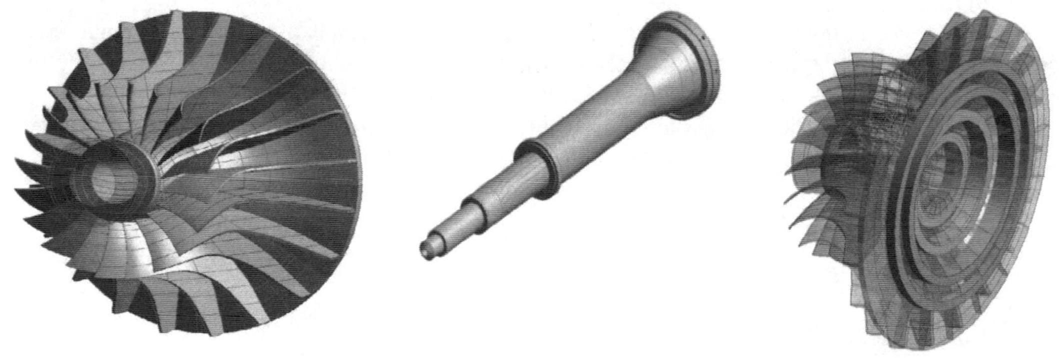

Figure 2. Three-dimensional model of centrifugal compressor impeller part.

It is known that the efficiency of the Brighton cycle directly depends on the degree of increase in gas pressure in the compressor [6]; accordingly, makes sense to minimize gas leakage on cooling of engine elements; however, it is assumed that a reduction in gas consumption for cooling can cause heating of the rear surface of the impeller to temperatures significantly exceeding the glass transition temperature of the binder, which in turn can lead to melting of the binder and the destruction of the part. The task is to achieve the minimum required flow rate of gas leakage for cooling the rear surface of the impeller of a centrifugal compressor, nozzle unit and rear bearing.

When calculating the air parameters on the back of the rotor, a program was used for iterative calculation of gas-dynamic parameters. The graphical dependencies (figure 3) confirm the assumption that the reduction in leakage rate Gy can cause significant heating of the rear surface of the compressor rotor. In view of the restrictions on the glass transition temperature GTT = 450 degrees Celsius of the binder PN-3M, the choice of leakage rate G_y = 0.0102 kg/s is preferable; accordingly, gas-dynamic parameters of air will be used to calculate the thermal state of the impeller of a centrifugal compressor at a leakage rate of G = 0.0102 kg/s.

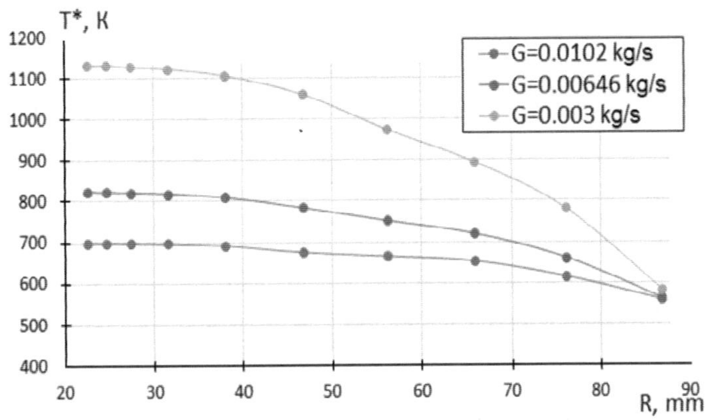

Figure 3. Dependence of the temperature T^* in the sections on the rear surface on the gas flow rate for cooling G_y.

2.1. Setting boundary conditions

In the ANSYS Mechanical module, the three-dimensional model was divided into annular regions according to the cross-sections for calculating boundary conditions (figure 4).

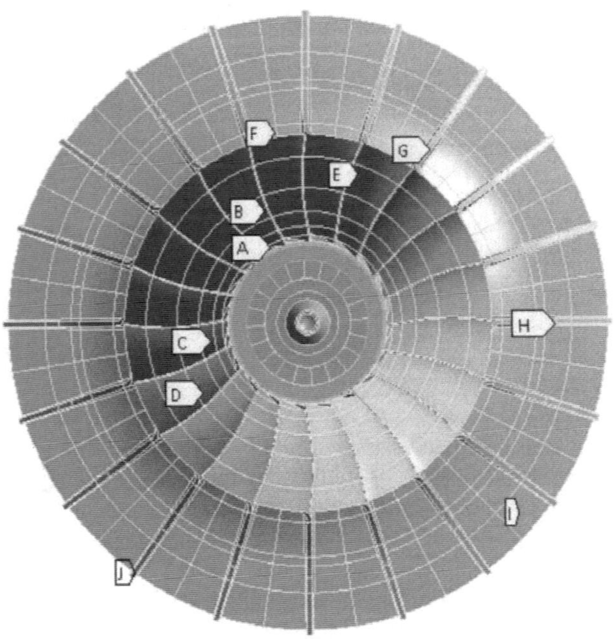

A	T=350.0 K	h=64754 W/m²·K
B	T=373.1 K	h=62721 W/m²·K
C	T=396.2 K	h=59154 W/m²·K
D	T=419.5 K	h=56852 W/m²·K
E	T=442.5 K	h=57402 W/m²·K
F	T=456.6 K	h=63512 W/m²·K
H	T=488.5 K	h=72692 W/m²·K
G	T=511.8 K	h=79527 W/m²·K
I	T=535.0 K	h=86572 W/m²·K
J	T=558.0 K	h=10632 W/m²·K

Figure 4. Setting boundary conditions - Convection.

To determine the thermal state of the impeller of a centrifugal compressor from composite material using ANSYS Meshing, a finite element mesh was generated containing 263787 nodes and 140344 elements (figure 5).

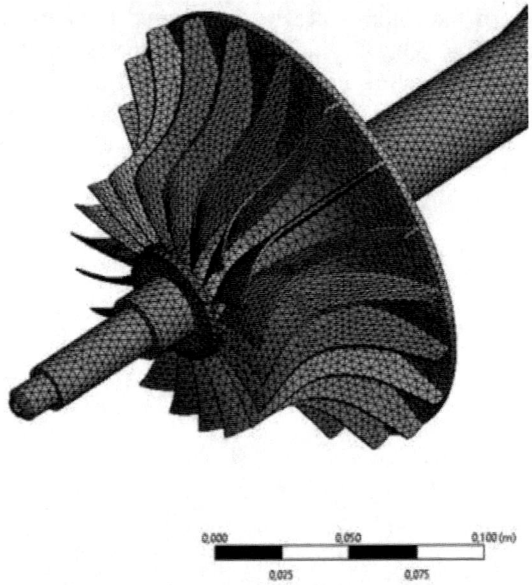

Figure 5. Mesh model of a centrifugal compressor.

2.2. Thermal state of the structure

As a result of the calculation, the thermal state of the power rings, the compressor segment, and the entire structure (figure 6) is determined. It is obvious that the back surface is heated to temperatures close to the binder destruction temperature, but not exceeding it.

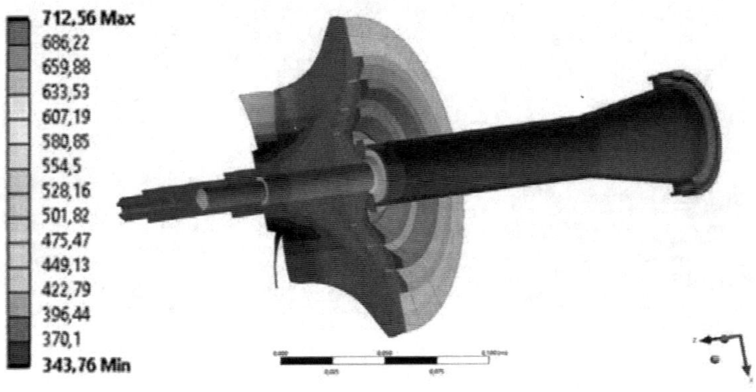

Figure 6. Thermal state of the structure (Temperature in K).

3. Results and discussion of centrifugal compressor manufacturing

Special attention is paid to the technological solutions used in this design. To manufacture the impeller segments of a centrifugal compressor, the technology of automated reinforcing filler patch was applied, which can be implemented using specialized sewing (or embroidery) machines equipped with CNC, which allow stitching of various types of roving (straight) fibres onto a flat fabric substrate, including carbon [7].

Such two-coordinate machines allow you to lay the fibre on the table plane along a complex path, while simultaneously (or locally) sewing it to the base with an auxiliary thread (figure 7). It is possible to sew the fibres end-to-end to each other with high accuracy and reproducibility, forming a monolayer of the future preform. In each layer, the fibres can be located in any given direction and, therefore, this method allows you to fully realize the strength properties of the fibres by manufacturing a preform with an optimal reinforcement scheme.

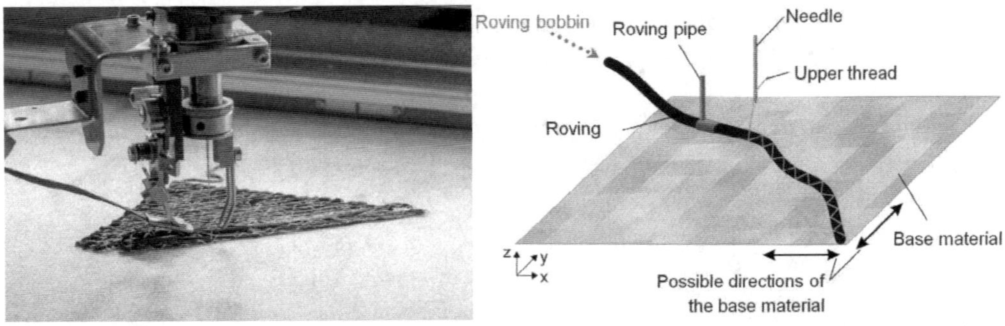

Figure 7. Laying principle and fiber fixation and stitching reinforcing material kevlar threads.

3.1. Separation of centrifugal compressor impeller segments into separable package

Due to the limitation on the thickness of the sewn preform, which is determined by the strength of the auxiliary (sewing) threads and the need for further processing of the received sewn packages, the sewed object is divided into separate packages.

For strips of a centrifugal compress rotor impeller segment, the most rational option for splitting was division into blades (figure 8). In this case, the feather of the blade is divided into two symmetrical packages and the disk part into 10 packages. The control programs for the embroidery machine were created in the GIS BasePac 10 software package taking into account the direction of fibre laying for each individual monolayer, the laying step (which affects the final density and, accordingly, the thickness of the bag). In this paper carbon fibre based on UMT49 fibres and binder PN-3M was used as the material for centrifugal compressor impeller segments.

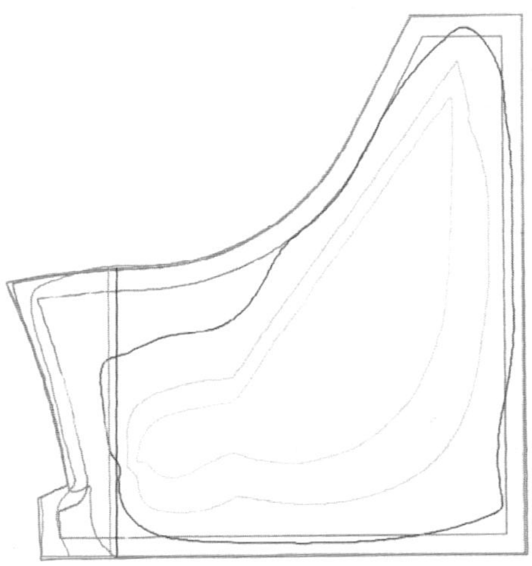

Figure 8. The model for the embroidery of the blade.

3.2. Centrifugal compressor impeller segment pressing technology

Figure 9 shows the preparation of a segment of the impeller of a centrifugal compressor, made by the method of directional laying of reinforcing material, which will be further impregnated by the RPM method and pressed in technological equipment. After manufacturing 20 segments of the impeller of the centrifugal compress and five power rings, the set of segments is fixed in the assembly tool, filled with molten paraffin and after curing the paraffin grooves are grooved under the power rings. Then, in the same snap-in, the power rings are pasted.

Studies have shown that the use of composite materials for the manufacture of the impeller of a centrifugal compressor is a promising direction for today, especially with the use of the described manufacturing technology, but the further development of this subject is impossible without the development of new types of Binders, which would increase the operating temperature and pressure ratio, thereby enabling to make high-performance centrifugal compressors.

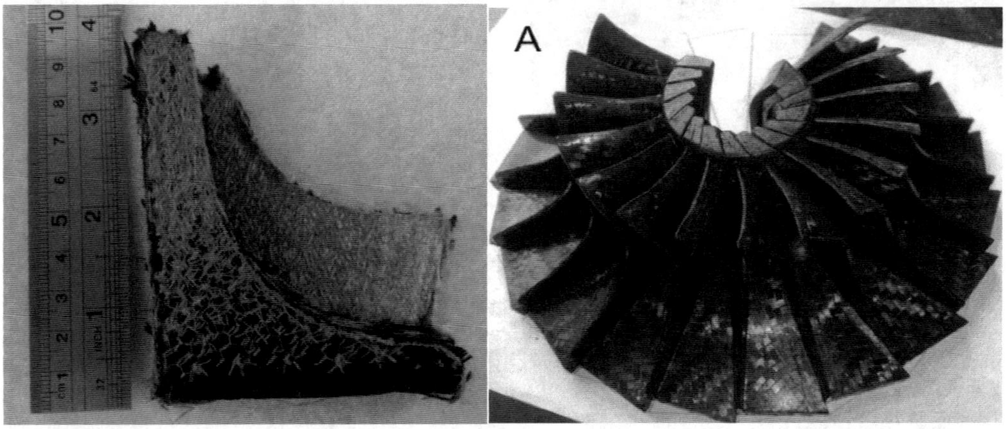

Figure 9. "Dry" workpiece of the blade segment of the impeller of a centrifugal compressor, directional styling reinforcing material and centrifugal compressor impeller segments.

4. Conclusion

A method has been developed for calculating the thermal state of the impellers of centrifugal compressors made of composite materials, taking into account the anisotropy of the thermophysical properties of materials and heating of the rear surface of the compressor.

As a result of the calculation, a temperature distribution is obtained for each individual structural element and the assembly as a whole. It was determined that the temperatures on the product do not exceed the permissible glass transition temperature HDT of the binder PN-3M of 450 C. The calculations showed that carbon fibre based on reinforcing fibres UMT-49S and a binder PN-3M can be used to make the impeller of a centrifugal compressor.

The use of this material will reduce the weight of the impeller by 45% compared with the aluminium prototype. The proven technology of directional laying of the reinforcing material allows to reduce in 3 times the manufacturing time of "dry preforms" in comparison with the manual laying of monolayers, and also reduces the frequency of revealing production defects of products from polymer composite materials.

References

[1] Pastuszak D P and Muc A 2013 Application of Composite Materials in Modern Constructions *Key Eng. Mater.* **542** 119–129.

[2] Windhorst T and Blount G 1997 Carbon-carbon composites: a summary of recent developments and applications *Mater. Des.* **18**(1) 11–15.

[3] Manocha L M 2003 High performance carbon–carbon composites *Sadhana* **28** 349–358

[4] Ezhov A, Bykov L and Artemchuk N 2019 Related problems of designing heat-loaded structures from ceramic matrix *IOP Conf. Ser.: Mater. Sci. Eng.* **683** 012083

[5] Zheng X, Jin L, Du T, Gan B, Liu F and Qian, H. 2012 Effect of temperature on the strength of a centrifugal compressor impeller for a turbocharger. *Proc. Inst. Mech. Eng. C J. Mech. Eng. Sci.* **227** 896–904

[6] Avduevskii V S 1992 *The basics of heat transfer in aviation and rocket and space technology* Mashinostroenie, Moscow 528 (in Russian)

[7] Spickenheuer A, Scheffler C, Bittrich L, Haase R, Weise D, Garray D and Heinrich G 2017 Tailored Fiber Placement in Thermoplastic Composites *Technologies for Lightweight Structures* **1**(2) 114–127.

IOP Conf. Series: Materials Science and Engineering 1060 (2021) 012027 doi:10.1088/1757-899X/1060/1/012027

Working efficiency analysis of space mirror antenna reflector made of composite materials

Y A Azhevsky

Bauman Moscow State Technical University, 2-ya Baumanskaya, 5-1, Moscow, Russian Federation, 105005

E-mail: azhevsky.yaroslav@mail.ru

Abstract. The study is devoted to the possibility of using composite materials, composite sandwich panels and composite mesh structures as constructions for space mirror antennas using at space satellites. The modelling of the thermal and stress-strain states of the structure, which are characteristic of radiation heat transfer in open space operating conditions, has been carried out. Based on the results obtained, a design with the best indicators of mass-dimensional stability under operating conditions was selected.

1. Introduction

Space mirror antennas already were highly demanded at the beginning of the satellite communication era. Numerous research has been conducted in the framework of the designing the reflectors structures that are used mostly for space and ground applications. Space mirror antenna reflector (SMAR) have a relatively simple design and high reliability, and, at the same time, a wide transmission band of radio waves as well as a highly orientability. These qualities allow to use the mirror antennas as a part of the spacecraft of various purposes for organizing high-speed radio channels [1-4].

The demand for high-speed communications in the last decade has grown significantly, which in its turn leads to the development of new higher frequency ranges with a reliable communication channel that can transmit data with short delays. A promising solution of this problem is the usage of the V – range frequency spectrum (frequency band of 40-75 GHz). This frequency band will allow reducing the mass and size of satellite antennas, while increasing energy efficiency and data transfer rate. However, one of the main limitations for the mass production of the V-band RSMAs as the main means of communication for spacecraft is the high cost of SMAR [5-8]. The high frequency of radio band restricts the requirements for the quality of space satellite antenna structures. The accepted deviation of the surface of the SMAR from the calculated value should not exceed $\Delta = \Lambda/50$ (Λ is the working length of the radio wave of the satellite antenna). Thus, for the correct operation of satellite communications, it is necessary to keep the dimensional stability of the reflective surface of the SMAR at a level of 0.1 to 0.4 mm, depending on the selected operating frequency [10, 11].

Space satellites mass reducing allows us to decrease costs of spacecraft orbiting. This fact makes us to think about a more detailed study of satellites constructive scheme and whole construction engineering research. Therefore, there is a necessity to create modern approaches in mirror space antennas reflector design in order to achieve the best results in linear density and dimensional stability of structures. During the operation in the near-Earth orbit, the spacecraft is subject to direct and reflected solar radiation, the Earth's self-radiation. The mentioned factors coupled with periodic approaches to the shadow part of the orbit, can create a temperature difference from 100 to 180 degrees on the meter

IOP Conf. Series: Materials Science and Engineering 1060 (2021) 012027 doi:10.1088/1757-899X/1060/1/012027

surface of the SMAR. Such operating conditions could adversely affect the dimensional stability of the reflector structure, which in its turn negatively affects the quality of the transmitted (received) signal [12-15]. High performance requirements for satellite antennas gives one thinking about using of advanced materials. Composite materials, especially CFRPs, have great importance for improving the design of SMAR. This material has high specific mechanical characteristics, low coefficients of linear thermal expansion, and relatively high thermal conductivity, which allows not only to reduce the weight of the antenna, but also to increase the stability of its shape and size.

One of the possible solutions to this problem is to use a space antenna structural scheme in the form of a thin-walled shell made of high-modulus carbon fibre with stiffening system made of three-layer panels with a non-woven polyester material filler. Current design can be manufactured in one technological operation. This approach will not only allow the design of a reflector with the required dimensional stability, but also reduce the cost of its production.

In this research, we propose a number of designs with a stiffening variation for thin-walled shell. In the course of finite element modelling, we conducted a comparative analysis of performance under operating conditions (inertial loading during orbiting and radiation heat transfer in open space). Based on the results of our research, we have selected the best reflector structural scheme and then we will conduct a study of design cases with various fastening options and see how certain fastening methods affect the stability of the dimensions of the whole antenna complex.

(a) – The reflector of the mirror antenna developed by a company "High Performance Space".

(b) – The reflector with the CFRP honeycomb core developed by "High performance Space".

Figure 1. SMAR with three-layer-type DLS.

(a) – the experimental reflector developed in Bauman MSTU.

(b) – the reflector of the mirror space antenna of "Artemis" satellite.

Figure 2. SMAR with the DLS of ribbed reinforcement.

2. The design-layout schemes of SMAR

Since the beginning of the use of composite materials as the main materials for space satellite antennas, the design-layout scheme (DLS) in the form of a three-layer panel with thin-walled skins and a honeycomb core has been widely used (figure 1). As a rule, CFRP panels are used as skins, while aluminium foil, carbon or organic plastic is used as the material for the honeycomb core. Such an approach makes it possible to obtain a rigid, dimensional-stable structure with the areal density at the level of 3 kg/m2 [16,17]. Another solution for the SMAR structure is DLS in the form of a thin-walled composite shell with developed thin-walled spatial ribbing (figure 2). Such structures are produced of CFRP. Ribbing allows to develop rigid structure with a high dimensional stability and areal density at the level of 2.5 kg/m^2 [18].

3. Development of the Design-layout scheme of SMAR

If we consider antennas of the V - range frequency spectrum, then, as a rule, antenna complexes located on spacecraft have small geometric dimensions. I would like to try out my ideas for the design of this kind of structures on just such design options. For further research, the antenna parameters were selected for the considered frequency range. The linear dimensions and arrangement of the elements were calculated using the Antenna Magus software package (figure 3). The following parameters were used as input data: Gain = 45 dB, Frequency = 40-75 GHz, Antenna type – parabolic offset.

As a result of calculations, we obtained the following parameters of the elements D = 618 mm – the height of the reflector mirror, Θ = 44 ° – the angle of feed rotation, F = 488 mm – focal length, H = 123 mm – the coordinate of the mirror positioning.

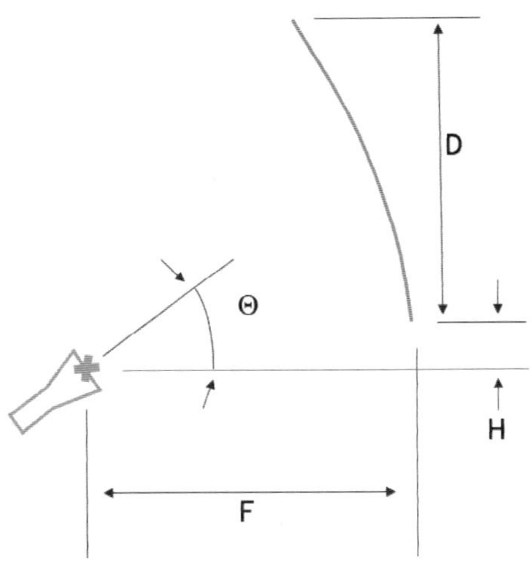

Figure 3. Geometric parameters of antenna construction

The geometry of the SMAR reinforcement set location could vary depending on the conditions for fixing the satellite dish and the mechanism of its deployment, while the design must comply with the requirements for stability of shape and size demanded for the operating conditions of orbital flight. Based on a domestic and foreign experience, several DLS of reflectors were designed with variation in the arrangement of reinforcement ribs that implies fastening to the spacecraft through the central zone (figure 4). A distinct feature of the presented designs is the original ribbing scheme. The reflector itself is made as a thin-walled CFRP shell, and the ribs are CFRP with an additional set of layers. T he width of the stiffeners of the samples under the study is 15 mm, and the thickness is 2.2 mm. The thickness of the reflector shell is 0.6 mm.

4. Refining the characteristics of the composite material

One of the important steps of the successful design of any structure is the creation of the correct mathematical model of the material. The current study deals with composite materials, which in its turn sufficiently complicated the task. Due to their specific nature, composites do not have their standardized libraries with the unambiguous defined characteristics. Since manufacturing and testing of structure in order to define the material characteristics cost a lot, it is necessary to use up-to-date mathematical modelling approaches in the preliminary design stages. Using of special software is one of such approaches. In current study MSC Digimat software package is used. This software complex allows to predict the orthotropic characteristics of a composite material based on the realistic representative volume element of the material (figure 4).

IOP Conf. Series: Materials Science and Engineering 1060 (2021) 012027 doi:10.1088/1757-899X/1060/1/012027

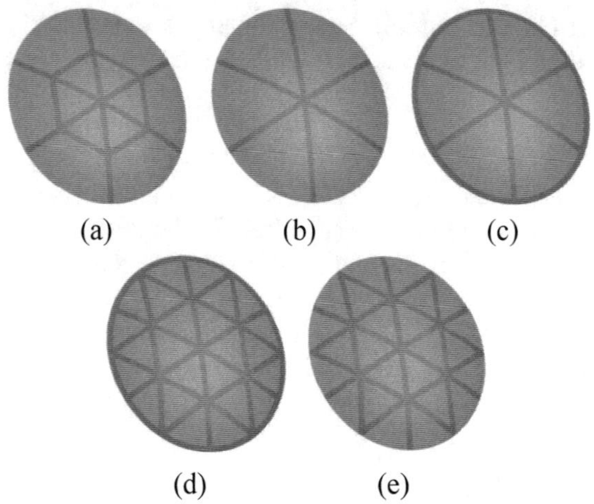

(a) – "Mark 20" scheme
(b) – "Mark 21" scheme
(c) – "Mark 22" scheme
(d) – "Mark 23" scheme
(e) – "Mark 24" scheme

Figure 4. The ribbing schemes of RSMA.

In current study, the characteristics of CFRP with the epoxy resin (Cytec 950-1) and high modulus carbon fibres (Cytec Thornel Carbon fibres P-75S) were calculated. The following data were used as input data for further definition of CFRP characteristics: the modulus of elasticity of the resin – 4.2 GPa; the shear modulus – 1.6 GPa; the Poisson's ratio – 0.3; the coefficient of linear thermal expansion of the resin – $4.5 \cdot 10^{-5}$ K^{-1}; the specific heat of the resin – 1000 J/(kg·K); the coefficient of thermal conductivity of the resin – 0.22 W/(m·K); the modulus of elasticity along the fibre – 517.1 GPa; the modulus of elasticity across the fibre – 9.1 GPa; the shear modulus – 13.1 GPa; the Poisson's ratio – 0.23; the coefficient of linear thermal expansion along the fibre – $-1.46 \cdot 10^{-6}$ K^{-1}; the coefficient of linear thermal expansion across the fibre – $12.5 \cdot 10^{-6}$ K^{-1}; the specific heat of carbon fibre– 660 J/(kg·K); the thermal conductivity coefficient along the fibre – 185 W/(m·K)); the thermal conductivity coefficient across the fibre – 2.4 W/(m·K).

As the result of homogenization using the MSC Digimat software package (figure 5), the following characteristics of CFRP with the layup scheme of [0/90/45/-45] were obtained: the thermal conductivity coefficient – 26 W/(m·K); the specific heat – 762 J/(kg·K); the density – 1 550 kg/m^3; the emissive factor – 0.85; the absorption factor – 0.735; the coefficient of linear thermal expansion – $6.1 \cdot 10^{-7}$ K^{-1}; the modulus of elasticity – 110 GPa; the Poisson's ratio – 0.3.

5. Thermal and thermal-mechanical analysis of the SMAR schemes under study

The most important stage in the design of an SMAR is to recreate operating conditions by the modelling of the temperature and stress-strain state of the structure that are typical for the radiation heat transfer in open space. The mathematical modelling of heat transfer process was conducted with the Siemens Simcenter 3D software by means of the Space System Thermal solver. The modelling results obtained for the finite-element model were used as input data for the definition of the stress-strain state of SMAR by means of the Siemens Simcenter Nastran solver. It was assumed that the studied variants of reflectors are operating as a part of the spacecraft in the geostationary Earth orbit.

A comparison of design variants was performed for moment of time of 108 000 seconds of operation in geostationary orbit, starting with a complete shading by the Earth. This moment of time corresponded to a rotation of the spacecraft by 150° relative to the Earth–Sun axis, when the temperature difference on the SMAR surface reaches its maximum value. Simulations of the orbital flight of the considered model and the following cases were carried out for the flight on the second orbital around the Earth. Since the temperature pattern on subsequent turns remains similar and the influence of the initial flight conditions is not levelled. According to the finite element modelling results, the best design from the position of dimensional stability is variant with the "Mark 1" identifying symbol (figure 6).

As a result of the antenna mirror temperature state, we calculated the temperature difference (maximum temperature = 37.19 °C, minimum temperature = 0.79 °C). Next, we exported the obtained temperature difference to a structural solver and the finite element analysis showed, the smallest displacement values correspond to the "Mark 21" structural scheme and amount to 0.037 mm, which satisfies the requirements for dimensional stability of the space satellite reflector design in orbital flight conditions. The mass of the antenna reflector is equal 1.622 kg, the areal density – 1.354 kg/m^2.

Figure 5. The stages of composite material characteristics homogenization by means of MSC Digimat software package.

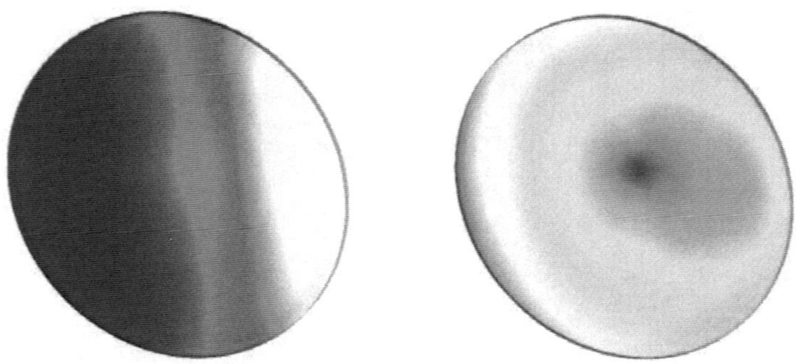

(a) – temperature distribution (b) – thermal displacement

Figure 6. The results of the modelling of radiation heat exchange in "Mark 21" variant.

6. Reflector's power Bracket designs

The power bracket used to attach the reflector to the hull of the spacecraft is another important design element of the satellite dish. Its dimensions are comparable to the dimensions of the reflector, but strict requirements are imposed on it in terms of dimensional stability and weight.

For this study, I have tried to use the types of structures that can be scaled and applied to larger ones. I settled on two types, radically different from each other. These are: sandwich panel construction and composite mesh construction. (figure 7).

These types of power brackets are often used for space applications. However, they had to take into account some features. These types of mounts are meant to be connected to the structure of the reflector mirror. Therefore, within the framework of structures, it is necessary to leave the possibility of making this connection. Therefore, for the mesh structure, a printed bracket will then be made, and for a sandwich panel, a mortgage is required to also insert the bracket or then attach it mechanically. Just for this, structural foam is used, which makes it possible to significantly lighten the structure in comparison with metal and prevent breakage as in the case of honeycombs in the attachment points.

For the structure obtained, a similar modelling was carried out, recreating the orbital flight in the geostationary orbit of the Earth. The modelling results are presented in figure 8. Although good results were obtained when simulating a reflector mirror, it is just as important to see how the whole structure works as an assembly. Therefore, we will repeat the thermal analysis for the entire antenna structure, taking into account the re-reflection from the surface of the spacecraft.

As a result of the antenna mirror temperature state, we calculated the temperature difference (maximum temperature = - 49.07 °C, minimum temperature = - 109.18 °C). Next, we exported the obtained temperature difference to a structural solver and the finite-element modelling results, the thermal displacements in the "Mark 21" reflector structure with power bracket made by composite mesh

IOP Conf. Series: Materials Science and Engineering 1060 (2021) 012027 doi:10.1088/1757-899X/1060/1/012027

design are 0.0943 mm. The total weight of full construction is 550 g. These indices fully satisfy the requirements previously presented to the design, which indicates its performance for orbital flight conditions.

The conducted research showed that the SMAR design made according to the "thin-walled shell with three-layer ribbing" scheme can meet the requirements for the stability of shapes and sizes when flying in the geostationary orbit of the Earth and is promising for further in-depth studies.

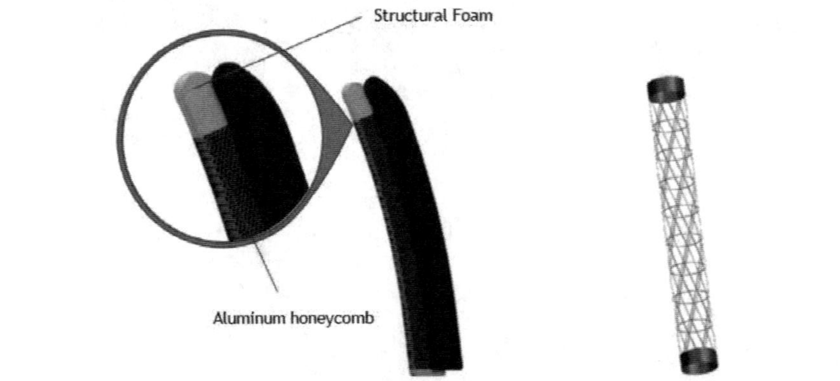

(a) – CFRP sandwich panel (b) – CFRP mesh construction

Figure 7. Variants of power brackets for SMAR.

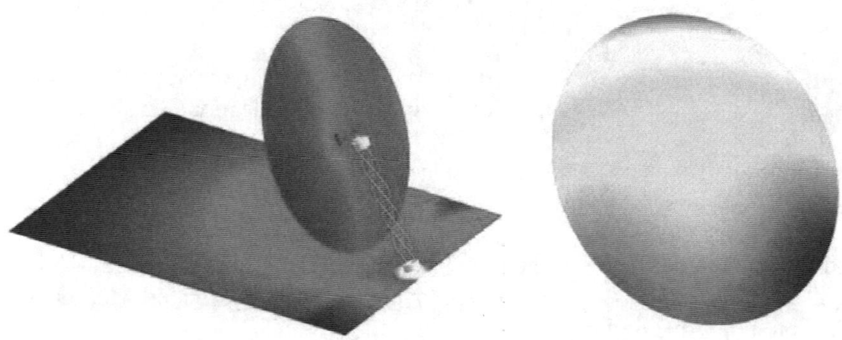

(a) – Temperature distribution (b) – Thermal displacements

Figure 8. The results of the modelling of radiation heat exchange in the sample "Mark 21" with power bracket.

7. The conditions for the antenna structure operability during the spacecraft launch into orbit

Another important point to determine the operability of a structure is the integrity analysis under transportation and orbital conditions. Each launch vehicle has its own loading parameters. Since I have no clear limitations on this point, I decided to conduct a small analysis and take the most loaded one. These are the Falcon family missiles with minimum load (figure 9). For my analysis, I will choose the point marked in the figure, which represents the highest launch load. This is 8.5g along and 2g across the rocket. As before, I will carry out my simulation in the Simcenter 3D software package with the Simcenter Nastran solver. I will do the fastening in the same way to prevent the rotation of the structure. As a result of the simulation, displacements on the mirror do not exceed 0.492 mm on the reflector mirror and 0.226 mm on the power mount. The stress also does not exceed the permissible.

8. Discussion

Although my work considers a sufficient number of design cases, it still does not affect those areas that may be important for the operation of the structure as a whole. First of all, all thermal characteristics for my materials are constants. This is not entirely true for the materials I have used, especially composites. More testing needs to be done to see if the property change is critical for the materials I have selected.

IOP Conf. Series: Materials Science and Engineering 1060 (2021) 012027 doi:10.1088/1757-899X/1060/1/012027

During operation, my design is exposed to significant temperature changes, and the surfaces of objects both heat up and cool down. Therefore, the correct solution would be to consider cyclic heating and take into account the parameters of durability for these structures. However, this work did not have such goals. The most important thing was to develop approaches that allow adapting the calculation methodology for more complex and important designs. Therefore, in the framework of this work, I consider these shortcomings not critical.

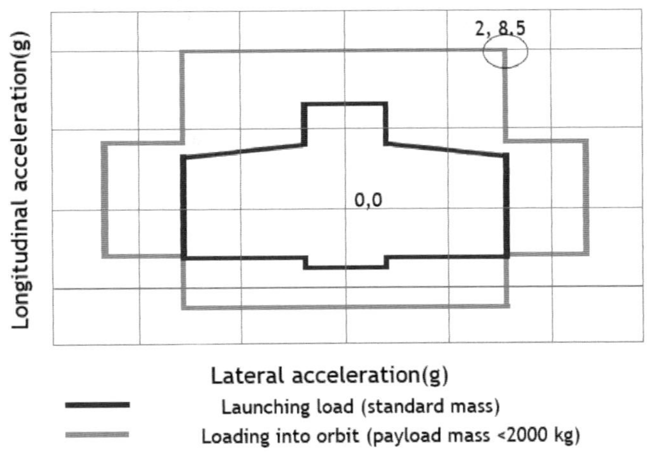

Figure 9. Payload loading scheme for the Falcon launch vehicle family

9. Conclusion

As part of my work, I went through several design stages that will further help me develop more complex structures for aerospace engineering. My research is not ideal, I have already described some of the shortcomings in the previous paragraph, but it is quite complex and allows us to speak with confidence about the applicability of a particular design to the conditions under consideration.

Firstly, the choice of the shape of the antenna mirror. Typically, these parameters are passed to the constructor as boundary conditions and this task is to improve the design in such a way that it satisfies the necessary constraints. In my work, I selected the necessary antenna parameters using the well-known expressions implemented in the Antenna Magus software product. For your research, you can also use this method to calculate the characteristics of your antennas. The choice of the power circuit of the reflector mirror is a creative task and there is no single approach. As a result, I got the geometric parameters of my future antenna and came up with finning schemes and then compared them under operating conditions.

Secondly, modelling the structure of the composite. This method allows you to easily calculate the required characteristics as thermal, mechanical and thermo-mechanical. This greatly facilitates the calculation process and allows both to simplify the design model and to reduce the number of tests. As a result, I obtained the thermal, mechanical and thermomechanical characteristics of materials and then used them for finite element analysis.

Thirdly, the simulation of the loading of our structures. In my research, I used the Simcenter 3D software package. It allows an integrated approach to the calculation process. I was able to run the analysis of orbital heating, determine the position in orbit with the largest temperature difference, determine the magnitude of the temperature movements. I was also able to calculate the loading conditions for the launch into orbit. As a result, I determined the parameters of my antenna. This antenna mirror is made of carbon fibre reinforced plastic with a thickness of 0.6 mm and fins made of CFRP in the form of 6 beams with a thickness of 2.2 mm. I also determined the shape of the power bracket in the form of a carbon fibre mesh structure. This engineering solution made it possible to achieve the stability of the mirror dimensions at the level of 0.0943 mm in orbital flight conditions and 0.492 mm from rocket launch.

References

[1] Gao S, Rahmat-Samii Y and Hodges R 2018 *Advanced antennas for small satellites*, *Proc. IEEE* **106**(3) 64–9

[2] Rao S, Shafai L, and Sharma S 2013 *Handbook of Reflector Antennas and Feed Systems Vol III Applications of Reflectors* (Boston: Artech House)

[3] Rao S K 2015 Advanced Antenna Technologies for Satellite Communications Payloads *Proc. IEEE Transactions on Antennas and Propagation* **63**(4) 169–75

[4] Rao S K 2003 Parametric design and analysis of multiple-beam reflector antennas for satellite communications *Proc. IEEE Antennas & Propagation Magazine* **45** 26–34

[5] Keen K M 1976 Surface efficiency measurements on a high-modulus carbon fibre composite reflector antenna at L- and S-band frequencies *Electronics Letters* **12**(7) 160–1

[6] Nicholson K J, Rowe W S T, Callus P J and Ghorbani K 2011 Split-ring resonator loading for the slotted waveguide antenna stiffened structure *IEEE Antennas and Wireless Propagation Letters* **10** 1524–1527

[7] Galehdar A 2012 Capacitively fed cavity-backed slot antenna in carbon-fiber composite panels *IEEE Antennas and Wireless Propagation Letters* **11** 1028–1031

[8] Galehdar A 2011 The effect of ply orientation on the performance of antennas in or on carbon fiber composites *Progress In Electromagnetics Research* **116** 123–136

[9] Reznik S V 2018 Thermal regimes of space composite structures. Part I MATEC Web of Conferences 194 01048

[10] Reznik S V 2018 Thermal regimes of space composite structures. Part II MATEC Web of Conferences 194 01049

[11] Reznik S V, Prosuntsov P V and Mikhaylovskii K V 2018 Thermal regime of large space structure with transformable elements from hybrid composite *J. Phys.: Conf. Ser.* 1134 012048

[12] Mikhailovskii K V, Reznik S V and Prosuntsov P V 2019 Method for modeling the interaction between transformable shells of spacecrafts and small space debris objects *AIP Conference Proceedings* **2171** 030017 DOI: 10.1063/1.5133183.

[13] Reznik S V, Prosuntsov P V and Novikov A D. 2018 Prospects of Increasing the Dimensional Stability and the Weight Efficiency of Mirror Space Antenna Reflectors Made of Composite Materials in Proc. Higher Educational Institutions. Machine Building.(Moscow: HEIMB) 71–83 (in Russian)

[14] Bitkina E V, Denisov A V and Bitkin V E. 2012 Design-Engineering Methods of Creating of Dimensionally Stable Space Structures of Integrated Type Made of Composite Materials *Izvestia of Samara Scientific Center of the Russian Academy of Sciences* (Moscow) 555–60. (in Russian)

[15] Manach L, Castel X and Himdi M 2012 Performance of a Lozenge monopole antenna made of pure composite laminate *Progress In Electromagnetics Research Letters* **35** 115–23

[16] Artner G and Langwieser R 2016 Performance of an automotive antenna module on a carbonfiber composite car roof *Proc. European Conf. Antennas and Propagation* (Davos) 243–8.

[17] De Assis R R and Bianchi I 2012 Analysis of microstrip antennas on carbon fiber composite material *J. Microwaves, Optoelectron. Electromagn. Appl.* **11**(1) 154–61

[18] Artner G, Gentner P K, Nicolics J and Mecklenbräuker F 2017 Carbon fiber reinforced polymer with shredded fibers: Quasi-isotropic material properties and antenna performance *Int. J. Antennas Propag.* **2017** 154-161

[19] Mehdipour A 2013 Mechanically reconfigurable antennas using an anisotropic carbon-fibre composite ground *IET Micro. Antennas Propag.* **7**(13) 1055–63

AUTHOR INDEX